Practical Recording Techniques

Sixth Edition

Practical Recording Techniques

The Step-by-Step Approach to Professional Audio Recording

Sixth Edition

Bruce Bartlett
Jenny Bartlett

Focal Press
Taylor & Francis Group

NEW YORK AND LONDON

Please visit the Companion Website at
booksite.focalpress.com/companion/9780240821535/dev/index.php

First published 2013
by Focal Press
225 Wyman Street, Waltham, MA 02451

Simultaneously published in the UK
by Focal Press
2 Park Square, Milton Park, Abingdon, Oxon OX14 4RN

Focal Press is an imprint of the Taylor & Francis Group, an informa business

Library of Congress Cataloging in Publication Data
A catalog record has been requested for this book.

ISBN: 978-0-240-82153-5 (pbk)
ISBN: 978-0-240-82463-5 (ebk)

Typeset in Giovanni
by MPS Limited, Chennai, India
www.adi-mps.com

SFI Certified Sourcing
www.sfiprogram.org
SFI-00453

Printed and bound in the United States of America
by Edwards Brothers, Inc.

To family, friends, and music.

Contents

Preface

Recording is a highly skilled craft combining art and science. It requires technical knowledge as well as musical understanding and critical listening ability. By learning these skills, you can capture a musical performance and reproduce it with quality sound for the enjoyment and inspiration of others.

Your recordings will become carefully tailored creations of which you can be proud. They will be a legacy that can bring pleasure to many people for years to come.

This book is intended as a hands-on, practical guide for beginning recording engineers, producers, musicians—anyone who wants to make better music recordings by understanding recording equipment and techniques. I hope to prepare the reader for work in a home studio, a small professional studio, or an on-location recording session.

Practical Recording Techniques offers up-to-date information on the latest music recording technology, such as hard-disk and flash-memory recorders, computer recording, loop-based recording, keyboard and digital workstations, MIDI, Web audio, and online collaboration. But it also guides the beginner through the basics, showing how to make quality recordings with the new breed of inexpensive home-studio equipment.

The first chapter answers the question, "Why do we record?" Next, the book overviews the recording process to instill a system concept. The basics of sound, signals, and studio acoustics are explained so that you'll know what you're doing when you adjust the controls on a piece of recording equipment or build a studio. Then advice is given on equipping a home studio for any style of recording, with examples of equipment manufacturers.

Studio setup is covered next, including cables and connections, choosing monitor speakers, and preventing hum.

Each piece of recording equipment is explained in detail, as well as the control-room techniques you'll use during actual sessions. Special attention is given to microphones (including detailed microphone techniques), effects and signal processors, and mixers.

A special chapter explains how to judge recordings and improve them. The engineer must know not only how to use the equipment, but also how to tell good sound from bad.

A chapter is devoted to the technology of digital recording, followed by a detailed chapter on computer recording, which includes a section on audio for video. The text continues with a chapter that covers the latest ways of creating and recording music and then a chapter discussing MIDI sequencing. The next two chapters cover on-location recording techniques for both popular and classical music.

The latest developments in recording are Web audio and online collaboration. These topics are covered in detail in the final chapter.

Finally five appendices explain the decibel, suggest how to optimize your computer for multitrack recording, explain impedance, provide insights on phantom power, and suggest further education.

On the book's website at *www.elsevierdirect.com/companions/9780240811444* are audio samples that demonstrate various topics explained in the book. Throughout the text, references to specific tracks on the website guide the reader to relevant audio demonstrations.

Based on my work as a professional recording engineer, the book is full of tips and shortcuts for making great-sounding recordings, whether in a professional studio, project studio, on-location, or at home. You'll find many topics not covered in similar texts:

Loop-based recording
Hum prevention tips
The latest monitoring methods
Examples of mic models by type
Microphone selection guide
Tonal effects of microphone placement
Glossary of sound-quality descriptions
The latest types of digital recorders
Up-to-date coverage of computer recording
Optimizing your computer for multitrack recording
Documenting the recording session
Audio-for-video techniques
On-location recording
Troubleshooting bad sound; guidelines for good sound
Audio on the Web
Selling your music online
File sharing for online collaboration

Acknowledgments

Thank you to Nick Batzdorf of Recording magazine for giving me permission to draw from my "Take One" series. For my education, thank you to the College of Wooster, Crown International, Shure Brothers Inc., Astatic Corporation, and all the studios I've worked for.

Thank you to Carlin Reagan, Sarah Hudson, Anais Wheeler, Michele Cronin, and Catharine Steers at Focal Press for their fine work and support.

My deepest thanks to Jenny Bartlett for her many helpful suggestions as a layperson consultant and editor. She made sure the book could be understood by beginners.

A note of appreciation goes to Sting, the Pat Metheny Group, and Samuel Barber ("Adagio for Strings"), among many others, whose music inspired the chapter "Music: Why We Record."

We appreciate the following manufacturers who provided photos of their products: Avid, Edirol, Boss, Korg, Alesis, Mackie, Frontier Design, Crown International, Audio-Technica, M-Audio, TASCAM, Zoom, Eventide, Rane, Lexicon, CamelAudio, iZotope, Line 6, RML Labs, E-MU, PreSonus, Cakewalk, Propellerhead Software, CD Baby, TuneCore, and imeem (Snocap, Inc.).

Finally, to the musicians I've recorded and played with, a special thanks for teaching me indirectly about recording.

CHAPTER 1
Music: Why We Record

As you learn about recording techniques for music, it's wise to remember that music is a wonderful reason for recording.

Music can be exalting, exciting, soothing, sensuous, and fulfilling. It's marvelous that recordings can preserve it. As a recording engineer or recording musician, it's to your advantage to better understand what music is all about.

Music starts as musical ideas or feelings in the mind and heart of its composer. Musical instruments are used to translate these ideas and feelings into sound waves. Somehow, the emotion contained in the music—the message—is coded in the vibrations of air molecules. Those sounds are converted to electricity and stored magnetically or optically. The composer's message manages to survive the trip through the mixing console and recorders; the signal is transferred to disc or computer files. Finally, the original sound waves are reproduced in the listening room, and miraculously the original emotion is reproduced in the listener as well.

Of course, not everyone reacts to a piece of music the same way, so the listener may not perceive the composer's intent. Still, it's amazing that anything as intangible as a thought or feeling can be conveyed by tiny magnetic patterns on a hard disk or by pits on a compact disc.

The point of music lies in what it's doing now, in the present. In other words, the meaning of "Doo wop she bop" is "Doo wop she bop." The meaning of an Am7 chord followed by an Fmaj7 chord is the experience of Am7 followed by Fmaj7.

INCREASING YOUR INVOLVEMENT IN MUSIC

Sometimes, to get involved in music, you must relax enough to lie back and listen. You have to feel unhurried, to be content to sit between your stereo speakers or wear headphones, and listen with undivided attention. Actively analyze or feel what the musicians are playing.

Music affects people much more when they are already feeling the emotion expressed in the song. For example, hearing a fast Irish reel when you're in a party mood, or hearing a piece by Debussy when you're feeling sensuous, is more moving because your feelings resonate with those in the music. When you're falling in love, any music that is meaningful to you is enhanced.

If you identify strongly with a particular song, that tells you something about yourself and your current mood. And the songs that other people identify with tell you something about them. You can understand individuals better by listening to their favorite music.

DIFFERENT WAYS OF LISTENING

There are so many levels on which to listen to music—so many ways to focus attention. Try this. Play one of your favorite records several times while listening for these different aspects:

- Overall mood and rhythm
- Lyrics
- Vocal technique
- Bass line
- Drum fills
- Sound quality
- Technical proficiency of musicians
- Musical arrangement or structure
- Reaction of one musician to another musician's playing
- Surprises versus predictable patterns

By listening to a piece of music from several perspectives, you'll get much more out of it than if you just hear it as background. There's a lot going on in any song that usually goes unnoticed. Sometimes when you play an old familiar record and listen to the lyrics for the first time, the whole meaning of the song changes for you.

Most people react to music on the basic level of mood and rhythmic motivation. But as a recording enthusiast, you hear much more detail because your focus demands sustained critical listening. The same is true of trained musicians focusing on the musical aspects of a performance.

It's all there for anyone to hear, but you must train yourself to hear selectively and to focus attention on a particular level of the multidimensional musical event. For example, instead of just feeling excited while listening to an impressive lead-guitar solo, listen to what the guitarist is actually playing. You may hear some amazing things.

Here's one secret of really involving yourself in recorded music: imagine yourself playing it! For example, if you're a bass player, listen to the bass line in a particular record, and imagine that you're playing the bass line. You'll hear the part as never before. Or respond to the music visually—see it as you do in the movie *Fantasia*.

Follow the melody line and see its shape. Hear where it reaches up, strains, and then relaxes. Hear how one note leads into the next. How does the musical expression change from moment to moment?

There are times when you can almost touch music: some music has a prickly texture (many transients, emphasized high frequencies); some music is soft and sinuous (sine-wave synthesizer notes, soaring vocal harmonies); and some music is airy and spacious (lots of reverberation).

WHY RECORD?

Recording is a service and a craft. Without it, people would be exposed to much less music. They would be limited to the occasional live concert or to their own live music, played once and forever gone.

With recordings, you can preserve a performance for thousands of listeners. You can hear an enormous variety of musical expressions whenever you want. Unlike a live concert, a record can be played over and over for analysis. Tapes or discs are also a way to achieve a sort of immortality. The Beatles may be gone, but their music lives on.

Records can even reveal your evolving consciousness as you grow and change. A computer audio file or CD stays the same physically, but you hear it differently over the years as your perception changes. Recordings are a constant against which you can measure change in yourself.

Be proud that you are contributing to the recording art—it is done in the service of music.

CHAPTER 2

The Recording Process

Welcome to the brave new world of 21st-century recording! The digital technologies of the past few years have given us possibilities undreamed of just 10 years ago. This book will show you an overview of current recording technology, help you choose the equipment that best suits your needs, and guide you in using it to create great recordings. And it will explain the technical jargon in plain English.

Thanks to the shift from analog to digital technology, the excitement and satisfaction of recording are accessible to more people than ever before. It used to take a whole roomful—or truckful—of expensive equipment to produce a good recording. But the new generation of smaller, cheaper gear means you may be able to tuck your studio into a corner of your bedroom or the backseat of your Toyota. As a result, many more people are involved in the process of recording—as musicians recording their own albums, or as engineers offering services to others.

As a recording engineer, you are a key player. Your skills help artists realize their visions in sound. Your miking techniques capture the vibrancy of the performance, whether it's the shimmering overtones of a string quartet or the sonic assault of an electric blues band. Your "post" work in the studio—adding effects, tweaking levels, etc.—will take the raw material of the performance and shape and blend it into a polished musical statement. As you master the technology and become fluent with the audio tools at hand, you will produce exciting recordings that will delight your clients and give you a real sense of pride and achievement.

Be sure to practice what you learn in this book. There's no substitute for hands-on experience. You might offer to record a band's rehearsal for free while you experiment and master the gear. Be patient, let yourself make mistakes, and above all, listen to how the sound changes when you move a mic or tweak a knob.

TYPES OF RECORDING

Let's get started. Currently there are six main ways to record music:

- Live stereo recording
- Live-mix-to-2-track recording
- Multitrack recorder and mixer
- Standalone digital audio workstation (recorder-mixer)
- Computer digital audio workstation (DAW)
- MIDI sequencing

Live Stereo Recording

This method uses a stereo mic technique to record with a stereo microphone or two microphones into a recorder. It is most commonly used to record an orchestra, symphonic band, pipe organ, small ensemble, quartet, or soloist. The microphones pick up the overall sound of the instruments and the concert-hall acoustics from several feet away. You might use this minimalist technique to record a folk group, rock group, or acoustic jazz group in a good-sounding room.

Figure 2.1 shows the stages of this method—the links in the recording chain. From left to right:

1. The musical instruments or voices make sound waves.
2. The sound waves travel through the air and bounce or reflect off the walls, ceiling, and floor of the concert hall. These reflections add a pleasing sense of spaciousness.
3. The sound waves from the instruments and the room reach the microphones, which convert the sound into electrical signals.
4. The sound quality is greatly affected by mic technique (microphone choice and placement).

FIGURE 2.1
The recording chain for live stereo recording.

5. The signals from the microphones go to a 2-track recorder. It may be a hard-drive recorder, CD-R burner, DVD-R burner, flash memory recorder, or computer hard drive. The signal changes to a pattern stored on a medium, such as magnetic patterns on a hard disk. During playback, the patterns on the medium are converted back into a signal.

During recording, signals are stored along a track—a path or channel on the medium containing a recorded signal. A single medium can record one or more tracks. For example, a 2-track hard-disk recording stores two tracks on hard disk, such as the two different audio signals required for stereo recording.

6. To hear the signal you're recording, you need a monitor system: headphones or a stereo power amplifier and loudspeakers. You use the monitors to judge how well your mic technique is working.

The speakers or headphones convert the signal back into sound. This sound resembles that of the original instruments. Also, the acoustics of the listening room affect the sound reaching the listener.

Live-Mix-to-2-Track Recording

This method is used mainly for live broadcasts or recordings of PA mixes and some orchestral recordings. A PA is a public address or sound reinforcement system. Using a mixer, you set up a mix of several microphones and record the mixer's output signal on a 2-track recorder (flash memory recorder or computer hard drive). Each mic is close to its sound source. Figure 2.2 shows the recording chain for this method.

Multitrack Recorder and Mixer

For this method, you record with several mics into a mixer, which is connected to a multitrack hard-disk recorder. You record the signal of each microphone on its own

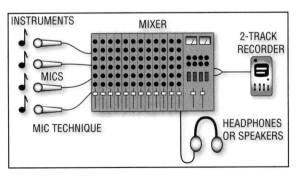

FIGURE 2.2
The recording chain for live-mix-to-2-track recording.

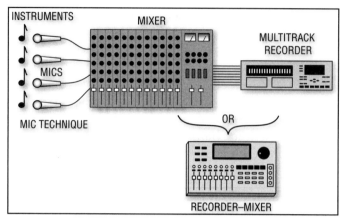

FIGURE 2.3
The recording chain for multitrack recording.

track, then mix these recorded signals after the performance is done. You can also record different groups of instruments on each track. Figure 2.3 shows the stages in this method; they are:

1. Place microphones near the instruments.
2. Plug the mics into a mixing console (a big, sophisticated mixer). During multi-track recording, the mixing console amplifies the weak microphone signals up to the level needed by the recorder. The console also sends each microphone signal to the desired track.
3. Record the amplified mic signals on the multitrack recorder.

You can record more instruments later on unused tracks—a process called overdubbing. Wearing headphones, the performer listens to the recorded tracks and plays or sings along with them. You record the performance on an unused track.

After the recording is done, you play all the tracks through the mixing console to mix them with a pleasing balance (Figure 2.4). Here are the steps:

1. Play back the multitrack recording of the song several times, adjusting the track volumes and tone controls until the mix is just the way you want it. You can add effects to enhance the sound quality. Some examples are echo, reverberation, and compression (explained in Chapter 9). Effects are made by signal processors that connect to your mixer, or by software applications (plug-ins) that are part of a recording program.
2. Record or export your final stereo mix onto your computer hard drive or to an external 2-track recorder such as the Tascam DV-RA1000HD.

FIGURE 2.4
The recording chain for a multitrack mixdown.

FIGURE 2.5
TASCAM DP-03, an example of a recorder-mixer.

Standalone Digital Audio Workstation (Recorder-Mixer)

This is a multitrack recorder and a mixer combined in one portable chassis (Figure 2.5). It's relatively easy to use. The recording medium is a hard drive or a flash memory card. Other names for this device include "digital multitracker," "personal digital studio," or "portable studio." Most recorder-mixers have built-in effects.

FIGURE 2.6
Computer with a choice of audio interface and recording/editing software.

Computer Digital Audio Workstation

This low-cost system includes a computer, recording software, and a sound card or audio interface that gets audio into and out of your computer (Figure 2.6). The computer's hard drive records the audio.

Using the recording software, you perform these operations:

1. Record music on the computer's hard drive.
2. Edit the tracks to fix mistakes, delete unwanted material, or copy/move song sections.
3. Mix the tracks with a mouse or controller by adjusting virtual controls that appear on your computer screen.

You might also assemble a song from samples or from loops. Samples are recordings of single notes of various instruments. Loops are repeating musical patterns.

MIDI Sequencing

With this recording method, a musician performs on a MIDI controller, such as a piano-style keyboard or drum pads. The controller puts out a MIDI signal, a series

FIGURE 2.7
MIDI sequencing system.

of numbers that indicates which keys were pressed and when they were pressed. The MIDI signal is recorded and stored by a standalone sequencer or sequencer program in a computer. When you play back the MIDI sequence, it plays the tone generators in a synthesizer or sound module. The synthesizer can be hardware or software (as a "soft synth," also called a "virtual instrument"). A MIDI sequencer also can play samples (digital recordings of musical notes played by real instruments). Like a player piano, MIDI sequencing records your performance gestures rather than audio. Figure 2.7 shows the process and Chapter 16 gives a detailed explanation of MIDI.

MIDI/digital-audio software lets you record MIDI sequences and digital audio onto a hard disk. First record a few tracks of MIDI sequences, then add audio tracks: lead vocal, sax solo, or whatever. All these elements stay synchronized.

PROS AND CONS OF EACH METHOD

Live stereo recording is simple, cheap, and fast. But it usually sounds too muddy with rock music, and you must adjust balances by moving musicians. It can work well with classical music, and sometimes with folk or acoustic jazz music.

Live-to-2-track recording is fairly simple and quick. However, loud instruments might sound distant in the recording because their sound "leaks" into other instruments' mics. You can't correct mistakes in the mix or performance unless you rerecord. Also, the live sound of the band can make it hard to hear the monitored sound clearly.

Multitrack recording has many advantages. You can punch-in (fix a musical mistake on a track by recording a new, correct part over the mistake). You can overdub (record one instrument at a time). This reduces leakage and gives a tighter sound. Also, you can postpone mixing decisions until after the performance. Then you can monitor the mix in quiet surroundings. This method is more complex and expensive than live-mix recording.

If you use a separate multitrack recorder and mixer, each component can be used independently. For example, you might do a PA job with just the mixer. Or, if you already have a mixer, all you need to buy is a recorder. This system is a little difficult to set up because you need to connect cables between the mixer and recorder, and between the mixer and outboard effects units.

A standalone DAW (recorder-mixer) is easy to use because it is a single portable chassis that includes most of your studio equipment: recorder, mixer, effects, and often a CD burner. It doesn't require cables except for the mics, instruments, and monitor speakers. High-end units let you edit the music. They also have automated mixing: memory chips in the mixer remember your mixdown settings, and reset the mixer accordingly the next time you play back the recording.

A computer DAW is inexpensive, powerful, and flexible. It lets you do sophisticated editing and automated mixing. Several plug-in (software) effects are included, and you can purchase and install other plug-ins. Recording software can be updated at little cost. As for drawbacks, computers can crash and can be difficult to set up and optimize for audio work.

MIDI sequencing lets you record musical parts by entering notes slowly or one-at-a-time if you wish. After the sequencing is finished, you can edit notes to correct mistakes. You can even change the instrument sounds or the tempo. A huge variety of sounds are available in synthesizers, sound modules, and soft synths. However, you are limited to their sounds unless you use MIDI/digital-audio software, which lets you add miked instruments to the mix.

Play audio clips 1 and 2 at www.elsevierdirect.com/companions/9780240811444. Clip 2 demonstrates:

- Live stereo recording—orchestra
- Live stereo recording—rock group
- Live-to-2-track recording— jazz group
- Multitrack recording—pop group
- MIDI sequencing—synthesizer funk

RECORDING THE MIXES

No matter which recording method you use, eventually you'll mix each song and record the mix on a 2-track recorder, or record the mix as a stereo wave file on your hard drive. You can convert the wave file to MP3 or WMA format for uploading to the web (explained in Chapter 19).

You might want to assemble an album of your recorded mixes. You remove noises and count-offs between songs, put the songs in the desired order, and put a few seconds of silence between songs. This is done with a computer and editing software (a DAW, Figure 2.6). The last step is to copy the album to a blank CD. There's your final product, ready to duplicate or replicate.

No matter what type of recording you do, each stage contributes to the sound quality of the finished recording. A bad-sounding CD can be caused by any weak link: low-quality microphones, ineffective mic placement or mixer settings, and so on. You can strive for quality recordings by optimizing each stage. This book will help you reach that goal.

CHAPTER 3

Sound, Signals, and Studio Acoustics

When you make a recording, you deal with at least two kinds of invisible energy: sound waves and electrical signals. For example, a microphone converts sound into a signal. A signal is a varying voltage that carries information. In our case, it's musical information.

This chapter covers some characteristics of sound and audio signals. These facts will help you work with room acoustics, and will help you know what's going on inside your mixer as you adjust the controls. With this knowledge you can make better recordings.

SOUND WAVE CREATION

To produce sound, most musical instruments vibrate against air molecules, which pick up the vibration and pass it along as sound waves. When these vibrations strike your ears, you hear sound.

To illustrate how sound waves are created, imagine a vibrating speaker cone in a guitar amp. When the cone moves out, it pushes the adjacent air molecules closer together. This forms a compression. When the cone moves in, it pulls the molecules farther apart, forming a rarefaction. As shown in Figure 3.1, the compressions have a higher pressure than normal atmospheric pressure; the rarefactions have a lower pressure than normal.

These disturbances pass from one molecule to the next in a springlike motion—each molecule vibrates back and forth to pass the wave along. The sound waves travel outward from the sound source at 1130 feet per second (344 meters per second), which is the speed of sound in air at room temperature.

At some receiving point, such as an ear or a microphone, the air pressure varies up and down as the disturbances pass by. Figure 3.2 is a graph showing how sound pressure

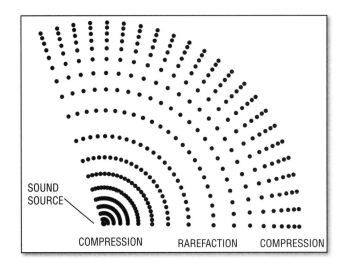

FIGURE 3.1
A sound wave.

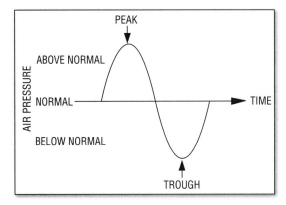

FIGURE 3.2
Sound pressure vs. time of one cycle of a sound wave.

varies with time, like a wave. The high point of the graph is called a peak; the low point is called a trough. The horizontal center line of the graph is normal atmospheric pressure.

Sound waves tend to spread out as they travel away from the source. The compressions and rarefactions move out as expanding spheres. As the spherical waves expand, the sound pressure spreads over a larger area, so the pressure becomes weaker with distance from the source. That means, the farther you are from a sound source, the quieter is its sound. Specifically, the sound pressure halves (drops 6 decibels [dB]) each time the distance from the source doubles. This phenomenon is called the inverse square law.

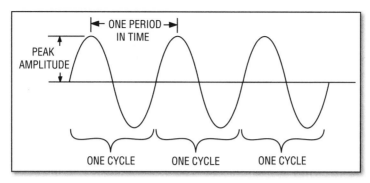

FIGURE 3.3
Three cycles of a wave.

CHARACTERISTICS OF SOUND WAVES

Figure 3.3 shows three waves in succession. One complete vibration from normal to high to low pressure and back to the starting point is called one cycle. The time it takes to complete one cycle—from the peak of one wave to the next—is called the period of the wave. One cycle is one period long.

Amplitude

The height of the wave is its amplitude. Loud sounds have high amplitudes (big pressure changes); quiet sounds have low amplitudes (small pressure changes). *Play audio clip 3 at www.elsevierdirect.com/companions/9780240811444 to hear an example.*

Frequency

The sound source (in this case, the guitar-amp loudspeaker) vibrates back and forth many times a second. The number of cycles completed in one second is called frequency. The faster the speaker vibrates, the higher the frequency of the sound. Frequency is measured in hertz (Hz), which means cycles per second. One thousand hertz is called "one kilohertz," abbreviated kHz.

The higher the frequency, the higher the perceived pitch of the sound. Low-frequency tones have a low pitch (like low E on a bass, which is 41 Hz). High-frequency tones have a high pitch (like four octaves above middle-C, or 4186 Hz). *Audio clip 4 at www. elsevierdirect.com/companions/9780240811444 illustrates this.* Doubling the frequency raises the pitch one octave.

Children can hear frequencies from 20 Hz to 20 kHz, and most adults with good hearing can hear up to 15 kHz or higher. Each musical instrument produces a range of frequencies, say, 41 Hz to 9 kHz for a string bass, or 196 Hz to 15 kHz for a violin.

Wavelength

When a sound wave travels through the air, the physical distance from one peak (compression) to the next is called a wavelength (refer to Figure 3.1). Low-pitched sounds have long wavelengths (several feet); high-pitched sounds have short wavelengths (a few inches or less). Wavelength is the speed of sound divided by frequency. So the wavelength of a 1000-Hz wave is 1.13 feet (0.344 m), 100 Hz is 11.3 feet (3.44 m), and 10 kHz is 1.35 inches (3.45 cm).

Phase and Phase Shift

The phase of any point on the wave is its degree of progression in the cycle—the beginning, the peak, the trough, or anywhere between. Phase is measured in degrees, with 360 degrees being one complete cycle. The beginning of a wave is 0 degrees; the peak is 90 degrees (one-quarter cycle), and the end is 360 degrees. Figure 3.4 shows the phases of various points on the wave.

If there are two identical waves traveling together, but one is delayed with respect to the other, there is a phase shift between the two waves. The more delay, the more phase shift. Phase shift is measured in degrees. Figure 3.5 shows two waves separated by 90 degrees (one-quarter of a cycle) of phase shift. The dashed wave lags the solid wave by 90 degrees.

If you combine two identical sound waves, such as a sound and its reflection off a wall, the peaks of the two waves add together at certain points in the room. This doubles the sound pressure or amplitude, creating areas of louder sound at certain frequencies.

Phase Interference

When there is a 180-degree phase shift between two identical waves, the peak of one wave coincides with the trough of another (Figure 3.6). If these two waves are combined, they cancel out. This phenomenon is called phase cancellation or interference.

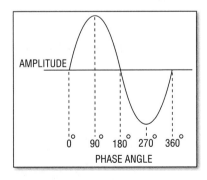

FIGURE 3.4
The phases of various points on a wave.

Suppose you have a signal with a wide range of frequencies, such as the singing voice. If you delay this signal and combine it with the original undelayed signal, some frequencies will be 180 degrees out of phase and will cancel. This creates a hollow, filtered tone quality.

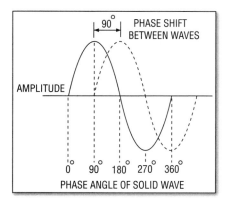

FIGURE 3.5
Two waves that are 90 degrees out of phase.

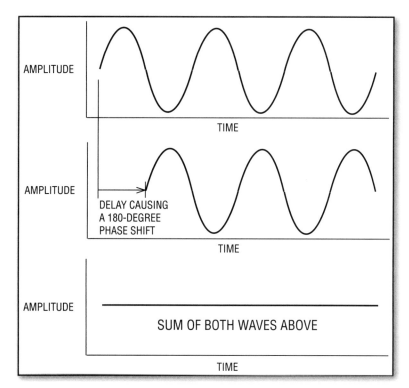

FIGURE 3.6
Phase interference: adding two waves that are out of phase cancels the sound at that frequency.

Here's an example of how this can happen. Suppose you're recording a singer/guitarist with one mic near the singer and another mic near the guitar. Both mics pick up the singer. The singer's mic is close to the mouth, and you hear it with no delay in the signal. The guitar mic is farther from the mouth, so its voice signal is delayed. When you mix the two mics, you often hear a colored tone quality caused by phase cancellations between the two mics.

Suppose you're recording a stage play with a mic on a short stand on the floor. The mic picks up the direct sound from the actors, but it also picks up delayed reflections off the floor. Direct and delayed sounds combine at the mic, causing phase cancellations. You hear it as a hollow, filtered sound that changes when the actor walks while talking.

Harmonics

The type of wave shown in Figure 3.2 is called a sine wave. It is a pure tone of a single frequency, like a signal from a tone generator. In contrast, most musical tones have a complex waveform, which has more than one frequency component. All sounds are combinations of sine waves of different frequencies and amplitudes. Figure 3.7 shows sine waves of three frequencies combined to form a complex wave.

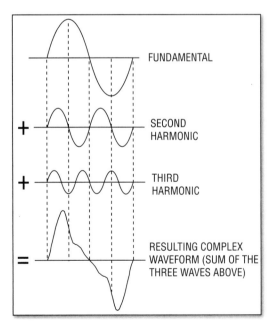

FIGURE 3.7
Adding fundamental and harmonic waveforms to form a complex waveform.

The lowest frequency in a complex wave is called the fundamental frequency. It determines the pitch of the sound. Higher frequencies in the complex wave are called overtones or upper partials. If the overtones are multiples of the fundamental frequency, they are called harmonics. For example, if the fundamental frequency is 200 Hz, the second harmonic is 400 Hz, and the third harmonic is 600 Hz.

Harmonics and their amplitudes partly determine the tone quality or timbre of a sound. They help you identify the sound as a drum, piano, organ, voice, etc. *Play audio clip 5 at www.elsevierdirect.com/companions/9780240811444.* In general, instruments with few or weak harmonics—such as a flute—tend to sound pure and smooth. Instruments with many or strong harmonics—such as a trumpet or distorted guitar—tend to sound bright and edgy.

As explained in Chapter 9, equalization can change the tonal balance of a recorded instrument by boosting or cutting its harmonics and fundamental frequencies. Boosting the fundamentals tends to make the sound warm, while cutting the fundamentals makes the sound thin; boosting the harmonics makes the sound bright, defined, or trebly, while cutting the harmonics makes the sound dark or muffled.

Playing an instrument louder usually increases its harmonics. So a piano played loudly sounds brighter than the same piano played softly.

Noise (such as tape hiss) contains a wide band of frequencies and has an irregular, nonrepeating waveform. Some musical sounds with a hissy or noise-like character are a cymbal, snare drum, or a singer's "s" sounds (sibilance).

Envelope

Another characteristic that identifies a sound is its envelope. When a note sounds, it doesn't last forever—it rises in volume, lasts a short time, then falls back to silence. This rise and fall in volume of one note is called the note's envelope. The envelope connects the peaks of successive waves that make up a note. Each musical instrument has a different envelope.

Most envelopes have four sections: attack, decay, sustain, and release (see Figure 3.8). During the attack, a note rises from silence to its maximum volume. Then it decays from maximum to some midrange level. This middle level is the sustain portion. During release, the note falls from its sustain level back to silence.

Percussive sounds, such as drum hits, are so short that they have only a rapid attack and decay. Other sounds, such as organ or violin notes, last longer. They have slower attacks and longer sustains. Guitar plucks and cymbal crashes have quick attacks and slow releases. They hit hard then fade out slowly. *Play audio clip 6 at www.elsevierdirect. com/companions/9780240811444.*

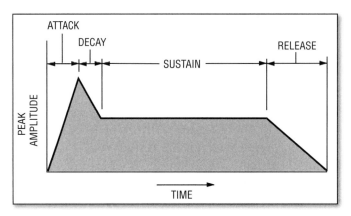

FIGURE 3.8
The four sections of the envelope of a note.

You can shorten a guitar string's decay or ringing by damping the string with the side of your hand. You can press a blanket against a kick drum head to damp the decay and get a tighter sound.

Harmonics usually change during a note's envelope. For example, if an instrument has a percussive attack—like a guitar pluck or tom hit—the harmonics are strongest at the attack then become weaker during the decay.

BEHAVIOR OF SOUND IN ROOMS

Because most music is recorded in rooms, you need to understand how room surfaces affect sound.

Echoes

Musical instruments generate sound waves that travel outward in all directions. Some of the sound travels directly to your ears (or to a microphone) and is called direct sound. The rest strikes the walls, ceiling, floor, and furnishings of the recording room. At those surfaces, some of the sound is absorbed, some is transmitted through the surface, and the rest is reflected back into the room.

Because sound waves take time to travel (moving at about 1 foot per millisecond), the reflected sound reaches you after the direct sound. The reflection repeats the original sound after a short delay. If the sound is delayed about 50 milliseconds or more, we call it an echo (Figure 3.9). In some concert halls we hear single echoes; in small rooms we often hear a short, rapid succession of echoes called flutter echoes. You can detect them by clapping your hands next to a wall. Flutter echoes happen when sound bounces back and forth between two parallel walls.

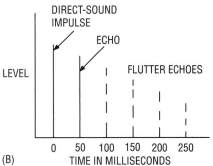

FIGURE 3.9
Echoes. (A) Echo formation. (B) Intensity vs. time of direct sound and its echoes.

Reverberation

Sound reflects many times from all the surfaces in the room. These reflections sustain the sound of each note the musician plays. This persistence of sound in a room after the original sound has stopped is called reverberation (reverb). For example, reverberation is the sound you hear just after you shout in an empty gymnasium. The sound of your shout stays in the room and gradually dies away (decays). *Play audio clip 10 at www.elsevierdirect.com/companions/9780240811444 to hear echoes and reverberation.*

Reverb is hundreds of echoes that gradually get quieter. The echoes follow each other so rapidly that they merge into a single continuous sound. Eventually, the room surfaces completely absorb the echoes. The timing of the echoes is random, and the echoes increase in number as they decay. Figure 3.10 shows how reverberation develops in a recording room.

Reverberation is a continuous fade-out of sound ("HELLO-O-O-o-o"), while an echo is a discrete repetition of a sound ("HELLO hello hello hello").

Reverberation time (RT60) is the time it takes for reverb to decay 60 dB. Too long a reverb time makes a recording sound distant, muddy, and washed-out. That's why

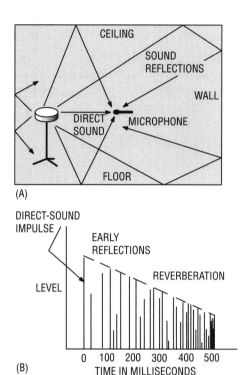

FIGURE 3.10
Reverberation. (A) Multiple sound reflections create reverberation. (B) Intensity vs. time of direct sound, early reflections, and reverberation.

pop-music recordings are usually made in a fairly "dead" or nonreverberant studio, which has an RT60 of about 0.4 second or less. In contrast, classical music is recorded in "live," reverberant concert halls (RT60 about 1 to 3 seconds) because we want to hear reverb with classical music—it's part of the sound.

Reverberation comes to you from every direction because it is a pattern of many sound reflections off the walls, ceiling, and floor. Because we can tell where sounds come from, we can distinguish between the direct sound of an instrument coming to us from a single location and the reverberation coming to us from everywhere else. So we can ignore the reverb and concentrate on the sound source. In fact, we normally are not aware of reverberation.

But suppose you put a mic next to your ears, record an instrument in that room, and play back the recording. You'll hear a lot more reverb than you heard live. What's going on? The reverb you recorded and played back is not all around you. Instead, it's all up front between the speakers. So it's more audible; you can't discriminate against the reverb spatially. To reduce the amount of reverb in your recordings, you need to place mics close to instruments, and maybe add some sound-absorbing materials to the room.

Diffusion

When sound waves strike and reflect off a very bumpy or convex surface, they spread out or diffuse. This diffusion can be used to weaken sound reflections. Sound waves also spread out when they travel through a small opening.

HOW TO TAME ECHOES AND REVERB

Echoes and reverb can make your recordings sound mushy and distant. There are two ways to prevent these problems: with recording techniques and with acoustic treatment.

Controlling Room Problems with Recording Techniques

Sometimes you can make clean recordings in an ordinary room, such as a club, living room, or basement, if you follow these suggestions:

- Mike close. Place each mic 1 to 6 inches from each instrument or voice. Then the mics will pick up more of the instruments and less of the room reflections. You might want to use mini mics, which attach directly to instruments.
- Use directional mics—cardioid, supercardioid, or hypercardioid—which reject room acoustics.
- Record bass guitar and synth directly with a guitar cord or a direct box. By eliminating the microphone, you remove room acoustics from the recording. To get a good sound when recording electric guitar direct, record off the effects boxes or use a guitar-amp simulator.

See also the "Leakage" section in Chapter 7 for tips on reducing leakage.

Controlling Room Problems with Acoustic Treatments

When should you add acoustic treatments?

- You clap your hands next to a wall and you hear flutter echoes (a fluttering sound). These are caused by sounds bouncing back and forth between hard parallel walls.
- Your studio is a very live environment, such as a garage or concrete-block basement, so you hear too much room reverberation.
- Your studio is very small.
- You hear outside noises in your recordings.
- Bass-guitar amps and monitor speakers sound boomy.
- You want the freedom to mike several feet away without picking up noise or excess room reverb.
- You hear a lot of leakage in the mic signals. Examples of leakage are drums picked up by the guitar mic, or electric guitar picked up by the cymbal mics.

FIGURE 3.11
Acoustic treatments.

If those conditions apply, upgrade the acoustics of your studio. Here are some suggestions.

Reverb and echoes are caused by sound reflections off room surfaces. So any surface that is highly sound-absorbent helps to reduce those problems.

To absorb high frequencies, use porous materials such as convoluted (bumpy) acoustic foam. Foam that is not coated with flame-retardant treatment can be flammable. Natfire.com or flamestop.com offer treatments. Nail or glue the acoustic foam to the walls, or mount it on frames. Thick foam works better than thin. Four-inch foam on the wall absorbs frequencies from about 200 to 800 Hz and up, depending on the angle at which sound strikes the foam. Leave some space between the foam panels (Figure 3.11); this helps to diffuse or spread out the sound in the room. Don't overdo the foam padding. A stuffed, dead room is uncomfortable to play in. Keep some reflections because they add "air" and liveliness to the sound.

Other high-frequency absorbers are sleeping bags, moving blankets, carpeting, curtains, and dense fiberglass insulation covered with muslin or burlap. If possible, space these materials several inches from the wall. The spacing helps them absorb mid-bass frequencies. A wide-range absorber is 4-inch pressed fiberglass board (Owens-Corning Type 703, $3\,\text{lb/ft}^3$) covered in muslin or burlap.

Start with just a little absorption behind or above the musician you're recording. Add more absorbers a few at a time, until your recordings sound as dead as you wish—typically covering 50-60% of the surfaces. Place absorption where you see mirror images of the monitors from the mix position. Put an absorber on the wall behind the monitors, and also hang an absorbent panel centered halfway between the mix position and monitors, suspended on hooks and wire.

Another absorber is a stand-mounted acoustic panel placed near a microphone. Examples are the sE Reflexion Filter, the Auralex MudGuard, and the RealTraps Portable Vocal Booth.

To absorb low frequencies, you can make bass traps. Here are four types:

1. Resonant tube trap: Take a 35- to 55-gallon rubber trashcan, stuff it with fiberglass insulation (wear a dust mask and gloves), and cover the open end with muslin or canvas. This tube trap absorbs frequencies near 1130/2 H, where H is the height of the trashcan in feet. For example, a 3-foot-high trashcan absorbs 188 Hz. Placement is not critical.
2. Frictional tube trap: Make a 2-foot-diameter canvas bag, 8 feet tall, and fill it with fiberglass insulation. Hang one bag a few feet out from each room corner (Figure 3.11). The distance should be 1130/4 F, where F is the frequency in Hz that you want to absorb. For example, to absorb 80 Hz, hang the bag 3.5 feet out from a room corner. (Thanks to David Moulton for these ideas.)
3. Lattice: Build a 3-foot-wide flat lattice frame of wood slats, the same height as the ceiling (Figure 3.11). Cover the frame with muslin or burlap. Put the frame diagonally across a room corner and fill the corner with R-30 insulation. Leave the foil on, foil side out toward the room. Put one assembly in each corner. (Thanks to Chips Davis for this idea.)
4. Insulation panel: Get 8 pieces of rigid fiberglass insulation, type 705, 2 × 4 feet and 4 inches thick, from an insulation supplier. Cover each piece with muslin or burlap to contain the fibers. Place a piece across each room corner with the 2-foot edge touching the floor. Stack two panels to make them 8 feet high. (Thanks to Ethan Winer for this idea.)

There are other ways to absorb bass. Wood paneling works well. It also helps to open closet doors and place couches and books a few inches from the walls. In a basement studio, nail acoustic tile to the ceiling joists, with fiberglass insulation in the air space between the tiles and ceiling.

You may not need any bass traps if you don't put any bass into the room. For example, don't turn up the bass guitar amp—just record the bass direct and have the musicians wear headphones to hear the bass.

Controlling Standing Waves

Let's look at another acoustic problem: standing waves. If you play an amplified bass guitar through a speaker in a room, and do a bass run up the scale, you may hear some

Table 3.1	Room dimensions in feet to reduce standing waves	
Height	**Width**	**Length**
8	9.1	11.1
8	9.4	11.8
8	10.1	11.3
8	10.2	12.3
8	11.6	16.8
8	11.8	13.6
8	12.8	18.6
8	13.0	21.0
10	11.4	13.9
10	11.7	14.7
10	12.6	14.1
10	12.8	15.4
10	14.5	21.0
10	14.7	17.0
10	16.0	23.3
10	16.2	26.2

notes that stand out in the room. The room is resonating at those frequencies. These resonance frequencies, which are strongest below 300 Hz, occur in patterns called "standing waves." They can give a tubby or boomy coloration to musical instruments and monitor speakers.

Room resonances are worst in a cubical room. They are less of a problem if the room's length, width, and height are not multiples of each other. Table 3.1 shows several room dimensions in feet that reduce boomy-sounding standing waves.

For example, if the room width is 9.1 feet, and the ceiling is 8 feet high, the length should be 11.1 feet for best reduction of boominess.

Try to record in a large room because the room resonance frequencies are likely to be below the musical range. Use bass traps to absorb room resonances. Contrary to popular opinion, nonparallel walls don't prevent standing waves.

MAKING A QUIETER STUDIO

Ideally the noise level in a studio should be 23–28 dB A-weighted as measured on an SPL meter. The following tips will keep noises out of your recordings:

FIGURE 3.12
Staggered-stud construction reduces noise transmission.

- Consider having the studio in a basement, because the earth blocks noises from the outside. The furnace or air conditioning might be a problem, however.
- Turn off appliances, air conditioning, and telephones while recording.
- Schedule your recordings when road and air traffic is light, or pause for ambulances and airplanes to pass.
- Close windows. Consider covering them with thick plywood.
- Close doors and seal with towels.
- Remove small objects that can rattle or buzz.
- Weather-strip doors all around, including underneath.
- Replace hollow doors with solid doors.
- Block openings in the room with thick plywood and caulking.
- Put several layers of plywood and carpet on the floor above the studio, and put insulation in the air space between the studio ceiling and the floor above.
- Place microphones close to instruments and use directional microphones. This won't reduce noise in the studio, but it will reduce noise picked up by the microphones.

When building a new studio, you might want to make the walls of plastered concrete block because massive walls reduce sound transmission, or make the walls of gypsum board and staggered studs. Nail gypsum board to 2 × 4 staggered studs on 2 × 6 footers as shown in Figure 3.12. Staggering the studs prevents sound transmission through the studs. Fill the air space between walls with insulation.

The ideal home-recording room for pop music is a large, well-sealed room with optimum dimensions. This room is in a quiet neighborhood. It should have some soft surfaces (carpet, acoustic-tile ceiling, drapes, couches), and some hard vibrating surfaces (wood paneling or gypsum board walls on studs).

Your home studio may not need acoustic treatment. Do some trial recordings to find out. But if your room could stand some improvement, the suggestions above should point you in the right direction.

For better results and a more professional appearance, consider buying some acoustic treatments from the following companies. Their websites are atsacoustics.com, realtraps.com, acousticsciences.com, acousticalsolutions.com, primacoustic.com, auralex.com, acousticsfirst.com, wallmate.net, pinta-acoustic.com, and rpginc.com. Atsacoustics.com has a free online room acoustics analysis, auralex.com and atsacoustics. com offer low-cost absorbers, and primacoustic.com offers acoustic "London Kits" for rooms of various sizes.

SIGNAL CHARACTERISTICS OF AUDIO DEVICES

When a microphone converts sound to electricity, this electricity is called the signal. It has the same frequencies and the same amplitude changes as the incoming sound wave.

When this signal passes through an audio device, the device may alter the signal. It might change the level of some frequencies, or add unwanted sounds that are not in the original signal. Let's look at some of these effects.

Frequency Response

Suppose you have an audio device—a mic, mixer, effects unit, recorder, or speaker. You send a musical signal through the device. Usually the music contains some high and some low frequencies.

The device might respond differently to different frequencies. It might amplify the low notes and weaken the high notes. You can graph how the device responds to different frequencies by plotting its output level versus frequency. This graph is called a frequency response (Figure 3.13). The level in the graph is measured in dB, while frequency is measured in Hz. Generally, 1 dB is the smallest change in loudness that we can hear.

Suppose the level is the same at all frequencies. Then the graph forms a horizontal straight line and is called a "flat frequency response" (Figure 3.14). All the frequencies are reproduced at an equal level. In other words, the device passes all the frequencies

FIGURE 3.13
An example of a frequency response.

without changing their relative levels. You get out the same amount of bass and treble that went in. A flat response does not affect the tonal balance of the incoming sound.

Many audio devices do not have a flat response across the audio band from 20 to 20,000 Hz. They have a limited range of frequencies that can be reproduced at an equal level (within a tolerance, such as ±3 dB). In Figure 3.14, the frequency response shown by the solid line is 50 to 12,000 Hz ±3 dB. That means the audio device passes all frequencies from 50 to 12,000 Hz at a nearly equal level—within 3 dB. It reproduces low sounds and high sounds equally well. The response is down 3 dB at 50 and 12,000 Hz, and is up 3 dB at 5000 Hz.

Usually, the more extended or "wide" the frequency range is, the more natural and real the recording sounds. A wide, flat response gives accurate reproduction. A frequency response of 200 to 8000 Hz (±3 dB) is narrow (poor fidelity); 80 to 12,000 Hz is wider (better fidelity), and 20 to 20,000 Hz is widest (best fidelity). *Play audio clip 7 at www. elsevierdirect.com/companions/9780240811444.*

Also, the flatter the frequency response, the greater the fidelity or accuracy. A response deviation of ±3 dB is good, ±2 dB is better, and ±1 dB is excellent. There are exceptions to this statement, which we'll look at in Chapter 9 in the section on equalizers.

When you turn a bass or treble knob on your guitar amp, mixer EQ, or stereo, you're changing the frequency response. If you turn up the bass, the low frequencies rise in level. If you turn up the treble, the high frequencies are emphasized. The ear interprets these effects as changes in tone quality—warmer, brighter, thinner, duller, and so on.

Figure 3.13 shows a nonflat frequency response. Toward the right side of this line, the response at high frequencies "rolls off" or declines. This shows that the upper harmonics are weak. The result is a dull sound. Toward the left side, the response at low frequencies rolls off. This means the fundamentals are weakened and the result is a thin sound.

The frequency response of an audio device might be made nonflat on purpose. For example, you might cut low frequencies with an equalizer to reduce breath pops from

FIGURE 3.14
Flat frequency response.

a microphone. Also, a microphone may sound best with a nonflat response, such as boosted high frequencies that add presence and sizzle.

Noise

Noise is another characteristic of audio signals. Every audio component produces a little noise—a rushing sound like wind in trees. Noise in a recording is undesirable unless it's part of the music.

You can make noise less audible by keeping the signal level in a device relatively high. If the level is low, you have to turn up the listening volume in order to hear the signal well. Turning up the volume of the signal also turns up the volume of the noise, so you hear noise along with the signal. But if the signal level is high, you don't have to turn up the listening level as much. Then the noise remains in the background.

Distortion

If you turn up the signal level too high, the signal distorts and you hear a gritty, grainy sound or clicks. This type of distortion is sometimes called "clipping" because the peaks of the signal are clipped off so they are flattened. To hear distortion, simply record a signal at a very high recording level (with the meters going well into the red area) and play it back. *Play audio clip 9 at www.elsevierdirect.com/ companions/9780240811444.* Digital recorders also produce "quantization" distortion at very low signal levels.

Optimum Signal Level

You want the signal level high enough to cover up the noise, but low enough to avoid distortion. Every audio component works best at a certain optimum signal level, and this is usually indicated on a level meter built into the device. This is discussed in more detail in Appendix A on the decibel.

Figure 3.15 shows the range of signal levels in an audio device. At the bottom is the noise floor of the device—the level of noise it produces with no signal. At the top is the distortion level—the point at which the signal distorts and sounds grungy. In between is a range in which the signal sounds clean. The idea is to maintain the signal around normal operation level on the average. Generally you want the signal level to be as high as possible without clipping or distorting.

With digital recorders (such as computer recording software), "0 dBFS" (0 decibels full scale) on the meter is the maximum undistorted signal. In that case, an average level of -20 dBFS and a peak level of -6 dBFS is the normal operation level. In a digital audio file that is used to master a CD, an average level of -14 dBFS is a good goal to aim for. There's more on digital signal levels in Chapter 13, Digital Recording.

FIGURE 3.15
The range of signal levels in an audio device.

Signal-to-Noise Ratio

The level difference in decibels between the signal level and the noise floor is called the "signal-to-noise ratio" or S/N (see Figure 3.15). The higher the S/N, the cleaner the sound. An S/N of 60 dB is fair, 70 dB is good, and 80 dB or greater is excellent. *Play audio clip 8 at www.elsevierdirect.com/companions/9780240811444.*

To illustrate S/N, imagine a person yelling a message over the sound of a train. The message being yelled is the signal; the noise is the train. The louder the message, or the quieter the train, the greater the S/N. And the greater the S/N, the clearer the message.

Headroom

The level difference in dB between the normal signal level and the distortion level is called "headroom" (see Figure 3.15). The greater the headroom, the greater the signal level the device can pass without running into distortion. If an audio device has a lot of headroom, it can pass high-level peaks without clipping them.

You want to set your mixer controls so that the signal has some headroom, is well above the noise floor, and is below distortion. Tips on doing that are in Chapter 13, Digital Recording, under the heading "Digital Recording Level." Also see Chapter 11, Mixer Operation, under the headings "Set Recording Levels" and "Set Levels."

With these settings, the signal levels in the mixer should be just about right, with no audible noise or distortion. And the mixer should have enough headroom so that loud peaks won't distort.

CHAPTER 4

Equipping Your Studio

You want to set up a recording system that's affordable, easy to use, and sonically excellent. With today's wide array of user-friendly sound tools, you can do just that. This chapter is a guide to equipment for a recording studio: what it does, what it costs, and how to set it up.

In this chapter we'll examine:

- Equipment
- Cables and connectors
- How to hook up your studio gear
- Preventing hum and radio frequency interference

What is the bare-bones equipment you need to crank out a decent demo CD? How much does it cost? Thanks to the new breed of affordable equipment, you can put together a complete home recording studio for as little as $1200. That includes powered speakers, two mics and mic stands, recording software, an audio interface, headphones, and cables.

EQUIPMENT

Let's examine each piece of equipment in a recording studio. You'll need some sort of recording device, headphones, cables, mics, direct box, monitor speakers, audio interface, and effects.

Recording Device

Six types of recording devices to choose from are a 2-track recorder, an iOS recording system, a recorder-mixer, a separate multitrack recorder and mixer, a computer, and a keyboard workstation. We'll look at each one.

PORTABLE 2-TRACK RECORDER

This device records in stereo on two tracks, either with two external mics or two built-in mics. Prices are $200 to $1800. Three types of 2-track recorder are:

- Handheld flash-memory recorder (Figure 4.1)
- Apple iOS device with recording software
- Laptop computer with recording software

A 2-track recorder works well for on-location stereo recording of classical music: an orchestra, symphonic band, string quartet, pipe organ, or soloist. Connect two high-quality mics and record in stereo as described in Chapter 7.

You can also use a 2-track recorder to capture practice sessions or gigs of pop, folk, or jazz groups. Recordings made this way sound sort of distant, as if you are listening from out in the audience. They might have background noise and not the best balance. But the recording process is very simple: just place the recorder, set a level, and hit Record.

Some handheld recorders are the Sony PCM-D50 and PCM-M10; Alesis PalmTrack; Zoom H4n; H2N, and H1; Tascam DR-07mkII, DR-05, DR-100, and DR-08; Yamaha

FIGURE 4.1
Roland R-09HR, an example of a portable 2-track recorder.

PocketTrak; Korg MR-2 and SOS; Roland R05 and R26; Marantz PMD620 and PMD661; Edirol (Roland) R-09HR (Figure 4.1); M-Audio MicroTrack II 24/96; and Olympus LS-7 and LS-100.

iOS RECORDING SYSTEM

Apple's iOS devices including the iPad have loads of recording app options. On the iPad, simply tap the App Store Icon and search for recording apps. Check out GarageBand, MultiTrack DAW by Harmonic Dog, Sonoma Wire Works' StudioTrack (Figure 4.2), StudioMini XL, Wizdom Music Morphwiz, and n-Track Studio (http://ntrack.com). On the iPhone and iPod touch, be sure to check out Audiofile Engineering's FiRe.

Also available are audio and audio/MIDI interfaces that plug into iOS devices. An example is the Tascam iU2. The Alesis iO Dock acts as an interface for an iPad, providing multiple I/O options.

Some stereo microphones are designed to plug into an iPad or other iOS devices. Three examples are the IK Multimedia iRig Mic, Blue Mikey Digital, and Tascam iM2.

More details on 2-track recorders are in Chapter 13 on digital recording.

MINI STUDIO

Also called a portable studio, notebook studio, personal studio, or pocket studio, this piece of gear combines a 4-track or 8-track recorder with a mixer—all in one portable

FIGURE 4.2
Sonoma Wire Works StudioTrack app for the Apple iPad.

FIGURE 4.3
Fostex R16, an example of a mini studio.

chassis (Figure 4.3). A similar option is an Apple iPad with recording software, mentioned earlier.

A mini studio records in MP3 or WAV format to a flash memory card. It lets you record several instruments and vocals, then mix them to stereo. That is, you "bounce" the tracks to a stereo file in the memory card. You can copy the stereo mix file to your computer via USB (Universal Serial Bus, an industry standard that defines certain cables and communications protocols for high-speed data transfer). Pocket studios sound good enough to make demos, or to use as a musical scratchpad for ideas, but not to release commercial albums. Some examples are the Boss BR-800 and Micro BR, TASCAM DP-004, and Zoom R16. Costing about $200 to $599, a mini studio can be a good choice for beginning recordists.

Some features include

- An internal MIDI sound module (synthesizer) that plays back MIDI sequences and includes MIDI files or rhythm patterns to jam with.
- Built-in effects.
- Built-in mic (in some models).
- Autopunch (automatically "punch" into and out of record mode to correct mistakes).
- Battery or AC-adapter powered.
- Virtual tracks let you record multiple takes of a single performance, then select your favorite take during mixdown.
- Guitar-amp modeling simulates various guitar amps; mic modeling simulates mic models.
- 2 mic inputs; records up to 2 tracks at a time.
- Plays back 4 or 8 tracks at once.

FIGURE 4.4
Korg D3200, an example of a digital multitracker (recorder-mixer).

DIGITAL MULTITRACKER (RECORDER-MIXER)

Other names for this device are standalone digital audio workstation (DAW), portable digital studio, or personal digital studio.

Like the mini studio, the digital multitracker combines a multitrack recorder with a mixer in a single chassis (Figure 4.4). It's convenient and portable. Offering better sound than a mini studio, the digital multitracker provides CD sound quality because it can record wave files. It records 8 to 32 tracks onto a built-in hard drive or flash memory card. The mixer includes faders (volume controls) for mixing, EQ or tone controls, and aux sends for effects (such as reverb). An LCD screen displays recording levels, waveforms for editing, and other functions.

The price for this device ranges from $700 to $3600. Some manufacturers of mini studios or digital multitrackers are TASCAM, Boss, Fostex, Korg, Roland, Yamaha, and Zoom (Samson).

FIGURE 4.5
Alesis HD24XR, an example of a multitrack hard-disk recorder.

Chapter 13 (under the heading "Recorder-Mixer") lists some features to look for in a recorder-mixer.

SEPARATE MULTITRACK RECORDER AND MIXER

Ideal for on-location recording, a multitrack hard-disk recorder (HD recorder) records up to 48 tracks reliably on a built-in hard drive (Figure 4.5), just like the hard drive in your computer. The recorder does not have mic preamps, so it must be used with a mixer or mic preamps. Multiple recorders can be linked to get more tracks. Some examples are the Alesis ADAT HD24XR, TASCAM X-48MKII, iZ Technology RADAR V, and Fostex D2424L Mk2. Prices range from $1,500 to $20,000.

Current units record with 24-bit resolution and up to 96 kHz sampling rate. Some record 24 tracks of 24-bit programs at sample rates of 44.1 and 48 kHz. They record 12 tracks of 24-bit recording at 88.2 and 96 kHz (these terms are explained in Chapter 13 on digital recording). Some HD recorders allow track editing, either on a built-in LCD screen or on a plug-in computer monitor. A few models have built-in removable hard drives for backing up projects, while others include network or high-speed data connections for data transfer.

A mixer (Figure 4.6) is an electronic device with several mic preamplifiers, volume and tone controls for each mic, monitor volume controls, and outputs for effects devices. To use it, plug mics and electric instruments into the mixer, which amplifies their signals. Connect the output signal of each mic preamp to a multitrack recorder. While recording, you use the mixer to send those signals to the desired recorder tracks, and to set recording levels. During mixdown, the mixer combines (mixes) the tracks to stereo. It also lets you adjust each track's volume, tone quality, effects, and stereo position. A large, complex mixer is called a mixing console (board or desk) and costs upwards of $3000. Chapter 10 explains mixing consoles in detail.

The normal price range for mixers is about $180 to $1500. Mixers are made by such companies as TASCAM, Alesis, Yamaha, Mackie, Allen & Heath, Soundcraft, M-Audio, Tapco, Peavey, Samson, and many others.

FIGURE 4.6
Mackie Onyx 1620i, an example of a mixer. It has a FireWire port for connecting to a computer.

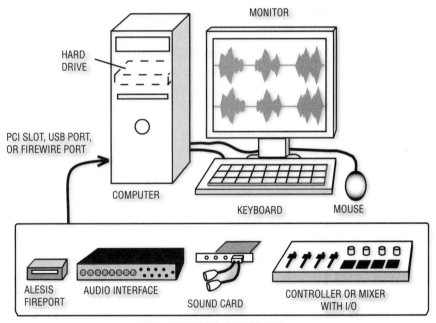

FIGURE 4.7
Computer with recording software and a choice of audio interface.

COMPUTER DIGITAL AUDIO WORKSTATION (DAW)

This low-cost recording setup has three parts: a personal computer, an audio interface, and recording software (Figure 4.7). Some software records MIDI data (explained later) as well as audio. The audio interface converts audio from your mics, mic preamps, or mixer into a signal that is recorded on the computer's hard drive as a magnetic pattern. Dozens of tracks can be recorded with professional quality. You mix the tracks with a mouse by adjusting the "virtual" controls that appear on your computer screen. Then you record the mix on your computer's hard drive. Details are in Chapter 14 on computer recording.

A computer studio costs about the same as a mini studio and is more powerful. It's a bargain. But since software requires computer skills, it's a little harder to learn and use than a hardware multitracker. Software recordings are at least CD quality—better than the MP3 recordings you get with a mini studio.

Recording software costs anywhere from $0 (freeware) to $1800, with $150-$500 being typical prices. Software examples include Cakewalk Home Studio, Sonar Studio, and X1 Producer; Avid Pro Tools 10 and HD Native; Apple Logic Studio and Logic Pro; Reaper; Steinberg Cubase and Nuendo; PreSonus Studio One and Studio One Pro; FL Studio Producer; Propellerhead Record; Sony Vegas and Sound Forge; BIAS Deck; Pro Tracks Plus; Magix Samplitude and Sequoia; SADiE; Pyramix; Mackie Tracktion 3; N Track Studio; RML Labs SAW Studio; Adobe Audition; and MOTU Digital Performer.

Free multitrack recording programs are available for download. Although they lack extensive features, they offer a chance to practice your skills at no cost. Examples are Audacity (from http://audacity.sourceforge.net) and GarageBand (supplied free with any new Apple computer or $79 with an iLife software upgrade).

Related to recording software is **music creation software**, which comes with samples of musical instruments. You set up loops (repeating drum riffs and musical patterns), play along with them using a MIDI keyboard, and record that MIDI sequence. Examples include Ableton Live, Propellerhead Reason, Sony ACID Pro, and Spectrasonics Stylus RMX.

A **sound card** (PCIe audio interface) costs $100 to $400. Some makers of sound cards include Echo, Event, Frontier Design, M-Audio, RME, Lynx, and E-MU.

An **audio interface** is a chassis with 2 to 16 mic preamps, A/D convertors, and a USB, FireWire, or Thunderbolt port that sends digital audio on a single cable to your computer. Some units have MIDI ins and out, and a high-impedance instrument input for an electric guitar, electric bass, acoustic guitar pickup, or synth. More details are in Chapter 14 on Computer Recording under the heading "FireWire, USB, or Thunderbolt Audio Interface."

Audio interface prices: 2-channel costs $150 to $500, 4-channel costs about $150 to $450 (Figure 4.8), and 8-channel starts at $400. If you link (connect together) two 8-channel interfaces, you can record up to 16 mic signals at once.

Manufacturers of audio interfaces include Akai, Avid, Metric Halo, MOTU, PreSonus, Apogee, Focusrite, M-Audio, TC Electronic, ART, Prism, Mackie, E-MU, Alesis, Edirol, TASCAM, and many others.

FIGURE 4.8
M-Audio Fast Track Ultra, an example of a 4-channel USB 2.0 audio interface. It offers low-latency direct monitoring with reverb, and can create two headphone mixes.

An audio interface usually has built-in mic preamps. Advanced users might prefer to use separate mic preamps connected to an **A/D converter (analog-to-digital converter)**. Some converter manufacturers are Lynx, Mitek, Crane Song, Millenia Media, RME, Universal Audio, and Apogee.

Another type of audio interface is a **mixer** that connects to a computer via a FireWire or USB connection. It sends the output signal of each mic preamp to a separate track in your recording software, and returns the stereo mix back to the mixer for monitoring. This can be a good system for recording a band because it offers easy setup of monitor mixes with effects. Some examples are the Mackie Onyx 1640i (Figure 4.6) with 16 recording channels, PreSonus StudioLive (16 recording channels and 16 mic inputs), Alesis Multimix 16 USB 2.0 (18 recording channels and 8 mic inputs), and Behringer Xenyx X2442USB (2 recording channels). Generally these mixers are not controllers for recording software—you still need to use a mouse.

A **USB microphone** has a USB interface built in, so you just plug the mic directly into your computer's USB port. No audio interface is needed. Examples are the Audio Technica AT2020 USB and AT2005 USB; Samson C01U; Blue Snowball, Yeti Pro, and Spark Digital; Rode Podcaster USB microphone; Apogee ONE and MicC; and MXL USB.009, 990 USB, and 990 USB Stereo mics.

An alternative to a USB microphone is a standard microphone plugged into an **XLR-to-USB mic adapter**. The MXL MicMate, Shure X2U, and the Blue Icicle connect to any mic and convert its output to USB. Phantom power and headphone monitoring are included. Its digital-audio format is 16-bit, 44.1 or 48 kHz. The Centrance MicPort Pro is similar but records up to 24-bit/96 kHz audio.

Using a mouse can be fatiguing and can lead to repetitive stress syndrome. An alternative to the mouse is a **control surface** or controller. It looks like a mixer with faders, but it adjusts the virtual controls you see on the computer screen. That way you can use

FIGURE 4.9
Frontier Design AlphaTrack, an example of a controller for recording software.

your computer for recording, and control the software with knobs and faders instead of a mouse. Two controllers with one fader are the Frontier Design AlphaTrack ($200, Figure 4.9) and PreSonus FaderPort ($150). Multifader controllers are made by such companies as Mackie, TASCAM, M-Audio, Avid, and Alesis ($350 to $1300).

You might want to get a wireless QWERTY keyboard so you can control your DAW from a mic location in the studio. Or add a USB extender cable to the keyboard.

KEYBOARD WORKSTATION

Here's another device to record music. This is a synthesizer/sampler with a piano-style keyboard, a built-in multitrack sequencer (MIDI recorder) and effects. Examples include the Korg Kronos and M50; Kurzweil PC3K8 and PC3LE; Yamaha Motif XF8; and Roland Fantom-G8 and Juno-Gi. The workstation lets you create a tune with several instrument sounds. Then you record the audio from the keyboard into a computer. If you want to add a vocal, use MIDI/audio recording software. An arranger keyboard workstation automatically creates backing tracks (drums, bass, and chords) by following your left hand's notes and right hand's melodies. Examples include the Yamaha Tyros4 and PSR-S910.

A similar device is a beat box or groove box, which is a sample player with pads that you tap to generate sounds and grooves. It lets you assemble a stereo music track from drum and synth samples. Copy the music from the beat box into your computer and add vocals. Examples include the Dave Smith Tempest; Korg Monotribe and Electribe; Akai MPC series and XR20; Native Instruments Maschine 1.6; Arturia Spark; Boss DR-880; Roland TD-20X; and Zoom RT-223. Beat boxes are available as software, too, such as MOTU BPM and Korg iElectribe for the Apple iPad.

LEGACY RECORDING DEVICES

Some recording equipment has become almost obsolete, but you may need to work with it if a client brings older gear into your studio to transfer audio tracks. Some of these formats are listed below.

Analog tape recorder: This records 2 to 24 tracks of analog audio on reels of magnetic tape from 1/4 inch to 2 inches wide. Compared to digital recorders, analog recorders have more noise, distortion, frequency-response errors, and unstable pitch. Their electronics and tape heads need frequent alignment. Still, many people love the sound of them. Some have mic inputs; all have analog line inputs and line outputs.

DAT (Digital Audio Tape) recorder: This records 2 tracks of audio digitally on a small DAT tape, a cassette about half the size of a standard analog cassette. The DAT also writes absolute time (hours, minutes, and seconds) on tape as it records. Inputs are analog mic/line level and digital; outputs are analog line level and digital.

Modular Digital Multitrack (MDM): This records 8 digital tracks on a videocassette, using a rotating drum like a DAT recorder. Two popular models are the Alesis ADAT, which records on S-VHS tape, and the TASCAM DA-88, DA-78, or DA-38, which record on Hi-8 mm tape. ADAT records up to 40 minutes on a single tape; DA-88 records up to 1 hour 48 minutes. With both types, you can sync several 8-track units with a cable to add more tracks. Both have analog line inputs and outputs. ADAT has Lightpipe-format digital outputs; TASCAM models have TDIF-format digital outputs.

Microphone

So far we covered recording devices. Let's move on to other studio equipment.

A microphone converts the sound of a voice or instrument into an electrical signal that can be recorded. Microphone sound quality varies widely, so be sure to get one or more good mics costing at least $100 each. Your ears should tell you if a mic's fidelity is adequate for your purpose. Some people are happy to get any sound recorded; others settle for nothing less than professional sound quality.

Probably the most useful mic types for home recording are the cardioid condenser mic and cardioid dynamic mic. The cardioid pickup pattern helps reject room acoustics for a tighter sound. The condenser type is commonly used on cymbals, acoustic instruments, and studio vocals; dynamics are used typically on drums and guitar amps. You'll also need at least one mic cable, mic stand, and boom costing about $25 each. For more information on microphones, see Chapter 6.

Do you want to record solo instruments or classical musical ensembles in stereo with two mics at a distance? You'll need either a stereo mic or a matched pair of high-quality condenser or ribbon mics of the same model number, plus a stereo mic-stand adapter. See Chapter 18 for details.

Phantom Power Supply

A phantom power supply powers the circuits in condenser mics. It uses the same cable that carries the mic's audio signal. You don't need the supply if your condenser mic has a battery, or if your mixer or audio interface supplies phantom power at its mic connectors (most do). Some makers of phantom supplies are Behringer, ART, Audio-Technica, Rolls, and Whirlwind. The price ranges from $20 to $139.

Mic Preamp

This device amplifies a mic signal up to a higher voltage called "line level," which is required by mixers and recorders. A standalone mic preamp provides a slightly cleaner, smoother, or more colorful sound than a mic preamp built into a mixer or audio interface, but costs much more. A 2-channel preamp costs $120 to $2000, while 8-channel preamps range from $600 to $6000. Studios on a budget can do without a standalone preamp.

Some manufacturers are Manley, True Systems, Focusrite, Universal Audio, GLM, Chandler, A Designs, Millennia Media, Avalon, Great River, John Hardy, Benchmark, AEA, Apogee, Vintech, Grace, Groove Tubes, PreSonus, Mackie, Summit Audio, Studio Projects, Samson, dbx, Aphex, and M-Audio.

Direct Box (DI)

A direct box (Figure 4.10) is used to connect an electric instrument (guitar, bass, synth) to an XLR-type mic input of a mixer, recorder-mixer, or audio interface. It lets you record electric instruments directly into your mixer without a microphone. A direct box picks up a very clean sound, which may be undesirable for electric guitar. If you want to pick up the distortion of the guitar amp, use a microphone instead. Or use a guitar-amp modeling device or modeling plug-in.

Some recorder-mixers and audio interfaces have "high-impedance" or "instrument" inputs meant for electric guitars and synthesizers. In this case, simply use a short guitar

FIGURE 4.10
A direct box.

cord between the instrument and the mixer high-impedance input. If the cable run is over 15 feet, you'll get higher quality with a direct box plugged into an XLR mic input.

Direct boxes sell for $50 to $700. Some manufacturers are Behringer, Radial, Countryman, Whirlwind, Manley, Groove Tubes, Pro Co, BSS, Tapco, Aphex, and ART.

Monitor System

Another important part of your studio is the monitor system, a pair of quality headphones or loudspeakers. The monitor system lets you hear what you're recording and mixing—what you're doing to the recorded sound. The sound you hear over the monitors is approximately what the final listener will hear.

Very good headphones are available for $100 and up from Sony, AKG, Sennheiser, Beyerdynamic, Audio Technica, Ultrasone, Shure, and others. A headphone amp can power several headphones at once, and some units let the musician adjust the headphone mix.

If you're recording only yourself, one set of headphones is enough. But if you're recording another musician, you both need headphones. Many recorder-mixers have two headphone jacks for this purpose.

If you want to record or overdub several people at once, you need headphones for all of them. For example, if you're overdubbing three harmony vocalists, each one needs headphones to hear previously recorded tracks to sing with. To connect all these headphones, you could build a headphone junction box, an aluminum or plastic box that contains several headphone jacks. These are wired to a cable coming from your mixer's headphone jack. Or you could use a splitter cable, which makes two jacks out of one. Another option is to purchase an affordable Rolls or Behringer headphone amplifier. Feed the amp from aux sends or the stereo mix.

FIGURE 4.11
An effects unit.

Powered speakers with built-in amplifiers provide accurate, high-fidelity monitoring. Described in Chapter 5, these Nearfield studio monitors are small, bookshelf-type speakers that are placed about 3 feet apart and 3 feet from you as you sit at your mixer. Some monitor companies include Genelec, Dynaudio, JBL, Event, Quested, Focal, Lipinski, KRK, Blue Sky, Yamaha, Adam, Mackie, Alesis, Tannoy, M-Audio, Behringer, Edirol, and Samson. Prices range from $200 to $3860 a pair.

Effects

A recording without effects sounds rather dead and plain. Effects such as reverberation, echo, and chorus can add sonic excitement to a recording. They are produced by devices called signal processors (see Figure 4.11) or by plug-ins, which are software effects used in a computer recording program. See Chapter 9 for more information on effects.

Although effects are built into most recording programs and recorder-mixers, most analog mixers require external effects units. On the mixer is a set of connectors (labeled "send" and "return") for hooking up an external effects unit, such as a reverb or delay device. A unit with one effects send lets you add one type of effect; a unit with two effects sends lets you add two different effects to create more sonic interest.

Miscellaneous Equipment

Other equipment for your home studio includes audio cables, USB or FireWire cables, power outlet strips, lighting, tables or studio furniture, mic pop filters, masking tape and a pen to label inputs and cables, contact cleaning fluid, MIDI equipment stands, music stands, session forms, connector adapters, pen and paper, a flashlight, and so on.

MIDI Studio Equipment

MIDI is covered in detail in Chapter 16. It describes the components in a typical MIDI studio, such as shown in Figure 4.12. Also see Chapter 2 under the heading "MIDI Sequencing."

We've looked at several types of recording setups. All can help you create quality demos. As you go higher in price you get more features and better sound. For example, if you want to record an entire band playing all at once, with each instrument having its own mic, you'll need a system with more microphones, more tracks, and more headphones.

FIGURE 4.12
One type of MIDI studio.

As we've seen, putting together a home studio or project studio needn't cost much. As technology develops, better equipment is available at lower prices. That dream of owning your own studio is within reach.

SETTING UP YOUR STUDIO

Once you have your equipment, you need to connect it together with cables, and possibly install equipment racks and acoustic treatment. Let's consider each step.

Cables

Cables carry electric signals from one audio component to another. They are usually made of one or two insulated conductors (wires) surrounded by a fine-wire mesh shield that reduces hum. Outside the shield is a plastic or rubber insulating jacket. On both ends of the cable are connectors. See www.bartlettmics.com/newsletter/ newsletter%204-11.pdf.

Cables are either balanced or unbalanced. A balanced line is a cable that uses two wires (conductors) to carry the signal, surrounded by a shield (see Figure 4.13). Each wire has equal impedance to ground. An unbalanced line has a single conductor surrounded by a shield (see Figure 4.14). The conductor and shield carry the signal. A balanced line rejects hum better than an unbalanced line, but an unbalanced line less than 10 feet long usually provides adequate hum rejection and costs less.

A cable carries one of these five signal levels or voltages:

- Mic level: about 2 mV (0.002 volt) to about 1 V depending on how loud the sound source is, and how sensitive the mic is
- Instrument level: typically 0.1 V to 1 V for passive pickups; up to 1.75 V for active pickups
- Semipro or consumer line level: −10 dBV (0.316 volt)
- Pro line level: +4 dBu (1.23 volts)
- Speaker level: about 20 volts.

Note that "instrument level" can overlap mic level and line level.

Some manufacturers use dBu instead of volts (for an explanation, see Appendix A on the decibel):

Mic level: about −52 dBu to +2.2 dBu
Instrument level: −17.7 dBu to +7 dBu
Semipro or consumer line level: −7.8 dBu
Pro line level: +4 dBu

FIGURE 4.13
A 2-conductor shielded, balanced line.

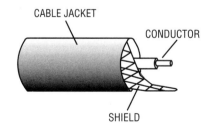

FIGURE 4.14
A 1-conductor shielded, unbalanced line.

Equipment Connectors

Recording equipment also has balanced or unbalanced connectors built into the chassis. Be sure your cable connectors match your equipment connectors.

Balanced equipment connectors:

- 3-pin (XLR-type) connector—Figure 4.15
- 1/4-inch TRS (tip-ring-sleeve) phone jack—Figure 4.16

Unbalanced equipment connectors:

- 1/4-inch TS (tip-sleeve) phone jack—Figure 4.16
- Phono jack (RCA connector)—Figure 4.17

A jack is a receptacle; a plug inserts into a jack.

FIGURE 4.15
A 3-pin XLR-type connector used in balanced equipment.
Left: male output connector. Right: female input connector.

FIGURE 4.16
A 1/4-inch phone jack used in balanced and unbalanced equipment.

FIGURE 4.17
A phono (RCA) jack used in unbalanced equipment.

Connectors are confusing because a single connector can have several functions (usually not at the same time). Here are some examples:

- XLR: balanced line input at +4 dBu, balanced mic input at 2 mV to 1 V, or balanced line output at +4 dBu
- TRS (stereo 1/4-inch phone jack): balanced mic input, insert send/return connectors (line level), instrument input, or stereo headphones
- TS (mono 1/4-inch phone jack): unbalanced mic input, unbalanced line-level input or output (+4 dBu or −10 dBV), instrument input, or low-cost speaker connector
- Combi connector: An XLR mic input plus a TRS input (instrument level or line level)
- RCA (phono): home stereo line-level input or output at −10 dBV, composite video input/output, or SPDIF digital-audio input/output
- 1/8-inch (3.5 mm) mini phone jack: headphone jack, low-cost stereo mic input, line output in a portable recorder at −10 dBV; or a computer sound card's line in, line out, or speaker out

Equipment connectors are labeled according to their function. If you see an XLR connector with the label "MIC," you know it's a balanced mic input. If it's a 1/8-inch connector on a sound card, look at the icon near the connector. It's either a mic input, line input, line output, or speaker output. You could download the manual for the sound card, which should describe the function of each connector.

Cable Connectors

Several types of cable connectors are used in audio. Figure 4.18 shows a 1/4-inch mono phone plug (or TS phone plug), used with cables for unbalanced microphones, synthesizers, and electric instruments. The tip terminal is soldered to the cable's center conductor; the sleeve terminal is soldered to the cable shield.

Figure 4.19 shows an RCA or phono plug, used to connect unbalanced line-level signals. The center pin is soldered to the cable's center conductor; the cup terminal is soldered to the cable shield.

FIGURE 4.18
A mono (TS) 1/4-inch phone plug.

FIGURE 4.19
An RCA (phono) plug.

FIGURE 4.20
A 3-pin pro audio connector (XLR-type). (A) female. (B) male.

FIGURE 4.21
A stereo (TRS) phone plug.

Figure 4.20 shows a 3-pin pro audio connector (XLR-type). It is used with cables for balanced mics and balanced recording equipment. The female connector (with holes; Figure 4.20A) plugs into equipment outputs. The male connector (with pins; Figure 4.20B) plugs into equipment inputs. Pin 1 is soldered to the cable shield, pin 2 is soldered to the "hot" red or white lead, and pin 3 is soldered to the remaining lead. This wiring applies to both female and male connectors.

Figure 4.21 shows a stereo (TRS) phone plug, used with stereo headphones and with some balanced line-level cables. For headphones, the tip terminal is soldered to the left-channel lead, the ring terminal is soldered to the right-channel lead, and the sleeve terminal is soldered to the common lead. For balanced line-level cables, the sleeve terminal is soldered to the shield, the tip terminal is soldered to the hot red or white lead, and the ring terminal is soldered to the remaining lead.

Some mixers have insert jacks that are stereo phone jacks; each jack accepts a stereo phone plug. The tip is the send signal to an audio device input, the ring is the return signal from the device output, and the sleeve is the ground.

Cable Types

Cables are also classified according to their function. In a studio, you'll use several types of cables: power, mic, MIDI, speaker, USB, FireWire, S/PDIF, TASCAM TDIF, Alesis Lightpipe, guitar cords, and patch cords.

A power cable, such as an AC extension cord or a power cord on a device, is made of three heavy-gauge wires surrounded by an insulating jacket. The wires are thick to handle high current without overheating.

A mic cable is usually 2-conductor shielded. It has two wires to carry the signal, surrounded by a fine-wire cylinder or shield that reduces hum pickup. On one end of the cable is a connector that plugs into the microphone, usually a female XLR-type. On the other end is either a 1/4-inch phone plug or a male XLR-type connector that plugs into your mixer or audio interface.

Rather than running several mic cables to your mixer or interface, you might consider using a snake, which is a box with multiple mic connectors, all wired to a thick multiconductor cable (Figure 4.22). A snake is especially convenient if you're running long cables to recording equipment from another room. It's essential for most on-location recording.

Professional balanced equipment is interconnected with mic cable: 2-conductor shielded cable having a female XLR on one end and a male XLR on the other. Professional patch bays (described later) use balanced cables with tip-ring-sleeve phone plugs.

A MIDI cable uses a 5-pin DIN connector on each end of a 2-conductor shielded cable. The cable connects MIDI OUT to MIDI IN, or MIDI THRU to MIDI IN.

A speaker cable connects a power amp to each loudspeaker. To avoid wasting power, speaker cables should be as short as possible and should be heavy gauge (between 12 and 16 gauge). They can even be made from lamp cord (zip cord). Number 12 gauge is thicker than 14; 14 is thicker than 16.

A USB cable or a FireWire cable connects a peripheral device (like an audio interface) to a computer. USB and FireWire are covered in detail in Chapter 14.

An S/PDIF cable transfers a digital signal from one device's S/PDIF output to another device's S/PDIF input. It uses a shielded unbalanced cable (ideally a 75-ohm RG59 video cable) with an RCA plug on each end.

STAGE BOX

CONNECTORS TO
MIC CABLES

SNAKE

CONNECTORS TO MIXER

FIGURE 4.22
A stage box and snake.

A TASCAM TDIF cable is a multiconductor cable with a 25-pin D-sub connector on both ends. It's used to connect multiple digital-audio signals from TASCAM multitrack recorders to digital mixers or computer TDIF interfaces.

An Alesis Lightpipe cable is a fiber optic cable with a TOSLINK connector on both ends. This cable is used to connect 8 channels of digital-audio signals from an Alesis multitrack recorder to a digital mixer or computer Lightpipe interface.

A guitar cord is constructed of a 1-conductor shielded cable with a 1/4-inch phone plug on each end. It connects a guitar amp to an electric instrument (electric guitar, electric bass, synthesizer, or an acoustic guitar with a pickup). Also, it connects an electric instrument to a direct box or to an instrument input.

Patch cords connect your recorder-mixer to external devices: an effects unit, a 2-track recorder, and a power amplifier. They also connect an analog mixer to the analog inputs and outputs of a multitrack recorder, usually as a snake that combines several cables. An unbalanced patch cord is made of 1-conductor shielded cable with either a 1/4-inch phone plug or a phono (RCA) connector on each end. A stereo patch cord is two patch cords joined side-by-side.

Rack/Patch Bay

You might want to mount your signal processors in a rack, an enclosure made of wood or metal with rails of screw holes for mounting equipment (Figure 4.23). You also

FIGURE 4.23
A rack and patch panel.

might want to install a patch panel or patch bay, a group of connectors that are wired to equipment inputs and outputs. Using a patch bay and patch cords, you can change equipment connections easily. You also can bypass or patch around defective equipment. Note that a patch bay increases the chance of hum pickup slightly because of the additional cables and connectors. Racks and patch bays are not essential, but they are convenient. Figure 4.24 shows some typical patch-bay assignments.

Equipment Connections

The instruction manuals of your equipment tell you how to connect each component to the others. In general, use cables that are as short as possible to reduce hum, but that are long enough to let you make changes.

Be sure to label all your cables on both ends according to what they plug into—for example, MIXER CH1 MONITOR OUT or ALESIS 3630 IN. If you change connections temporarily, or the cable becomes unplugged, you'll know where to plug it back in. A piece of masking tape folded over the end of the cable makes a stay-put label.

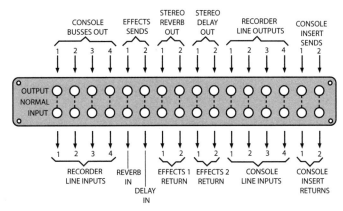

FIGURE 4.24
Some typical patch-bay assignments.

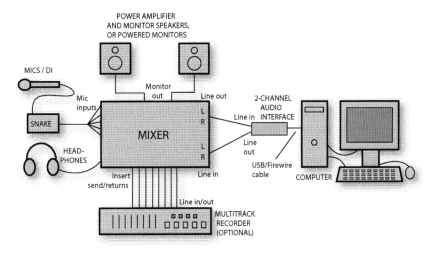

FIGURE 4.25
Typical connections for a mixer-based studio.

Let's say you have a hardware mixer in your recording setup. Here's a typical way to hook up the gear (see Figure 4.25):

1. Plug the AC power cords of audio equipment and electric musical instruments into AC outlet strips fed from the same circuit breaker. Make sure that the sum of the equipment current ratings does not exceed the breaker's amp rating for that outlet. Plug the power amplifier or powered speakers into their own outlet on the same breaker so that they receive plenty of current. Consider using an AC power conditioner such as made by Furman (www.furmansound.com). It provides clean, steady AC power to sensitive electronic equipment. Surge protection and noise filtering are included.

FIGURE 4.26
A female XLR to phone plug adapter, such as Radio Shack #274-016.

2. Connect mic cables to mics. Use mic cables with a male XLR connector on one end and a female XLR connector on the other end.

3. Connect mic cables to the female XLR connectors in either the snake junction box, or directly into mic inputs on a mixer or mic preamps. Plug the snake connectors into the mic inputs. If your mixer has phone-jack mic inputs, you may need to use an impedance-matching adapter (female XLR to phone) between the mic cable and the mic input jack (Figure 4.26).

4. Set the output volume of synthesizers and sound modules about three-quarters up. Using a guitar cord, connect their audio outputs to instrument or line inputs on your mixer. If this causes hum, use a direct box. Using a MIDI cable, connect the MIDI OUT of a MIDI controller to the MIDI IN of your audio interface or MIDI interface.

5. If you are recording a guitar direct, connect its cord either to (1) an instrument input on your mixer or audio interface (1/4-inch phone jack), or (2) a direct box. Connect the XLR output of the direct box to a mic input on your mixer or audio interface.

6. If the mixer is a standalone unit (not part of a recorder-mixer), connect the mixer's stereo line outputs to the inputs of an audio interface. Use a stereo RCA-to-RCA cable or two phone-to-phone cables. If the mixer has a USB or FireWire connector, connect that to the mating connector in your computer—you don't need an audio interface.

7. Connect the audio interface line outputs to the mixer's 2-track or tape inputs, or directly to powered speakers. Use a stereo RCA-to-RCA cable or two phone-to-phone cables. Again, if the mixer has a USB or FireWire connector, connect that to the mating computer connector and omit the audio interface.

8. Connect the mixer's monitor outputs to the power-amp inputs. Connect the power-amp outputs to loudspeakers. Or if you are using powered (active) monitors, connect the mixer monitor outputs to the monitor-speaker inputs.

9. If the mixer does not have internal effects, connect the mixer aux-send connectors to effects inputs (not shown). Connect the effects outputs to the mixer aux-return or bus-in connectors. Use phone-to-phone cables.

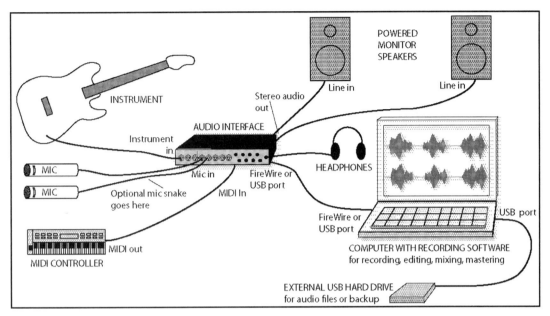

FIGURE 4.27
Typical connections in a DAW recording studio.

10. If you're using a separate mixer and multitrack recorder, connect mixer bus 1 to recorder track 1 IN; connect bus 2 to track 2, and so on. Also connect the recorder's track 1 OUT to the mixer's line input 1; connect the track 2 OUT to the mixer's line input 2, and so on. Use cables with RCA or phone connectors. As an alternative, connect insert jacks to multitrack inputs and outputs. At each insert plug, connect the tip (send) terminal to a track input and connect the ring (return) terminal to the same track's output. Use a TRS-to-2-TS cable (stereo phone plug to two mono phone plugs).

11. If you have several headphones for musicians, connect the mixer's headphone jack to a small amplifier to drive their headphones. Use a cable with a stereo phone plug on one end and two mono phone plugs on the other end (TRS-to-2-TS cable). Or if the mixer's headphone signal is powerful enough, connect it to a box with several headphone jacks wired in parallel.

Semipro studio equipment with unbalanced connectors (usually phone or RCA) operates at a level called −10 or −10 dBV. Pro studio equipment with balanced connectors (XLR or TRS) works at +4 or +4 dBu. Check your equipment manuals to determine their input and output levels. When you connect devices that run at different levels, set the +4/−10 switch on each unit to match the levels. If there is no such switch on either device, connect between them a +4/−10 converter box such as the Whirlwind LM2U

FIGURE 4.28
Typical layout of a DAW recording studio.

line level converter (www.whirlwindusa.com). Or try the cables shown in Appendix A, "dB or not dB," Figures A-3 and A-4.

Figure 4.27 shows typical connections in a DAW recording studio with a multichannel audio interface. Figure 4.28 shows a typical layout for a DAW recording studio.

As shown in Figure 4.27, you might connect the equipment like this:

1. Using a guitar cord, connect electric instruments to instrument inputs on the audio interface. If an instrument is more than about 15 feet from the interface, connect its output to a direct box (using guitar cords), and connect the direct box XLR output to a snake or to an audio interface mic input.

2. Using an XLR mic cable, connect each mic to a mic input on the audio interface. If the mics are more than about 15 feet from the interface, connect each mic to a snake box, and connect the snake XLR connectors to the interface mic inputs. If you prefer to use a separate mic preamp and A/D converter, plug the mic into the preamp, and connect the preamp's line output to the A/D converter's line input using an XLR or phone-to-phone cable.

3. Using a MIDI cable, connect the MIDI OUT of a MIDI controller to the MIDI IN of the audio interface.

4. Using two phone-to-phone cables (stereo or mono), connect the stereo output of the interface to two powered monitors. If your monitor speakers are passive,

connect the interface stereo outputs to the line inputs of a stereo power amp. Use speaker cable to connect the power-amp outputs to the speakers.

5. Plug headphones (or a headphone amplifier or junction box) into the headphone jack of the interface.

6. Using a USB or FireWire cable, connect the USB/FireWire port in the interface to the mating port in the computer.

7. Using a USB cable, connect the USB port of an external hard drive to a mating port in the computer. That hard drive can be used for audio files or for backup.

Hooking up a recorder-mixer studio can be quite simple. Plug mics into mic inputs, plug headphones into headphone jacks, and plug powered speakers into the mixer monitor outputs.

HUM PREVENTION

You patch in a piece of audio equipment, and there it is—HUM! It's a low-pitched tone or buzz. This annoying sound is a tone at 60 Hz (50 Hz in Europe) and multiples of that frequency.

Hum is caused mainly by

- Cables picking up magnetic and electrostatic hum fields radiated by power wiring—especially if the cable shield connection is broken.
- Ground loops. A ground loop is a conductive loop or circuit made of a cable shield and a power-ground wire. A ground loop occurs when two or more separated pieces of audio equipment are each connected to power ground through a 3-prong power cord, and are also connected to each other through a cable shield. The ground voltage may be slightly different at each piece of equipment, so a 50- or 60-Hz hum signal flows between the components along the cable shield.

These are the most important points to remember about hum prevention:

- To prevent ground loops, plug all equipment into outlet strips powered by the same AC outlet or AC circuit.
- Do not use an AC (electrical) 3-to-2 adapter to disconnect the power ground—it causes a safety hazard.
- Some power amps create hum if they don't get enough AC current. So connect the power amp (or powered speakers) AC plug to its own wall outlet socket—the same outlet that feeds the outlet strips for the recording equipment.
- If possible, use balanced cables going into balanced equipment. Balanced cables have XLR or TRS connectors and two conductors surrounded by a shield. At both ends of the cable, connect the shield to a screw in the chassis, not to XLR pin 1. Or use audio gear whose XLR connectors are wired with pin 1 to chassis ground, not to signal ground.
- Transformer-isolate unbalanced connections. If that is not an option, use the cable assemblies shown in Figures A-3 and A-4 in Appendix A.

- Don't use conventional SCR dimmers to change the studio lighting levels. Use Luxtrol® variable-transformer dimmers or multiway incandescent bulbs instead.

Even if your system is wired properly, a hum or buzz may appear when you make a connection. Follow these tips to stop the hum:

- If the hum is coming from a direct box, flip its ground-lift switch.
- Check cables and connectors for broken leads and shields.
- Unplug all equipment from each other. Start by listening just to the powered monitor speakers. Connect a component to the system one at a time, and see when the hum starts.
- Remove audio cables from your devices and monitor each device by itself. It may be defective.
- Lower the volume on your power amp (or powered speakers), and feed them a higher-level signal.
- Use a direct box instead of a guitar cord between instrument and mixer.
- To stop a ground loop when connecting two devices, connect between them a 1:1 isolation transformer, direct box, or hum eliminator (such as the Jensen CI-2RR, Behringer HD400, Rolls HE18, or Ebtech He2PKG). See Figures A-3 and A-4 in Appendix A.
- Make sure that the snake box is not touching metal.
- To prevent accidental ground loops, do not connect XLR pin 1 to the connector shell except for permanent connections to equipment inputs and outputs.
- Try another mic.
- If you hear a hum or buzz from an electric guitar, have the player move to a different location or aim in a different direction. You might also attach a wire between the player's body and the guitar strings near the tailpiece to ground the player's body.
- Turn down the high-frequency EQ on a buzzing bass guitar track.
- To reduce buzzing between notes on an electric-guitar track, apply a noise gate.
- Route mic cables and patch cords away from power cords; separate them vertically where they cross. Also keep recording equipment and cables away from computer monitors, power amplifiers, and power transformers.
- See Rane's excellent article on sound system interconnections at www.rane.com.

By following all these tips, you should be able to connect audio equipment without introducing any hum. Good luck!

REDUCING RADIO FREQUENCY INTERFERENCE (RFI)

RFI is heard as buzzing, clicks, radio programs, or "hash" in the audio signal. It's caused by CB transmitters, computers, lightning, radar, radio and TV transmitters, industrial machines, auto ignitions, stage lighting, and other sources. Many of the following techniques are the same used to reduce hum from other sources. To reduce RFI:

- If you think that a speaker cable, mic cable, or patch cord is picking up RFI, wrap the cable several times around an RFI choke (available at Radio Shack). Put the choke near the device that is receiving audio.
- Install high-quality RFI filters in the AC power outlets. The cheap types available from local electronics shops are generally ineffective.
- If the shield is disconnected in a balanced cable connector, solder it to pin 1.
- If a mic is picking up RFI, solder a 0.047 microfarad capacitor between pin 1 and 2, and between pin 1 and 3, in the female XLR connector of the mic cable. Caution: this might cause high-frequency distortion with some mics.
- At equipment XLR mic inputs, use a cable connector with a tab that is connected to the metal shell. Use a wide strip of metal braid to connect this tab to pin 1. Make sure the connector makes metal-to-metal contact with the chassis.
- Periodically clean connector contacts with Caig Labs DeoxIT, or at least unplug and plug them in several times.

This chapter briefly covered the equipment and connectors for a recording studio. The rest of this book explains each piece of equipment in detail and explains how to use it for best results.

CHAPTER 5
Monitoring

One of the most exciting moments in recording comes when the finished mix is played over the studio monitor speakers. The sound is so clear you can hear every detail, and so powerful you can feel the deep bass throbbing in your chest.

You use the monitor system to listen to the output signals of the console, audio interface, or recorders. It consists of the console monitor mixer, the power amplifiers, loudspeakers, and the listening room. The power amplifier boosts the electrical power of the console signal to a sufficient level to drive a loudspeaker. The speaker converts the electrical signal into sound, and the listening-room acoustics affect the sound from the speaker.

A quality monitor system is a must if you want your mixes to sound good. The power amp and speakers tell you what you're doing to the recorded sound. According to what you hear, you adjust the mix and judge your mic techniques. Clearly, the monitor system affects the settings of many controls on your mixer, as well as your mic selection and placement. And all those settings affect the sound you're recording. So, using inadequate monitors can result in a poor-sounding product coming out of your studio.

It's important to use accurate speakers that have a flat frequency response. If your monitors are weak in the bass, you will tend to boost the bass in the mix until it sounds right over those monitors. But when that mix is played over speakers with a flatter response, it will sound too bassy because you boosted the bass on your mixer. So, using monitors with weak bass results in bassy recordings; using monitors with exaggerated treble results in dull recordings, and so on. In general, colorations in the monitors will be inverted in your mixdown recording.

That's why it's so important to use an accurate monitor system—one with a wide, smooth frequency response. Such a system lets you hear exactly what you recorded.

SPEAKER REQUIREMENTS

The requirements for an accurate studio monitor are these:

- Wide, smooth frequency response. To ensure accurate tonal reproduction, the on-axis response of the direct sound should be ±4 dB or less from at least 40 Hz to 15 kHz. The low-frequency response of a small monitor speaker should extend to at least 70 Hz.
- Uniform off-axis response. The high-frequency output of a speaker tends to diminish off-axis. Ideally the response at 30 degrees off-axis should be only a few decibels down from the response on-axis. That way, a producer and engineer sitting side-by-side will hear the same amount of treble. Also, the treble will not change as the engineer moves around at the console.
- Good transient response. This is the ability of the speaker to accurately follow the attack and decay of musical sounds. If a speaker has good transient response, the bass guitar sounds tight, not boomy, and drum hits have sharp impact. Some speakers are designed so that the woofer and tweeter signals are aligned in time. This aids transient response.
- Clarity and detail. You should be able to hear small differences in the sonic character of instruments, and to sort them out in a complex musical passage.
- Low distortion. Low distortion is necessary because it lets you listen to the speaker for a long time without your ears hurting. A good spec might be total harmonic distortion under 3% from 40 Hz to 20 kHz at 90 dBSPL (sound pressure level).
- Sensitivity. Sensitivity is the sound pressure level a speaker produces at 1 meter (m) when driven with 1 watt (W) of pink noise. Pink noise is random noise with equal energy per octave. This noise is either band-limited to the range of the speaker or is a one-third-octave band centered at 1 kHz. Sensitivity is measured in dB/W/m (dB sound pressure level per 1 W at 1 m). A spec of 93 dB/W/m is considered high; 85 dB/W/m is low. The higher the sensitivity, the less amplifier power you need to get adequate loudness.
- High output capability. This is the ability of a speaker to play loudly without burning out. You often need to monitor at high levels to hear quiet details in the music. Plus, when you record musicians who play loudly in the studio, it can be a letdown for them to hear a quiet playback. So you may need a maximum output of 110 dBSPL.

This formula calculates the maximum output of a speaker (how loud it can play):

$$dBSPL = 10\log(P) + S$$

where dBSPL is the sound pressure level at 1 m, P is the continuous power rating of the speaker in watts, and S is the sensitivity rating in dB/W/m.

For example, if a speaker is rated at 100 W maximum continuous power, and its sensitivity is 94 dBSPL/W/m, its maximum output SPL is 10 log(100) + 94 = 114 dBSPL (at 1 m from the speaker). The level at 2 m will be about 4 to 6 dB less.

NEARFIELD™ MONITORS

Many professional recording studios use large monitor speakers that have deep bass. However, they are expensive, heavy, and difficult to install, and they are affected by the acoustics of the control room. All studios, large or small, need a pair of Nearfield monitor speakers (Figure 5.1). A Nearfield monitor is a small, wide-range speaker typically using a cone woofer and dome-shaped tweeter. You place a pair of them about 3 or 4 feet apart, on stands just behind the console, about 3 or 4 feet from you. Nearfields are far more popular than large wall-mounted speakers.

This technique, developed by audio consultant Ed Long, is called Nearfield monitoring. Because the speakers are close to your ears, you hear mainly the direct sound of the speakers and tend to ignore the room acoustics. Plus, Nearfield monitors sound very clear and provide sharp stereo imaging. Some units have bass or treble tone controls built in to compensate for the effects of speaker placement and room surfaces.

FIGURE 5.1
A Nearfield monitor speaker.

Nearfield monitors have enough bass to sound full when placed far from walls. Although most Nearfields lack deep bass, they can be supplemented with a subwoofer to reproduce the complete audio spectrum. Or you can check the mix occasionally with headphones that have deep bass.

Some Nearfields are in a satellite-subwoofer format. The two satellite speakers are small units, typically including a 4-inch woofer and 3/4-inch dome tweeter. The satellites are too small to produce deep bass, but that is handled by the subwoofer—a single cabinet with one or two large woofer cones. Typically, the subwoofer (sub) produces frequencies from 100 Hz down to 40 Hz or below. Because we do not localize sounds below about 100 Hz, all the sound seems to come from the satellite speakers. The sub-satellite system is more complicated to set up than two larger speakers, but offers deeper bass.

POWERED (ACTIVE) MONITORS

Most monitors have a power amplifier built in. You feed them a line-level signal (labeled MONITOR OUT or STEREO L/R) from your mixing console or audio interface. Most powered monitors are bi-amplified: they have one amplifier for the woofer and another for the tweeter. The advantages of bi-amplification include:

- Distortion frequencies caused by clipping the woofer power amplifier will not reach the tweeter, so there is less likelihood of tweeter burnout if the amplifier clips. In addition, clipping distortion in the woofer amplifier is made less audible.
- Intermodulation distortion is reduced.
- Peak power output is greater than that of a single amplifier of equivalent power.
- Direct coupling of amplifiers to speakers improves transient response—especially at low frequencies.
- Bi-amping reduces the inductive and capacitive loading of the power amplifier.
- The full power of the tweeter amp is available regardless of the power required by the woofer amp.

THE POWER AMPLIFIER

If your monitor speakers are not powered, you need a power amplifier (Figure 5.2). It boosts your mixer's line-level signal to a higher power in order to drive the speakers.

How many watts of power do you need? The monitor speaker's data sheet gives this information. Look for the specification called "Recommended amplifier power." A power amp of 50 W per channel continuous is about the minimum for Nearfield monitors; 150 W is better. Too much power is better than too little, because an underpowered system is likely to clip or distort. This creates high frequencies that can damage tweeters.

FIGURE 5.2
Crown D-75A, an example of a power amplifier.

A good monitor power amp has distortion under 0.05% at full power. It should have a high damping factor—at least 100—to keep the bass tight. The amp should be reliable. Look for separate level controls for left and right channels. The amplifier should have a clip or peak light that flashes when the amp is distorting.

SPEAKER CABLES AND POLARITY

When you connect the power amp to the speakers, use good wiring practice. Long or thin cables waste amplifier power by heating. So put the power amp(s) close to the speakers and use short cables with thick conductors—at least 16 gauge. The low resistance of these cables helps the power amplifier to damp the speaker motion and tighten the bass.

If you wire the two speakers in opposite polarity, one speaker's cone moves out while the other speaker's cone moves in. This causes vague stereo imaging, weak bass, and a strange sense of pressure on your ears. Be sure to wire the speakers in the same polarity as follows: in both channels, connect the amplifier positive (+ or red) terminal to the speaker positive (+ or red) terminal. Setting the correct polarity is also called "speaker phasing."

CONTROL-ROOM ACOUSTICS

The acoustics of the control room affect the sound of the speakers. At your ears, the sound waves reflected from the room surfaces combine with the direct sound from the speakers. Reflections that arrive within 20 to 65 milliseconds (msec) after the direct sound blend with the direct sound and affect the tonal balance you hear.

Chapter 3, under the heading "Controlling Room Problems with Acoustic Treatment," described how to treat the acoustics of studios and control rooms to reduce excessive reverb, echo, standing waves, and noise.

Using Nearfield monitors makes the room acoustics less important, but it still helps to put absorbent material on the wall behind the monitors. This treatment improves the monitors' sound. Stereo imaging and depth are greatly improved, the sound is clearer, and the frequency response is flatter. The treatment will reduce boominess and ringing, and make transients sharper. Also, your recordings will translate better to other speakers.

While sitting in front of your mixer, have someone slide a mirror along the walls at eye height. Note the spots where you can see the monitor speakers in the mirror, and put some absorbent material there to absorb sound reflections. Repeat this procedure with the mirror on the ceiling between the monitors and your mix position.

If your control room is separate from the studio, the control room should be built to keep out sound from the studio. You want to hear only the sound from the monitors, not the live sound from the musicians. In a home studio, you might be able to achieve isolation simply by putting the control-room equipment in a room far removed from the studio, with the doors closed.

A control room built next to the studio needs good isolation. Use staggered-stud construction, seal holes between rooms, and install a double-pane window (mounted in rubber) between the control room and studio.

In some home or project studios, the control room is the same room as the studio. Because no isolation is used, the cost of building the studio is much less. You record while listening with headphones, and do the critical monitoring with speakers during playback and mixdown.

SPEAKER PLACEMENT

Once you have acquired the speakers and worked on the room acoustics, you can install the speakers.

- Mount them at ear height so the mixer doesn't block their sound.
- To prevent sound reflections off the mixing console, place the speakers on stands behind the console's meter bridge, rather than putting them on top.
- For best stereo imaging, align the speaker drivers vertically and mount the speakers symmetrically with respect to the side walls.
- Place the two speakers as far apart as you're sitting from them; aim them toward you, and sit exactly between them (Figure 5.3). Some engineers recommend aiming the speakers 12–18 inches behind the listener.
- To get the smoothest low-frequency response, put the speakers near the shorter wall, and sit about $\frac{1}{3}$ of the way back from the front wall.
- Use a foam isolation device under each speaker to ensure deep, tight bass. Auralex MoPads and Primacoustics Recoil Stabilizers work well.

Try to position the monitors several feet from the nearest wall. Wall reflections can degrade the frequency response and stereo imaging. The closer to the wall the monitors are, the more bass you hear. In small rooms you might have to place the monitors against the wall, which will exaggerate the bass. But some monitors have a low-frequency attenuation switch to compensate.

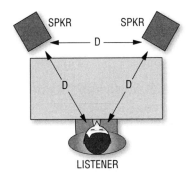

FIGURE 5.3
The recommended speaker/listener relationship for best stereo imaging.

KRK Systems ERGO is a monitor controller that corrects a room's phase and frequency-response errors, resulting in accurate monitoring—so mixes translate better to other systems.

USING THE MONITORS

You've treated the room acoustics, and you've connected and placed the speakers as described earlier. Now it's time to adjust the stereo balance.

1. Play a mono musical signal into an input channel on your mixer, and assign it to the stereo output channels 1 and 2.
2. Adjust the input channel's pan pot so that the signal reads the same on the stereo output channel 1 and 2 meters.
3. Place the two speakers the same distance from you.
4. Sit at the mixer exactly midway between the speakers. If you sit off-center, you will hear the image shifted toward one side. Listen to the image of the sound between the speaker pair. You should localize it midway between the monitors; that is, straight ahead.
5. If necessary, center the image by adjusting the left or right volume control on your power amp or powered monitors.

When you do a mixdown, try to keep the listening level around 85 dBSPL—a fairly loud home listening level. As discovered by Fletcher and Munson, we hear less bass in a program that is played quietly than in the same program played loudly. If you mix a program while monitoring at, say, 100 dBSPL, the same program will sound weak in the bass when heard at a lower listening level—which is likely in the home. So, programs meant to be heard at 85 dBSPL should be mixed and monitored at that level.

Loud monitoring also exaggerates the frequencies around 4 kHz. A recording mixed loud may sound punchy, but the same recording heard at a low volume will sound dull and lifeless.

Here's another reason to avoid extreme monitor levels: loud sustained sound can damage your hearing or cause temporary hearing loss at certain frequencies. If you must do a loud playback for the musicians (who are used to high SPLs in the studio), protect your ears by wearing earplugs or leaving the room.

You can get a low-cost sound level meter from Radio Shack. Play a musical program at 0 VU or 0 dB on the mixer meters and adjust the monitor level to obtain an average reading of 85 dBSPL on the sound level meter. Mark the monitor-level setting.

Before doing a mix, you may want to play some familiar commercial CDs over your monitors to remind yourself what a good tonal balance sounds like. Listen to the amount of bass, midrange, and treble, and try to match those in your mixes. But listen to several CDs because they vary.

While mixing, monitor the program alternately in stereo and mono to make sure there are no out-of-phase signals that cancel certain frequencies in mono. Also beware of center-channel buildup. Instruments or vocals that are panned to center in the stereo mix sound 3 dB louder when monitored in mono than they do in stereo. That is, the balance changes in mono—the center instruments are a little too loud. To prevent this, don't pan tracks hard left and hard right. Bring in the side images a little so they will be louder in mono.

You'll mix the tracks to sound good on your accurate monitors. But also check the mix on small inexpensive speakers to see whether anything is missing or whether the mix changes drastically. Make sure that bass instruments are recorded with enough edge or harmonics to be audible on the smaller speakers. It's a good idea to make a CD copy of the mix for listening in a car, boom box, or compact stereo.

HEADPHONES

Compared to speakers, headphones have several advantages:

- They cost much less.
- There is no coloration from room acoustics.
- The tone quality is the same in different environments.
- They are convenient for on-location monitoring.
- It's easy to hear small changes in the mix.
- Transients are sharper due to the absence of room reflections.

Headphones have several disadvantages:

- They become uncomfortable after long listening sessions.
- Cheap headphones have inaccurate tone quality.
- Headphones don't project bass notes through your body.
- The bass response varies due to changing headphone pressure.

- The sound is in your head rather than out front.
- You hear no room reverberation, so you may add too much or too little artificial reverb.
- It's difficult to judge the stereo spread. Over headphones, panned signals tend to sound closer to center than the same signals heard over speakers. The same is true of stereo recordings made with a coincident pair of mics.

Because speakers sound different from headphones, it's best to do mixes over speakers. But check your mixes on headphones too because so many listeners use them with MP3 players.

Focusrite's Saffire PRO 24 DSP Audio Interface with VRM technology simulates the sound of studio monitor loudspeakers over headphones. You can choose three room models and 15 different speaker emulations.

If your monitor speakers are in the same room as your microphones, the mics pick up the sound of the speakers. This causes feedback or a muddy sound. In this case you must monitor only with headphones while recording or overdubbing, then monitor with speakers during playback or mixdown.

If you monitor with headphones as the musicians are playing, external sounds leak through the headphones and mask the sounds you are trying to monitor. That makes it hard to judge the sound quality you are picking up. Check it during playback. Closed-cup (closed back) headphones or in-the-ear earphones provide the best isolation from outside sounds.

THE CUE SYSTEM

The cue system is a monitor system for musicians to use as they're recording. It includes some of the aux knobs in your mixer, a small power amplifier or headphone amp, a headphone connector box, and headphones. Musicians often can't hear each other well in the studio, but they can listen over headphones to hear each other in a good balance. Also, they can listen to previously recorded tracks while overdubbing.

Headphones for a cue system should be durable and comfortable. They should be closed-cup to avoid leakage into microphones—especially for a click track. Also, the cue "phones" should have a smooth response to reduce listening fatigue, and should play loud without burning out. Make sure they are all the same model so each musician hears the same thing. A built-in volume control is convenient.

A **headphone amplifier** is shown in Figure 5.4. Connect it to a headphone monitor jack on your mixer or audio interface. It drives up to eight headphones and lets you adjust the volume of each one. Examples include the PreSonus HP60 and HP4, ART HeadAmp 4 and HeadAmp 6 PRo, Samson S-Phone, and C-Que 8.

FIGURE 5.4
ART HeadAmp 4, an example of a headphone amplifier

Does your mixer or audio interface have a strong signal at its headphone jack? You could install four 1/4-inch stereo jacks in a small metal box wired in parallel (tip to tip, ring to ring, and sleeve to sleeve). Connect them by a single cable to the headphone jack of your mixer or audio interface.

Although some consoles can provide several independent cue mixes, the ideal situation is to set up a **personal monitor mixer** near each musician. Then they can set their own cue mix and listening level. The inputs of these mixers are fed from the console output buses. Other names for this device are headphone monitor mixer or cue mixer. Some examples are the Behringer PowerPlay System, Furman HR-6, Mackie HMX-56, Aviom A16II, Hear Technologies Hear-Back-Four-Pack, and Roland M-48.

Suppose a vocalist sings into a microphone and hears that mic's signal over the cue headphones. If the singer's voice and the headphone's sound are opposite in polarity, the voice partially cancels or sounds funny in the headphones. Make sure that the voice and headphones are the same polarity.

Here's how. While talking into a mic and listening to it on headphones, reverse the ground and signal leads to the headphones connector. The position that gives the fullest, most solid sound in the headphones is correct.

All the headphones in your studio should be the same model, so that everyone will hear with correct polarity.

CONCLUSION

Ultimately, what you hear from the monitors influences your recording techniques and affects the quality of your recordings. So take the time to plan and adjust the control-room acoustics. Choose and place the speakers carefully. Monitor at proper levels and listen on several systems. You'll be rewarded with a monitor system you can trust.

CHAPTER 6

Microphones

What microphone is best for recording an orchestra? What's a good snare mic? Should the microphone be a condenser or dynamic, omni or cardioid?

You can answer these questions more easily once you know the types of microphones and understand their specs. First, it always pays to get a high-quality microphone. The mic is a source of your recorded signal. If that signal is noisy, distorted, or tonally colored, you'll be stuck with those flaws through the whole recording process. Better get it right up front.

Even if you have a MIDI studio and get your sounds from samples or synthesizers, you still might need a good microphone for sampling, or to record vocals, sax, acoustic guitar, and so on.

A microphone is a transducer—a device that changes one form of energy into another. Specifically, a mic changes sound into an electrical signal. Your mixer amplifies and modifies this signal.

TRANSDUCER TYPES

Mics for recording can be grouped into three types depending on how they convert sound to electricity: dynamic, ribbon, or condenser.

A dynamic mic capsule, or transducer, is shown in Figure 6.1. A coil of wire attached to a diaphragm is suspended in a magnetic field. When sound waves vibrate the diaphragm, the coil vibrates in the magnetic field and generates an electrical signal similar to the incoming sound wave. Another name for a dynamic mic is moving-coil mic, but this term is seldom used.

In a ribbon mic capsule, a thin metal foil or ribbon is suspended in a magnetic field (Figure 6.2). Sound waves vibrate the ribbon in the field and generate an electrical signal.

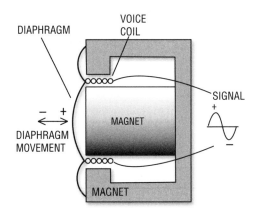

FIGURE 6.1
A dynamic transducer.

FIGURE 6.2
A ribbon transducer.

A condenser or capacitor mic capsule has a conductive diaphragm and a metal backplate placed very close together (Figure 6.3). They are charged with static electricity to form two plates of a capacitor. When sound waves strike the diaphragm, it vibrates. This varies the spacing between the plates. In turn, this varies the capacitance and generates a signal similar to the incoming sound wave. Because of its lower diaphragm mass and higher damping, a condenser mic responds faster than a dynamic mic to rapidly changing sound waves (transients).

Two types of condenser mic are true condenser and electret condenser. In a true condenser mic (externally biased mic), the diaphragm and backplate are charged with a voltage from a circuit built into the mic. In an electret condenser mic, the diaphragm and backplate are charged by an electret material, which is in the diaphragm or on the backplate. Electrets

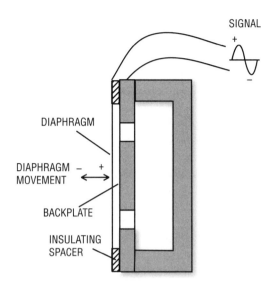

SIGNAL

DIAPHRAGM

DIAPHRAGM − +
MOVEMENT

BACKPLATE

INSULATING
SPACER

FIGURE 6.3
A condenser transducer.

and true condensers can sound equally good, although some engineers prefer true condensers, which tend to cost more.

A condenser mic needs a power supply to operate, such as a battery or phantom power supply. Phantom power is 12 to 48 volts DC applied to pins 2 and 3 of the mic connector through two equal resistors. The microphone receives phantom power and sends audio signals on the same two conductors. Ground for the phantom power supply is through the cable shield. Nearly all mixing consoles and audio interfaces supply phantom power at their mic input connectors. You simply plug the mic into the mixer to power it. Phantom power is explained in detail in Appendix D.

Dynamics and ribbons need no power supply. You can plug these types of mics into a phantom supply without damage because the voice coil or ribbon is not connected to ground (unless they are accidentally shorted to the mic housing, or one cable conductor is shorted to the shield). However, because damage is possible in theory, most ribbon mic manufacturers recommend that phantom power be turned off unless you're using one of the newer ribbon mics that have a built-in preamp that is phantom powered.

Figure 6.4 shows a cutaway view of a typical dynamic vocal mic and condenser instrument mic.

General Traits of Each Transducer Type

Condenser

- Wide, smooth frequency response
- Detailed sound, extended highs

FIGURE 6.4
Inside a typical dynamic vocal mic and condenser instrument mic.

- Omni type has excellent low-frequency response
- Transient attacks sound sharp and clear
- Preferred for acoustic instruments, cymbals, studio vocals
- Can be miniaturized

Dynamic

- Tends to have rougher response, but still quite usable
- Rugged and reliable
- Handles heat, cold, and high humidity
- Handles high volume without distortion
- Preferred for guitar amps and drums
- If flat response, can take the "edge" off woodwinds and brass

Ribbon

- Prized for its warm, smooth tone quality
- Delicate
- Complements digital recording
- Preferred for horns and guitar amps

There are exceptions to the tendencies listed above. Some dynamics have a smooth, wide-range frequency response. Some condensers are rugged and handle high SPLs. It depends on the specs of the particular mic. *Audio clip 13 at www.elsevierdirect.com/ companions/9780240811444 demonstrates the sound of each transducer type.*

POLAR PATTERN

Microphones also differ in the way they respond to sounds coming from different directions. An omnidirectional microphone is equally sensitive to sounds arriving from all directions. A unidirectional mic is most sensitive to sound arriving from one direction—in front of the mic—but softens sounds entering the sides or rear of the mic. A bidirectional mic is most sensitive to sounds arriving from two directions—in front of and behind the mic—but rejects sounds entering the sides.

There are three types of unidirectional patterns: cardioid, supercardioid, and hypercardioid. A mic with a cardioid pattern is sensitive to sounds arriving from a broad angle in front of the mic. It is about 6 dB less sensitive at the sides, and about 15 to 25 dB less sensitive in the rear. The supercardioid pattern is 8.7 dB less sensitive at the sides and has two areas of least pickup at 125 degrees away from the front. The hypercardioid pattern is 12 dB less sensitive at the sides and has two areas of least pickup at 110 degrees away from the front.

To hear how a cardioid pickup pattern works, talk into a cardioid mic from all sides while listening to its output. Your reproduced voice is loudest when you talk into the front of the mic, and softest when you talk into the rear. *Play audio clip 14 at www. elsevierdirect.com/companions/9780240811444.*

The super- and hypercardioid reject sound from the sides more than the cardioid. They are more directional, but they pick up more sound from the rear than the cardioid does.

A microphone's polar pattern is a graph of its sensitivity versus the angle at which sound approaches the mic. The polar pattern is plotted on polar graph paper. Sensitivity is plotted as distance from the origin. Figure 6.5 shows various polar patterns.

Traits of Each Polar Pattern

Omnidirectional

- All-around pickup
- Most pickup of room reverberation (*play audio clip 14 at www.elsevierdirect.com/ companions/9780240811444*)
- Not much isolation unless you mike close
- Low sensitivity to pops (explosive breath sounds)

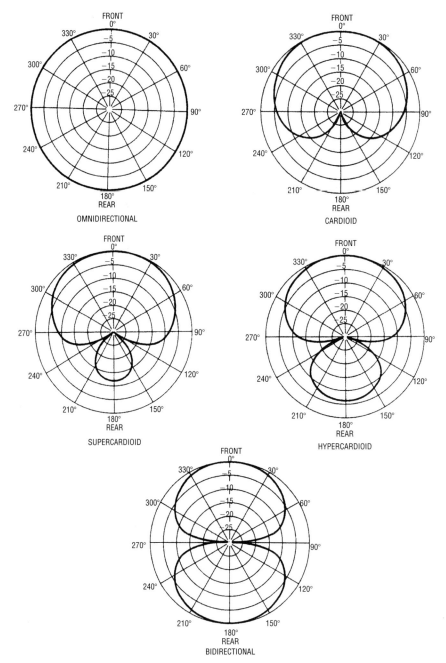

FIGURE 6.5
Various polar patterns. Sensitivity is plotted vs. angle of sound incidence.

- Low handling noise
- No up-close bass boost (proximity effect)
- Extended low-frequency response in condenser mics—great for pipe organ or bass drum in an orchestra or symphonic band
- Lower cost in general

Unidirectional (cardioid, supercardioid, hypercardioid)

- Selective pickup
- Rejection of room acoustics, background noise, and leakage
- Good isolation—good separation between tracks
- Up-close bass boost (except in mics that have holes in the handle)
- Better gain-before-feedback in a sound-reinforcement system
- Coincident or near-coincident stereo miking (explained in Chapter 7)

Cardioid

- Broad-angle pickup of sources in front of the mic
- Maximum rejection of sound approaching the rear of the mic
- Most popular pattern

Supercardioid

- Maximum difference between front hemisphere and rear hemisphere pickup (good for stage-floor miking)
- More isolation than a cardioid
- Less reverb pickup than a cardioid

Hypercardioid

- Maximum side rejection in a unidirectional mic
- Maximum isolation—maximum rejection of reverberation, leakage, feedback, and background noise

Bidirectional

- Front and rear pickup, with side sounds rejected (for across-table interviews or two-part vocal groups, for example)
- Maximum isolation of an orchestral section when miked overhead
- Blumlein stereo miking (two bidirectional mics crossed at 90 degrees)

In a good mic, the polar pattern should be about the same from 200 Hz to 10 kHz. If not, you'll hear off-axis coloration: the mic will have a different tone quality on and off axis. Small-diaphragm mics tend to have less off-axis coloration than large-diaphragm mics.

You can get either the condenser or dynamic type with any kind of polar pattern (except bidirectional dynamic). Ribbon mics are either bidirectional or hypercardioid.

FIGURE 6.6
A typical multipattern mic that is side-addressed.

Some condenser mics come with switchable patterns. Note that the shape of a mic does not indicate its polar pattern.

If a mic is end-addressed, you aim the end of the mic at the sound source. If a mic is side-addressed, you aim the side of the mic at the sound source. Figure 6.6 shows a typical side-addressed condenser mic with switchable polar patterns.

Boundary mics that mount on a surface have a pattern that is half-omni (hemispherical), half-supercardioid, or half-cardioid (like an apple sliced in half through its stem). The boundary mounting makes the mic more directional so it picks up less room acoustics.

FREQUENCY RESPONSE

As with other audio components, a microphone's frequency response is the range of frequencies that it will reproduce at an equal level (within a tolerance, such as ±3 dB).

The following is a list of sound sources and the microphone frequency response that is adequate to record the source with high fidelity. A wider range response works, too.

- Most instruments: 80 Hz to 15 kHz
- Bass instruments: 40 Hz to 9 kHz
- Brass and voice: 80 Hz to 12 kHz
- Piano: 40 Hz to 12 kHz
- Cymbals and some percussion: 300 Hz to 15 or 20 kHz
- Orchestra or symphonic band: 40 Hz to 15 kHz

If possible, use a mic with a response that rolls off below the lowest fundamental frequency of the instrument you're recording. For example, the frequency of the low-E string on an acoustic guitar is about 82 Hz. A mic used on the acoustic guitar should roll off below that frequency to avoid picking up low-frequency noise such as rumble from trucks and air conditioning. Some mics have a built-in low-cut switch for this purpose. Or you can filter out the unneeded lows at your mixer.

A frequency-response curve is a graph of the mic's output level in dB at various frequencies. The output level at 1 kHz is placed at the 0 dB line on the graph, and the levels at other frequencies are so many decibels above or below that reference level.

The shape of the response curve suggests how the mic sounds at a certain distance from the sound source. (If the distance is not specified, it's probably 2 to 3 feet.) For example, a mic with a wide, flat response reproduces the fundamental frequencies and harmonics in the same proportion as the sound source. So a flat-response mic tends to provide accurate, natural reproduction at that distance.

A rising high end or a "presence peak" around 5 to 10 kHz sounds more crisp and articulate because it emphasizes the higher harmonics (Figure 6.7). *Play audio clip 15 at www.elsevierdirect.com/companions/9780240811444*. Sometimes this type of response is called tailored or contoured. It's popular for guitar amps and drums because it adds punch and emphasizes attack. Some microphones have switches that alter the frequency response.

Most uni- and bidirectional mics boost the bass when used within a few inches of a sound source. You've heard how the sound gets bassy when a vocalist sings right into the mic. This low-frequency boost related to close mic placement is called the proximity effect, and it's often plotted on the frequency-response graph. Omni mics have no proximity effect; they sound tonally the same at any distance.

The warmth created by proximity effect adds a pleasing fullness to drums. In most recording situations, though, the proximity effect lends an unnatural boomy or bassy

FIGURE 6.7
An example of the frequency response of a microphone with proximity effect and a presence peak around 5 kHz.

sound to the instrument or voice picked up by the mic. Some mics—multiple-D or variable-D types—are designed to reduce it. These types have holes or slots in the mic handle. Some mics have a bass-rolloff switch to compensate for the bass boost. Or you can roll off the excess bass with your mixer's equalizer until the sound is natural. By doing so, you also reduce low-frequency leakage picked up by the microphone.

Note that mic placement can greatly affect the recorded tone quality. A flat-response mic does not always guarantee a natural sound because mic placement has such a strong influence. Tonal effects of mic placement are covered in Chapter 7.

IMPEDANCE (Z)

This spec is the mic's effective output resistance at 1 kHz. A mic impedance between 150 and 600 ohms is low; 1000 to 4000 ohms is medium; and above 25 kilohms is high.

Always use low-impedance mics. If you do, you can run long mic cables without picking up hum or losing high frequencies. The input impedance of a mixer mic input is about 1500 ohms. If it were the same impedance as the mic, about 250 ohms, the mic would "load down" when you plug it in. Loading down a mic makes it lose level, distort, or sound thin. To prevent this, a mic input has an impedance much higher than that of the microphone. But it's still called a low-Z input. More information on impedance is in Appendix C.

MAXIMUM SPL

To understand this spec, first we need to understand sound pressure level (SPL). It is a measure of the intensity of a sound. The quietest sound we can hear, the threshold of hearing, is 0 dBSPL. Normal conversation at 1 foot measures about 70 dBSPL; painfully loud sound is above 120 dBSPL.

If the maximum SPL spec is 125 dBSPL, the mic starts to distort when the instrument being miked is putting out 125 dBSPL at the mic. A maximum SPL spec of 120 dB is good, 135 dB is very good, and 150 dB is excellent.

Dynamic mics tend not to distort, even with very loud sounds. Some condensers are just as good. Some have a pad you can switch in to prevent distortion in the mic circuitry. Because a mic pad reduces signal-to-noise ratio (S/N), use it only if the mic distorts.

SENSITIVITY

This spec tells how much output voltage a mic produces when driven by a certain SPL. A high-sensitivity mic puts out a stronger signal (higher voltage) than a low-sensitivity mic when both are exposed to an equally loud sound.

A low-sensitivity mic needs more mixer gain than a high-sensitivity mic. More gain usually results in more noise. When you record quiet music at a distance (classical guitar, string quartet), use a mic of high sensitivity to override mixer noise. When you record loud music or mike close, sensitivity matters little because the mic signal level is well above the mixer noise floor. That is, the S/N is high. Listed below are typical sensitivity specs for three transducer types:

- Condenser: 5.6 mV/Pa (high sensitivity)
- Dynamic: 1.8 mV/Pa (medium sensitivity)
- Ribbon or small dynamic: 1.1 mV/Pa (low sensitivity)

The louder the sound source, the higher the signal voltage the mic puts out. A very loud instrument, such as a kick drum or guitar amp, can cause a microphone to generate a signal strong enough to overload the mic preamp in your mixer. That's why most mixers have pads or input-gain controls—to prevent preamp overload from hot mic signals.

SELF-NOISE

Self-noise or equivalent noise level is the electrical noise or hiss a mic produces. It's the dBSPL of a sound source that would produce the same output voltage that the noise does.

Usually the self-noise spec is A-weighted. That means the noise was measured through a filter that makes the measurement correlate more closely with the annoyance value. The filter rolls off low and high frequencies to simulate the frequency response of the ear.

An A-weighted self-noise spec of 14 dBSPL or less is excellent (quiet), 21 dB is very good, 28 dB is good, and 35 dB is fair—not good enough for quality recording.

A mic with a self-noise spec around 30 dB can sound noise-free if it is used very close to a sound source—for example, clipped onto an acoustic guitar or fiddle.

Because a dynamic mic has no active electronics to generate noise, it has very low self-noise (hiss). So most spec sheets for dynamic mics do not specify self-noise.

SIGNAL-TO-NOISE RATIO

This is the difference in decibels between the SPL in dB and a mic's self-noise in dB. The higher the SPL of the sound source at the mic, the higher the S/N. Given an SPL of 94 dB, an S/N spec of 74 dB is excellent; 64 dB is good. The higher the S/N ratio, the cleaner (more noise-free) the signal, and the greater the "reach" of the microphone.

Reach is the clear pickup of quiet, distant sounds due to high S/N. Reach is not specified in data sheets because any mic can pick up a source at any distance if the source is loud enough. For example, even a cheap mic can reach several miles if the sound source is a thunderclap.

POLARITY

The polarity spec relates the polarity of the electrical output signal to the acoustic input signal. The standard is "pin 2 hot." That is, the mic produces a positive voltage at pin 2 with respect to pin 3 when the sound pressure pushes the diaphragm in (positive pressure).

Be sure that your mic cables do not reverse polarity. On both ends of each cable, the wiring should be pin 1 shield, pin 2 red, pin 3 white or black. Or the wiring on both ends should be pin 1 shield, pin 2 white, pin 3 black. If some mic cables are correct polarity and some are reversed, and you mix their mics to mono, the bass may cancel.

MICROPHONE TYPES

The following sections describe several types of recording mics.

Large-Diaphragm Condenser Microphone (LDC)

This is a condenser microphone, usually side-addressed, with a diaphragm 1 inch or larger in diameter (Figure 6.6). It generally has very good low-frequency response and low self-noise. Common uses are studio vocals and acoustic instruments. Examples are the AKG C12VR; C414; Perception 100 and 200; C2000B and C3000B; Apex 530; Audio-Technica AT2020/3035/4040; Audix SCX25; Blue Blueberry; Blue Reactor; CAD Equitek Series, GXL Series and M177; DPA 4041; Karma Mics Unity; Lawson L47MP MKII and L251; Mojave Audio MA-301 fet; Manley Gold Reference; Neumann U87; U47, and TLM 103; Sennheiser MK 4; Soundelux Elux 251; ADK various models; SE Electronics various models; Shure KSM Series; Shure Beta 181; MXL V67G, V69, 900, 2001, 2003, and 2006; Rode NT1A; Studio Projects B and C Series; Samson CL7 and C01; Nady SCM 950 and 100; Violet Flamingo; M-Audio Solaris, Sputnik, Luna, and Nova; and Behringer B1 and B2.

Small-Diaphragm Condenser Microphone (SDC)

This is a stick-shaped or "pencil" condenser microphone, usually cardioid and end-addressed, with a diaphragm under 1 inch in diameter (Figure 6.4). It generally has very good transient response and detail, making it a fine choice for close miking acoustic instruments—especially cymbals, acoustic guitar, and piano. Examples are the AKG C 451 B; Audio-Technica AT 3031 and AT 4051a; Audix SCX1, ADX50, and ADX51; Berliner CM-33, CAD Equitek e60; M-Audio Pulsar; Samson C02; Crown CM-700; DPA 4006; Mojave MA-100 and MA-101SP; Neumann KM 184; Sennheiser e614 and

MKH50; Shure KSM109/SL, KSM137/Sl, and SM81; MXL 600 and 603S; Behringer B5 with cardioid capsule; and Studio Projects C4.

Dynamic Instrument Microphone

This is a stick-shaped dynamic microphone, end-addressed (Figure 6.4). Although it may have a flat response, it generally has a presence peak and some low-frequency roll-off to prevent boominess when used up close. It's often used on drums and guitar amps. Examples include the Shure SM57; AKG D112 (kick drum); Audio-Technica AT AE2500 (kick); Beyerdynamic M-99; Electro-Voice N/D868 (kick); Heil PR40 (kick); Audix D2, D3, D6, and i5; and Sennheiser MD421, e604 and e602 (kick). A drum microphone kit is a collection of microphones for a drum set, such as the Audix DP QUAD mic pack.

Live-Vocal Microphone

This unidirectional mic is shaped like an ice-cream cone because of its large grille used to reduce breath pops. It can be a condenser, dynamic, or ribbon type, and it usually has a presence peak and some low-frequency rolloff. Examples are the Shure SM58; Beta 58A, SM85, SM86, SM87A, and KSM9; AKG D3800; C 535 EB, C5, and D 5/D 5S; Audix OM5 and OM7; Beyerdynamic M88 TG and TG V70d; CAD D189; EV N-Dym Series; Sennheiser e945; and Neumann KMS104 and KMS105.

Ribbon Microphone

This mic can be side- or end-addressed. It generally is used wherever you want a warm, smooth tone quality (sometimes with reduced highs). Examples include models by Beyerdynamic, Coles, Royer, Cascade, Audio-Technica, Shure, Blue, and AEA.

Boundary Microphone

Boundary mics are designed to be used on surfaces. Tape them to the underside of a piano lid, or tape them to the wall for pickup of room ambience. They can be used on hard baffles between instruments, or on panels to make the mics directional. A boundary mic uses a mini condenser mic capsule mounted very near a sound-reflecting plate or boundary (Figure 6.8). Because of this construction, the mic picks up direct sound and reflected sound at the same time, in phase at all frequencies. So you get a smooth response free of phase cancellations. A conventional mic near a surface sounds colored; a boundary mic on a surface sounds natural. Other benefits are a wide, smooth frequency response free of phase cancellations, excellent clarity and reach, and the same tone quality anywhere around the mic.

Examples of half-omnidirectional or hemispherical boundary mics include the AKG CBL 99, Audio-Technica models, Beyerdynamic models, Crown PZM-30D and PZM-6D, and Shure Beta 91.

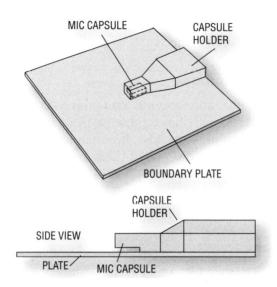

FIGURE 6.8
Typical PZM construction.

Some boundary mics have a half-cardioid or half-supercardioid polar pattern. They work great on a conference table, or near the front edge of a stage floor to pick up drama or musicals. Examples are the Crown PCC-170, and the Bartlett TM-125, TM-125C, and Floor Recording Mic.

Miniature Microphone

Mini condenser mics can be attached to drum rims, flutes, horns, acoustic guitars, fiddles, and so on. Their tone quality is about as good as larger studio microphones and the price is relatively low. With these tiny units you can mike a band in concert without cluttering the stage with boom stands (Figure 6.9), or you can mike a whole drum set with two or three of these. Although you lose individual control of each drum in the mix, the cost is low and the sound is quite good with some bass and treble boost. Compared to large mics, mini mics tend to have more noise (hiss) in distant-miking applications. A lavalier mic is a mini mic worn on the chest to pick up speech from a newscaster or a wandering lecturer. Examples are the AKG C 516, C 518, and C 519; Bartlett Spark Mini Mic, Guitar Mic, and Fiddle Mic; Shure Beta 98S; Audix M1245 and Micro-D; Countryman Isomax B6; DPA 4060; and Sennheiser e608.

Stereo Microphone

A stereo microphone combines two directional mic capsules in a single housing for convenient stereo recording (Figure 6.10). Simply place the mic a suitable distance and height from the sound source, and you'll get a stereo recording with little fuss. Examples include the Audio-Technica AT2022, AT4050ST, and AT8022; Neumann SM69; Shure VP88; AEA R88 MK2; and Royer SF-12.

FIGURE 6.9
A mini mic is the size of a penny.

FIGURE 6.10
Audio-Technica AT8022, an example of a stereo microphone.

Because there is no spacing between the mic capsules, there also is no delay or phase shift between their signals. Coincident stereo microphones are mono-compatible—the frequency response is the same in mono and stereo—because there are no phase cancellations if the two channels are combined.

Digital Microphone

This condenser microphone has a built-in analog-to-digital converter. It is usually side-addressed, has a large diaphragm, has a flat response, and very low self-noise. Its output is a digital signal, which is immune to picking up hum. An example is the Neumann Solution-D.

Headworn Microphone

This microphone is used for a live performance that might be recorded. It is a small condenser mic worn on the head, either omni- or unidirectional. The headworn mic allows the performer freedom of movement on stage. Some models provide excellent gain before feedback and isolation. Examples include the AKG C 520, Audio-Technica BP892 MicroSet, Samson SE50, Sennheiser ME 3-EW, Shure WBH53, Rode HS1-P, Countryman Isomax E6, Crown CM-311A, and DPA 4066F/4088F.

MICROPHONE SELECTION

Table 6.1 is a guide to choosing a mic based on your requirements.

Table 6.1 Mic Application Guide	
Requirement	**Characteristic**
Natural, smooth tone quality	Flat frequency response
Bright, present tone quality	Presence peak (emphasis around 5 kHz)
Extended lows	Omni condenser or dynamic with good low-frequency response
Extended highs (detailed sound)	Condenser
Reduced "edge" or detail	Dynamic or ribbon
Boosted bass up close	Directional mic
Flat bass response up close	Omni mic, or directional mic with sound holes in the handle
Reduced pickup of leakage, feedback, and room acoustics	Directional mic, or omni mic up close
Enhanced pickup of room acoustics	Omni mics
Miking close to a surface, even coverage of moving sources or large sources, inconspicuous mic	Boundary mic
Coincident or near-coincident stereo (see Chapter 18)	Stereo mic
Extra ruggedness	Dynamic mic
Reduced handling noise	Omni mic, or unidirectional with shock mount
Reduced breath popping	Omni mic, or unidirectional with pop filter
Distortion-free pickup of very loud sounds	Condenser with high maximum SPL spec, or dynamic
Low self-noise, high sensitivity, noise-free pickup of quiet sounds	Large-diaphragm condenser mic

Suppose you want to record a grand piano playing with several other instruments. You need the microphone to reduce leakage. Table 6.1 recommends a unidirectional mic or an omni mic up close. For this particular piano, you also want a natural sound, for which the table suggests a mic with a flat response. You want a detailed sound, so a condenser mic is the choice. A microphone with all these characteristics is a flat-response, unidirectional condenser mic. If you're miking close to a surface (the piano lid), a boundary mic is recommended.

Now suppose you're recording an acoustic guitar on stage, and the guitarist roams around. This is a moving sound source, for which the table recommends a mini mic attached to the guitar. Feedback and leakage might not be a problem because you're miking close, so you can use an omni mic. Thus, an omni condenser mic is a good choice for this application.

For a home studio, a suggested first choice is a cardioid condenser mic with a flat frequency response. This type of mic is especially good for studio vocals, cymbals, percussion, and acoustic instruments. Remember that the mic needs a power supply to operate, such as a battery or phantom power supply.

Your second choice of microphone for a home studio is a cardioid dynamic microphone with a presence peak in the frequency response. This type is good for drums and guitar amps. I recommend cardioid over omni for a home studio. The cardioid pattern rejects the leakage, background noise, and room reverb often found in home studios. An omni mic, however, can do that, too, if you mike close enough. Also, omni mics tend to provide a more natural sound at lower cost, and they have no proximity effect.

MIC ACCESSORIES

There are many devices used with microphones to route their signals or to make them more useful. These include pop filters, stands and booms, shock mounts, cables and connectors, stage boxes and snakes, and splitters.

Pop Filter

A much needed accessory for a vocalist's microphone is a pop filter or windscreen. It usually is a foam "sock" that you put over the mic. Some microphones have pop filters or ball-shaped grilles built in.

Why is it needed? When a vocalist sings a word starting with "p," "b," or "t" sounds, a turbulent puff of air is forced from the mouth. A microphone placed close to the mouth is hit by this air puff, resulting in a thump or little explosion called a pop. The windscreen reduces this problem.

The best type of pop filter is a nylon screen in a hoop, or a perforated-metal disk, placed a few inches from the mic.

You can also reduce pop by placing the mic above or to the side of the mouth, or by using an omni mic. *Audio clip 16 at www.elsevierdirect.com/companions/9780240811444 demonstrates how a pop filter or mic placement can prevent breath pops.*

Stands and Booms

Stands and booms hold the microphones and let you position them as desired. A mic stand has a heavy metal base that supports a vertical pipe. At the top of the pipe is a rotating clutch that lets you adjust the height of a smaller telescoping pipe inside the large one. The top of the small pipe has a standard 5/8-inch 27 thread, which screws into a mic stand adapter.

A boom is a long horizontal pipe that attaches to the vertical pipe. The angle and length of the boom are adjustable. The end of the boom is threaded to accept a mic stand adapter, and the opposite end is weighted to balance the weight of the microphone.

Shock Mount

A shock mount holds a mic in a resilient suspension to isolate the mic from mechanical vibrations, such as floor thumps and mic-stand bumps. Many mics have an internal shock mount which isolates the mic capsule from its housing; this reduces handling noise as well as stand thumps.

Cables and Connectors

Mic cables carry the electrical signal from the mic to the mixing console, mic preamp, or recorder. With low-impedance mics, you can use hundreds of feet of cable without hum pickup or high-frequency loss. Some mics have a permanently attached cable for convenience and low cost; others have a connector in the handle to accept a separate mic cable. The second method is preferred for serious recording because if the cable breaks, you have to repair or replace only the cable, not the whole microphone.

Mic cables are made of one or two insulated conductors surrounded by a fine-wire mesh shield to keep out electrostatic hum. If you hear a loud buzz when you plug in a microphone, check that the shield is securely soldered in place.

After acquiring a microphone, you may need to wire its 2-conductor shielded cable to a 3-pin XLR audio connector. Here are the solder connections:

Pin 1: shield
Pin 2: "hot" or "in-polarity" lead (usually red or white)
Pin 3: "cold" or "out-of-polarity" lead (usually black)

If the mic output is a 3-pin XLR, but your recorder or mixer mic input is an unbalanced phone jack, a different wiring is needed:

Phone-plug tip (the short terminal): hot lead
Phone-plug sleeve (the long terminal): shield and cold lead

Wind your mic cables onto a large spool, which can be found in the electrical section of hardware stores. Plug the cables together as you wind them.

Snake

It is messy and time-consuming to run mic cables from several mics all the way to a mixer. Instead, you can plug all your mics into a stage box with several connectors (Figure 6.11). The snake—a thick multiconductor cable—carries the signals to the mixer. At the mixer end, the cable divides into several mic connectors that plug into the mixer.

Splitter

When you record a band in concert, you might want to feed each mic's signal to your recording mixer and to the band's PA and monitor mixers. A mic splitter does the job. For each microphone channel, it has one XLR input for a microphone, a "direct" XLR output wired to the input, and one or more transformer-isolated XLR outputs with a ground-lift switch. The mixer that provides phantom power must be connected to the direct XLR output. There's more on splitters in Chapter 17 under the heading "Splitting the Mic Signals."

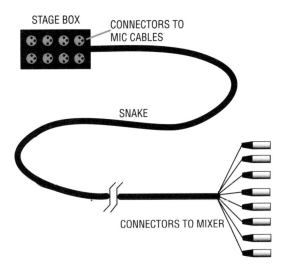

FIGURE 6.11
A stage box and snake.

SUMMARY

We talked about some mic types, specs, and accessories. You should have a better idea about what kind of microphone to choose for your own applications.

Mic manufacturers are happy to send you free catalogs and application notes, which are also available on mic company websites. Mic dealers also may have this literature.

Remember, you can use any microphone on any instrument if it sounds good to you. Just try it and see if you like it. To make high-quality recordings, though, you need good mics with a smooth, wide-range frequency response, low noise, and low distortion.

CHAPTER 7

Microphone Technique Basics

Suppose you're going to mike a singer, a sax, or a guitar. Which mic should you choose? Where should you place it?

Your mic technique has a powerful effect on the sound of your recordings. In this chapter we'll look at some general principles of miking that apply to all situations. Chapter 8 covers common mic techniques for specific instruments.

WHICH MIC SHOULD I USE?

Is there a "right" mic to use on a piano, a kick drum, or a guitar amp? No. Every microphone sounds different, and you choose the one that gives you the sound you want. Still, it helps to know about two main characteristics of mics that affect the sound: frequency response and polar pattern.

Most condenser mics have an extended high-frequency response—they reproduce sounds up to 15 or 20 kHz. This makes them great for cymbals or other instruments that need a detailed sound, such as acoustic guitar, strings, piano, and voice. Dynamic moving-coil microphones have a response good enough for drums, guitar amps, horns, and woodwinds. Loud drums and guitar amps sound dull if recorded with a flat-response mic; a mic with a presence peak (a boost around 5 kHz) gives more edge or punch.

Suppose you are choosing a microphone for a particular instrument. In general, the frequency response of the mic should cover at least the frequencies produced by that instrument. For example, an acoustic guitar produces fundamental frequencies from 82 Hz to about 1 kHz, and produces harmonics from about 1 to 15 kHz. So a mic used on an acoustic guitar should have a frequency response of at least 82 Hz to 15 kHz if you want to record the guitar accurately. Table 7.1 shows the frequency ranges of various instruments.

The polar pattern of a mic affects how much leakage and ambience it picks up. Leakage is unwanted sound from instruments other than the one at which the mic is aimed.

Table 7.1	Frequency Ranges of Various Musical Instruments	
Instrument	**Fundamentals (Hz)**	**Harmonics (kHz)**
Flute	261–2349	3–8
Oboe	261–1568	2–12
Clarinet	165–1568	2–10
Bassoon	62–587	1–7
Trumpet	165–988	1–7.5
French horn	87–880	1–6
Trombone	73–587	1–7.5
Tuba	49–587	1–4
Snare drum	100–200	1–20
Kick drum	30–147	1–6
Cymbals	300–587	1–15
Violin	196–3136	4–15
Viola	131–1175	2–8.5
Cello	65–698	1–6.5
Acoustic bass	41–294	1–5
Electric bass	41–300	1–7
Acoustic guitar	82–988	1–15
Electric guitar	82–1319	1–3.5 (through amp)
Electric guitar	82–1319	1–15 (direct)
Piano	28–4196	5–8
Bass (voice)	87–392	1–12
Tenor (voice)	131–494	1–12
Alto (voice)	175–698	2–12
Soprano (voice)	247–1175	2–12

Ambience is the acoustics of the recording room—its early reflections and reverb. The more leakage and ambience you pick up, the more distant the instrument sounds.

An omni mic picks up more ambience and leakage than a directional mic when both are the same distance from an instrument. So an omni tends to sound more distant. To compensate, you have to mike closer with an omni.

HOW MANY MICS?

The number of mics you need varies with what you're recording. If you want to record an overall acoustic blend of the instruments and room ambience, use just two

FIGURE 7.1
Overall miking of a musical ensemble with two distant microphones.

FIGURE 7.2
Individual miking with multiple close mics and a mixer.

microphones or a stereo mic (Figure 7.1). This method works great on an orchestra, symphonic band, choir, string quartet, pipe organ, small folk group, or a piano/voice recital. Stereo miking is covered in detail later in this chapter.

To record a pop-music group, you mike each instrument or instrumental section. Then you adjust the mixer volume control for each mic to control the balance between instruments (Figure 7.2).

FIGURE 7.3
Multiple miking with several sound sources on each microphone.

To get the clearest sound, don't use two mics when one will do the job. Sometimes you can pick up two or more sound sources with one mic (Figure 7.3). You could mike a brass section of four players with one mic on four players, or with two mics on every two players. Or mike a choir in a studio in four groups: put one mic on the basses, one on the sopranos, and so on.

Picking up more than one instrument with one mic has a problem: during mixdown, you can't adjust the balance among instruments recorded on the same track. You have to balance the instruments before recording them. Monitor the mic, and listen to see if any instrument is too quiet. If so, move it closer to the mic.

HOW CLOSE SHOULD I PLACE THE MIC?

Once you've chosen a mic for an instrument, how close should the mic be? Mike a few inches away to get a tight, present sound; mike farther away for a distant, spacious sound. (Try it to hear the effect.) *Play audio clip 17 at www.elsevierdirect.com/companions/9780240811444.* The farther a mic is from the instrument, the more ambience, leakage, and background noise it picks up. So mike close to reject these unwanted sounds. Mike farther away to add a live, loose, airy feel to overdubs of drums, lead-guitar solos, horns, etc. You'll need a room with good acoustics for best results.

Close miking sounds close; distant miking sounds distant. Here's why. If you put a mic close to an instrument, the sound at the mic is loud. So you need to turn up the mic gain on your mixer only a little to get a full recording level. And because the gain is low, you pick up very little reverb, leakage, and background noise (Figure 7.4A).

If you put a mic far from an instrument, the sound at the mic is quiet. You'll need to turn up the mic gain a lot to get a full recording level. And because the gain is high, you pick up a lot of reverb, leakage, and background noise (Figure 7.4B).

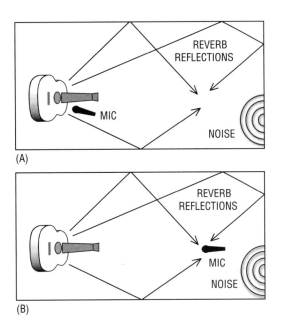

FIGURE 7.4
(A) A close microphone picks up mainly direct sound, which results in a close sound quality. (B) A distant microphone picks up mainly reflected sound, which results in a distant sound quality.

If the mic is very far away—maybe 10 feet—it's called an ambience mic or room mic. It picks up mostly room reverb. A popular mic for ambience is a boundary microphone taped to the wall. You mix it with the usual close mics to add a sense of space. Use two for stereo. When you record a live concert, you might want to place ambience mics over the audience, aiming at them from the front of the hall, to pick up the crowd reaction and the hall acoustics.

Classical music is always recorded at a distance (about 4 to 20 feet away) so that the mics will pick up reverb from the concert hall. It's a desirable part of the sound.

Leakage (Bleed or Spill)

Suppose you're close-miking a drum set and a piano at the same time (Figure 7.5). When you listen to the drum mics alone, you hear a close, clear sound. But when you mix in the piano mic, that nice, tight drum sound degrades into a distant, muddy sound. That's because the drum sound leaked into the piano mic, which picked up a distant drum sound from across the room. *Audio clip 11 at www.elsevierdirect.com/companions/9780240811444 is an example of leakage.*

There are many ways to reduce leakage:

- Mike each instrument closely. That way the sound level at each mic is high. Then you can turn down the mixer gain of each mic, which reduces leakage at the same time.

FIGURE 7.5
Example of leakage. The piano mic picks up leakage from the drums, which changes the close drum sound to distant.

- Overdub each instrument one at a time.
- Record direct. Record an acoustic guitar off its pickup during tracking, then overdub the guitar with a mic. Record an electric guitar off its pickup during tracking, then play the guitar signal through a guitar-amp modeling plug-in during mixdown. Or record the electric guitar through a Line 6 Pod, which is a guitar-amp emulator.
- Filter out frequencies above and below the range of each instrument.
- Use directional mics (cardioid, etc.) instead of omni mics.
- Record in a large, fairly dead studio. In such a room, leakage reflected from the walls is weak.
- Put portable walls (goboes) between instruments.
- Use noise gates on drum tracks (see Chapter 9 under the heading "Noise Gate").

Don't Mike Too Close

Miking too close can color the recorded tone quality of an instrument. If you mike very close, you might hear a bassy or honky tone instead of a natural sound. Why? Most musical instruments are designed to sound best at a distance, at least 1½ feet away. The sound of an instrument needs some space to develop. A mic placed a foot or two away tends to pick up a well-balanced, natural sound. That is, it picks up a blend of all the parts of the instrument that contribute to its character or timbre.

Think of a musical instrument as a loudspeaker with a woofer, midrange, and tweeter. If you place a mic a few feet away, it will pick up the sound of the loudspeaker accurately. But if you place the mic close to the woofer, the sound will be bassy. Similarly, if you mike close to an instrument, you emphasize the part of the instrument that the microphone is near. The tone quality picked up very close may not reflect the tone quality of the entire instrument.

Suppose you place a mic next to the sound hole of an acoustic guitar, which resonates around 80 to 100 Hz. A microphone placed there hears this bassy resonance, giving a boomy recorded timbre that does not exist at a greater miking distance. To make the guitar sound more natural when miked close to the sound hole, you need to roll off the excess bass on your mixer, or use a mic with a bass rolloff in its frequency response.

The sax projects highs from the bell, but projects mids and lows from the tone holes. So if you mike close to the bell, you miss the warmth and body from the tone holes. All that's left at the bell is a harsh tone quality. You might like that sound, but if not, move the mic out and up to pick up the entire instrument. If leakage forces you to mike close, change the mic or use equalization (EQ).

Usually, you get a natural sound if you put the mic as far from the source as the source is big. That way, the mic picks up all the sound-radiating parts of the instrument about equally. For example, if the body of an acoustic guitar is 18 inches long, place the mic 18 inches away to get a natural tonal balance. If this sounds too distant or hollow, move in a little closer.

WHERE SHOULD I PLACE THE MIC?

Suppose you have a mic placed a certain distance from an instrument. If you move the mic left, right, up, or down, you change the recorded tone quality. In one spot, the instrument might sound bassy; in another spot, it might sound natural, and so on. So, to find a good mic position, simply place the mic in different locations—and monitor the results—until you find one that sounds good to you.

Here's another way to do the same thing. Close one ear with your finger, listen to the instrument with the other ear, and move around until you find a spot that sounds good. Put the mic there. Then make a recording and see if it sounds the same as what you heard live. Don't try this with kick drums or screaming guitar amps! You could also move a mic around while monitoring its signal with good headphones.

Why does moving the mic change the tone quality? A musical instrument radiates a different tone quality in each direction. Also, each part of the instrument produces a different tone quality. For example, Figure 7.6 shows the tonal balances picked up at various spots near a guitar. *Audio clip 18 at www.elsevierdirect.com/companions/9780240811444 illustrates the effect of mic placement on guitar tonal balance. Audio clip 19 demonstrates close and distant stereo miking of the acoustic guitar.*

Other instruments work the same way. A trumpet radiates strong highs directly out of the bell, but does not project them to the sides. So a trumpet sounds bright when miked on-axis to the bell and sounds more natural or mellow when miked off to one side. A grand piano miked one foot over the middle strings sounds fairly natural, under the soundboard sounds bassy and dull, and in a sound hole it sounds mid-rangy.

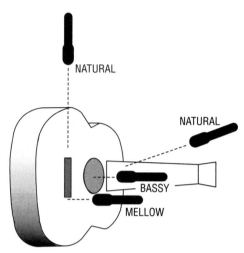

FIGURE 7.6
Microphone placement affects the recorded tonal balance.

It pays to experiment with all sorts of mic positions until you find a sound you like. There is no one right way to place the mics because you place them to get the tonal balance you want.

ON-SURFACE TECHNIQUES

Sometimes you're forced to place a mic near a hard reflecting surface. Examples include:

- Recording drama or opera with the mics near the stage floor
- Recording an instrument that has hard surfaces around it
- Recording a piano with the mic close to the lid

In these cases, you'll often pick up an unnatural, filtered tone quality. Here's why. Sound travels to the microphone via two paths: directly from the sound source, and reflected off the nearby surface (Figure 7.7). Because of its longer travel path, the reflected sound is delayed compared to the direct sound. The direct and delayed sound waves combine at the mic, which causes phase cancellations of various frequencies. The series of peaks and dips in the response is called a comb-filter effect, and it sounds like mild flanging.

Boundary mics solve the problem. In a boundary mic, the diaphragm is very close to the reflecting surface so that there is no delay in the reflected sound. Direct and reflected sounds add in phase over the audible range of frequencies, resulting in a flat response (Figure 7.8). *Play audio clip 20 at www.elsevierdirect.com/ companions/9780240811444.*

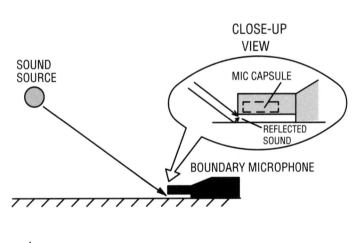

FIGURE 7.7
A mic placed near a surface picks up direct sound and a delayed reflection, which gives a comb-filter frequency response.

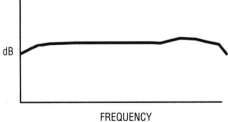

FIGURE 7.8
A boundary mic on the surface picks up direct and reflected sounds in phase.

You might tape an omni boundary mic to the underside of a piano lid, to a hard-surfaced panel, or to a wall for ambience pickup. A unidirectional boundary mic works great on a stage floor to pick up drama. A group of these mics will clearly pick up people at a conference table.

THE THREE-TO-ONE RULE

Suppose you're recording a singer/guitarist. There's a mic on the singer and a mic on the acoustic guitar. When you monitor the mix, something's wrong: the singer's voice sounds hollow or filtered. You're hearing the effect of phase interference.

In general, if two mics pick up the same sound source at different distances, and their signals are mixed to the same channel, this might cause phase cancellations. These are peaks and dips in the frequency response—a comb filter—caused by some frequencies combining out of phase. The result is a colored, filtered tone quality that sounds like mild flanging.

To prevent this problem, follow the three-to-one rule: space the mics at least three times the mic-to-source distance (as in Figure 7.9). For example, if two mics are 12 inches apart, they should be less than 4 inches from their sound sources to prevent phase cancellations. *Play audio clip 21 at www.elsevierdirect.com/companions/9780240811444.* The mics can be closer together than 3:1 if you use two cardioid mics aiming in opposite directions. The goal is to get at least 9 dB of separation between recorded tracks.

What if you pick up an instrument with two mics that are panned left and right? You don't get phase interference. Instead you get stereo imaging.

OFF-AXIS COLORATION

Some mics have off-axis coloration—a dull or colored effect on sound sources that are not directly in front of the mic. Try to aim the mic at sound sources that put out high frequencies, such as cymbals. When you pick up a large source such as an orchestra,

FIGURE 7.9
The three-to-one rule of microphone placement avoids phase interference between microphone signals.

use a mic that has the same response over a wide angle. Such a mic has similar polar patterns at middle and high frequencies. Most large-diaphragm mics have more off-axis coloration than smaller mics (under 1 inch).

STEREO MIC TECHNIQUES

Stereo mic techniques capture the sound of a musical group as a whole, using only two or three microphones. When you play back a stereo recording, you hear phantom images of the instruments in various spots between the speakers. These image locations—left to right, front to back—correspond to the instrument locations during the recording session.

Stereo miking is the preferred way to record classical-music ensembles and soloists. In the studio, you can stereo-mike a piano, drum set cymbals, vibraphone, harmony singers, or other large sound sources.

Goals of Stereo Miking

One goal is accurate localization. That is, instruments in the center of the group are reproduced midway between the two speakers. Instruments at the sides of the group are heard from the left or right speaker. Instruments halfway to one side are heard halfway to one side, and so on.

Figure 7.10 shows three stereo localization effects. Figure 7.10A shows some instrument positions in an orchestra: left, left-center, center, right-center, right. In Figure 7.10B, the

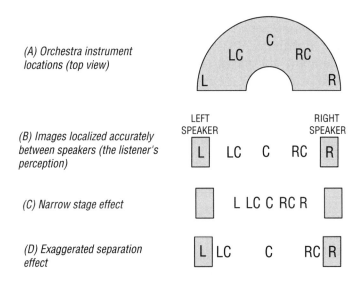

FIGURE 7.10
Stereo localization effects. (A) Orchestra instrument locations (top view). (B) Images localized accurately between speakers (the listener's perception). (C) Narrow stage effect. (D) Exaggerated separation effect.

reproduced images of these instruments are accurately localized between the speakers. The stereo spread, or stage width, extends from speaker to speaker. (You might want to record a string quartet with a narrower spread.)

If you space or angle the mics too close together, you get a narrow stage effect (Figure 7.10C). If you space or angle the mics too far apart, you hear exaggerated separation (Figure 7.10D). That is, instruments halfway to one side are heard near the left or right speaker.

To judge stereo effects, you have to sit exactly between your monitor speakers (the same distance from each). Sit as far from the speakers as the spacing between them. Then the speakers appear to be 60 degrees apart. This is about the same angle an orchestra fills when viewed from a typical ideal seat in the audience (say, tenth row center). If you sit off-center, the images shift toward the side on which you're sitting and are less sharp.

Types of Stereo Mic Techniques

To make a stereo recording, you use one of these basic techniques:

- Coincident pair (XY or MS)
- Spaced pair (AB)
- Near-coincident pair (ORTF, etc.)
- Baffled pair (sphere, OSS, SASS, PZM wedge, etc.)

Let's look at each technique.

COINCIDENT PAIR

With this method, you mount two directional mics with grilles touching, diaphragms one above the other, and angled apart (Figure 7.11). For example, mount two cardioid mics with one grille above the other, and angle them 120 degrees apart. You can use other patterns too: supercardioid, hypercardioid, or bidirectional. The wider the angle between mics, the wider the stereo spread. If the angle is too wide, center images will be weak (there will be a "hole in the middle").

How does this technique make images you can localize? A directional mic is most sensitive to sounds in front of the mic (on-axis) and progressively less sensitive to sounds arriving off-axis. That is, a directional mic puts out a high-level signal from the sound source it's aimed at, and produces lower-level signals from other sound sources.

The coincident pair uses two directional mics that are angled symmetrically from the center line (Figure 7.11). Instruments in the center of the group make the same signal from each mic. During playback, you hear a phantom image of the center instruments midway between your speakers. That's because identical signals in each channel produce an image in the center.

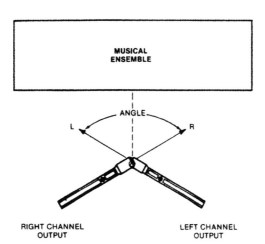

FIGURE 7.11
Coincident-pair technique.

If an instrument is off-center to the right, it is more on-axis to the right-aiming mic than to the left-aiming mic. So the right mic will produce a higher level signal than the left mic. During playback of this recording, the right speaker will play at a higher level than the left speaker. This reproduces the image off-center to the right—where the instrument was during recording.

The coincident pair codes instrument positions into level differences between channels. During playback, the brain decodes these level differences back into corresponding image locations. A pan pot in a mixing console works on the same principle. If one channel is 15 to 20 dB louder than the other, the image shifts all the way to the louder speaker.

Suppose we want the right side of the orchestra to be reproduced at the right speaker. That means the far-right musicians must produce a signal level 20 dB higher from the right mic than from the left mic. This happens when the mics are angled far enough apart. The correct angle depends on the polar pattern.

Instruments partway off-center produce interchannel level differences less than 20 dB, so you hear them partway off-center.

Listening tests have shown that coincident cardioid mics tend to reproduce the musical group with a narrow stereo spread. That is, the group does not spread all the way between speakers.

A coincident-pair method with excellent localization is the Blumlein array. It uses two bidirectional mics angled 90 degrees apart and facing the left and right sides of the group.

A special form of the coincident-pair technique is mid-side or MS (Figure 7.12). In this method, a cardioid or omni mic faces the middle of the orchestra. A matrix circuit

TO MUSICAL
ENSEMBLE

MID
MIC

–

+

SIDE MIC

RIGHT CHANNEL = MID + SIDE
LEFT CHANNEL = MID – SIDE

FIGURE 7.12
Mid-side (MS) technique.

sums and differences the cardioid mic with a bidirectional mic aiming to the sides. This produces left- and right-channel signals. You can remotely control the stereo spread by changing the ratio of the mid signal to the side signal. This remote control is useful at live concerts, where you can't physically adjust the mics during the concert. MS localization can be accurate. Chapter 18, under the heading "Stereo-Spread Control," explains how to replace the matrix with a computer DAW.

To make coincident recordings sound more spacious, boost the bass 4 dB (+2 dB at 600 Hz) in the L–R or side signal.

A recording made with coincident mics is mono compatible. That is, the frequency response is the same in mono or stereo. Because the mics occupy almost the same point in space, there is no time or phase difference between their signals. And when you combine them to mono, there are no phase cancellations to degrade the frequency response. If you expect that your recordings will be heard in mono (say, on TV), then you'll probably want to use coincident methods.

SPACED PAIR

Here, you mount two identical mics several feet apart and aim them straight ahead (Figure 7.13). The mics can have any polar pattern, but omni is most popular for this method. The greater the spacing between mics, the greater the stereo spread.

How does this method work? Instruments in the center of the group make the same signal from each mic. When you play back this recording, you hear a phantom image of the center instruments midway between your speakers. If an instrument is off-center, it is closer to one mic than the other, so its sound reaches the closer microphone

FIGURE 7.13
Spaced-pair technique.

before it reaches the other one. Both mics make about the same signal, except that one mic signal is delayed compared with the other.

If you send a signal to two speakers with one channel delayed, the sound image shifts off-center. With a spaced-pair recording, off-center instruments produce a delay in one mic channel, so they are reproduced off-center.

The spaced pair codes instrument positions into time differences between channels. During playback, the brain decodes these time differences back into corresponding image locations.

A delay of 1.2 msec is enough to shift an image all the way to one speaker. You can use this fact when you set up the mics. Suppose you want to hear the right side of the orchestra from the right speaker. The sound from the right-side musicians must reach the right mic about 1.2 msec before it reaches the left mic. To make this happen, space the mics about 2 to 3 feet apart. This spacing creates the correct delay to place right-side instruments at the right speaker. Instruments partway off-center make interchannel delays less than 1.2 msec, so they are reproduced partway off-center.

If the spacing between mics is, say, 12 feet, then instruments that are slightly off-center produce delays between channels that are greater than 1.2 msec. This places their images at the left or right speaker. I call this "exaggerated separation" or a "ping-pong" effect (Figure 7.10D).

On the other hand, if the mics are too close together, the delays produced will be too small to provide much stereo spread. Also, the mics will tend to emphasize instruments in the center because the mics are closest to them.

To record a good musical balance of an orchestra, you might need to space the mics about 10 or 12 feet apart. But then you get too much separation. You could place a third mic midway between the outer pair and mix its output to both channels. That way, you pick up a good balance, and you hear an accurate stereo spread.

The spaced-pair method tends to make off-center images unfocused or hard to localize. Why? Spaced-pair recordings have time differences between channels. Stereo images produced solely by time differences are unfocused. You still hear the center instruments clearly in the center, but off-center instruments are hard to pinpoint. Spaced-pair miking is a good choice if you want the sonic images to be diffuse or blended, instead of sharply focused.

Another flaw of spaced mics is that if you mix both mics to mono, you may get phase cancellations of various frequencies. This may or may not be audible.

Spaced mics, however, give a "warm" sense of ambience, in which the concert-hall reverb seems to surround the instruments and, sometimes, the listener. Here's why: the two channels of recorded reverb are incoherent, that is, they have random phase relationships. Incoherent signals from stereo speakers sound diffuse and spacious.

Because spaced mics pick up reverb incoherently, it sounds diffuse and spacious. The simulated spaciousness caused by this phasiness is not necessarily realistic, but it is pleasant to many listeners.

Another advantage of the spaced pair is that you can use omni mics. An omni condenser mic has deeper bass than a uni condenser mic.

NEAR-COINCIDENT PAIR

In this method, you angle apart two directional mics, and space their grilles a few inches apart horizontally (Figure 7.14). Even a few inches of spacing increases the stereo spread and adds a sense of ambient warmth or air to the recording. The greater the angle or spacing between mics, the greater the stereo spread. If the angle is too wide, center images will be weak (there will be a "hole in the middle").

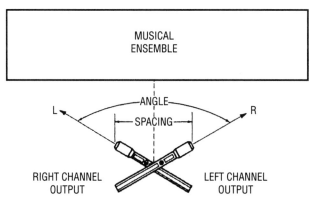

FIGURE 7.14
Near-coincident pair technique.

How does this method work? Angling directional mics produces level differences between channels. Spacing mics produces time differences. The level differences and time differences combine to create the stereo effect.

If the angling or spacing is too great, you get exaggerated separation. If the angling or spacing is too small, you'll hear a narrow stereo spread.

A common near-coincident method is the ORTF system, which uses two cardioids angled 110 degrees apart and spaced 7 inches (17 cm) horizontally. Usually this method gives accurate localization. That is, instruments at the sides of the orchestra are reproduced at or very near the speakers, and instruments halfway to one side are reproduced about halfway to one side.

BAFFLED OMNI PAIR

This method uses two omni mics, usually ear-spaced, and separated by either a hard or soft baffle (Figure 7.15). To create stereo, it uses time differences at low frequencies and level differences at high frequencies. The spacing between mics creates time differences. The baffle creates a sound shadow (reduced high frequencies) at the mic farthest from the source. Between the two channels, there are spectral differences—differences in frequency response.

Some examples of baffled-omni pairs are the Schoeps or Neumann sphere microphones, and the Jecklin Disk.

FIGURE 7.15
Baffled-omni technique.

Comparing the Four Techniques

Coincident pair:

- Uses two directional mics angled apart with grilles touching.
- Level differences between channels produce the stereo effect.
- Images are sharp.
- Stereo spread ranges from narrow to accurate.
- Signals are mono compatible.

Spaced pair:

- Uses two mics spaced several feet apart, aiming straight ahead.
- Time differences between channels produce the stereo effect.
- Off-center images are diffuse.
- Stereo spread tends to be exaggerated unless a third center mic is used, or unless spacing is under 2 to 3 feet.
- Provides a warm sense of ambience.
- Tends not to be mono compatible, but there are exceptions.
- Good low-frequency response if you use omni condensers.

Near-coincident pair:

- Uses two directional mics angled apart and spaced a few inches apart horizontally.
- Level and time differences between channels produce the stereo effect.
- Images are sharp.
- Stereo spread tends to be accurate. Provides a greater sense of air than coincident methods.
- Tends not to be mono compatible.

Play audio clip 22 at www.elsevierdirect.com/companions/9780240811444 to hear a comparison of the coincident, near-coincident, and spaced-pair techniques.

Baffled omni pair:

- Uses two omni mics, usually ear-spaced, with a baffle between them.
- Level, time, and spectral differences produce the stereo effect.
- Images are sharp.
- Stereo spread tends to be accurate but is not adjustable (except partly by panning).
- Good low-frequency response.
- Good imaging with headphones.
- Provides more air than coincident methods.
- Tends not to be mono compatible, but there are exceptions.

Hardware

A handy device is a stereo mic adapter or stereo bar (Figure 7.16). It mounts two mics on a single stand, and lets you adjust the angle and spacing. You might prefer to

FIGURE 7.16
Stereo mic adapter.

use a stereo mic instead of two mics. It has two mic capsules in a single housing for convenience.

How to Test Imaging

Here's a way to check the stereo imaging of a mic technique.

1. Set up the stereo mic array in front of a stage.
2. Record yourself speaking from various locations on stage where the instruments will be—center, half-right, far right, half-left, far left. Announce your position.
3. Play back the recording over speakers.

You'll hear how accurately the technique translated your positions, and you'll hear how sharp the images are.

We looked at several mic arrays to record in stereo. Each has its pros and cons. Which method you choose depends on the sonic compromises you're willing to make.

CHAPTER 8

Microphone Techniques

This chapter describes some ways to select and place mics for musical instruments and vocals. These techniques are popular, but they're just suggested starting points. Feel free to experiment.

Before you mike an instrument, listen to it live in the studio, so you know what sound you're starting with. You might want to duplicate that sound through your monitor speakers.

ELECTRIC GUITAR

Let's start by looking at the chain of guitar, effects, amplifier, and speaker. At each point in the chain where you record, you'll get a different sound (Figure 8.1).

1. The electric guitar puts out an electrical signal that sounds clean and clear.
2. This signal might go through some effects boxes, such as distortion, wah wah, compression, chorus, or stereo effects.

FIGURE 8.1
Three places to record the electric guitar.

3. Then the signal goes through a guitar amp, which boosts the signal and adds distortion. At the amplifier output (preamp out or external speaker jack), the sound is very bright and edgy.

4. The distorted amp signal is played by the speaker in the amp. Because the speaker rolls off above 4 kHz, it takes the edge off the distortion and makes it more pleasant.

You can record the electric guitar in many ways (Figure 8.1):

- With a mic in front of the guitar amp
- With a direct box
- Both miked and direct
- Through a signal processor or stomp box

The song you're recording will tell you what method it wants. Just mike the amp when you want a rough, raw sound with tube distortion and speaker coloration. Rock 'n' roll or heavy metal usually sounds best with a miked amp. If you record through a direct box, the sound is clean and clear, with crisp highs and deep lows. That might work for quiet jazz or R&B. Use whatever sounds right for the particular song you're recording.

First, try to kill any hum you hear from the guitar amp. Turn up the guitar's volume and treble controls so that the guitar signal overrides hum and noise picked up by the guitar cable. Ask the guitarist to move around, or rotate, to find a spot in the room where hum disappears. Flip the polarity switch on the amp to the lowest-hum position. To remove buzzes between guitar notes, try a noise gate, or ask the player to keep his or her hands on the strings.

Miking the Amp

Small practice amplifiers tend to be better for recording than large, noisy stage amps. If you use a small one, place it on a chair to avoid picking up sound reflections from the floor (unless you like that effect).

A common mic for the guitar amp is a cardioid dynamic type with a "presence peak" in its frequency response (a boost around 5 kHz). The cardioid pattern reduces leakage (off-mic sounds from other instruments). The dynamic type handles loud sounds without distorting, and the presence peak adds "bite." Of course, you can use any mic that sounds good to you.

As a starting point, try miking the amp about an inch from the grille near a speaker cone, slightly off-center—where the cone meets the dome.

The closer you mike the amp, the bassier the tone. The farther off-center the mic is, the duller the tone. Often, distant miking sounds great when you overdub a lead guitar solo played through a stack of speakers in a live room. Try a boundary mic on the floor or on the wall, several feet away.

Recording Direct

Now let's look at recording direct (also known as direct injection or DI). The electric guitar produces an electrical signal that you can plug into your mixer. You bypass the mic and guitar amp, so the sound is clean and clear. Just remember that amp distortion is desirable in some songs.

Mixer mic inputs tend to have an impedance (Z) around 1500 ohms. But a guitar pickup is several thousand ohms. So if you plug a high-Z electric guitar directly into a mic input, the input will load down the pickup and give a thin or dull sound.

To get around this loading problem, use a direct box between the guitar and your mixer (Figure 8.2). The DI box has a high-Z input and low-Z output, thanks to a built-in transformer or circuit. Some mixers and audio interfaces have a high-Z (instrument) input jack built in, so you can plug the electric guitar or bass directly into this jack.

The direct box should have a ground-lift switch to prevent ground loops and hum. Set it to the position where you monitor the least hum. You might try a mix of direct sound and miked sound. *Play audio clip 23 at www.elsevierdirect.com/companions/ 9780240811444 to hear demonstrations of electric-guitar recording methods.*

It's a good idea to record the guitar direct on its own track even if you mike the amp. Later during mixdown, you can run the DI track through a guitar-amp simulator plug-in, which might sound better than the real guitar amp did.

Electric Guitar Effects

If you want to record the guitarist's effects, connect the output of the effects boxes into the direct-box input. Many players have a rack of signal processors that creates their unique sound, and they just give you their direct feed. Be open to their suggestions, and be diplomatic about changing the sound. If they are studio players, they often have a better handle on effects than you might as the engineer.

FIGURE 8.2
Typical direct box.

You might want a "fat" or spacious lead-guitar sound. Here are some ways to get it:

- Send the guitar signal through a digital delay set to 20 to 30 msec. Pan guitar left, delay right. Adjust levels for nearly equal loudness from each speaker. (Watch out for phase cancellations in mono.)
- Send the guitar signal through a pitch-shifter, set for about 10 cents of pitch bending. Pan guitar left, pitch-shifted guitar right. (A cent is 1/100 of an equal-temperament semitone. There are 100 cents in a half-tone or semitone interval of pitch).
- Record two guitarists playing identical parts, and pan them left and right. This works great for rhythm-guitar parts in heavy metal.
- Double the guitar. Have the player rerecord the same part on an unused track while listening to the original part. Pan the original part left and pan the new part right.
- Add stereo reverb or stereo chorus.

Some guitar processors add many effects to an electric guitar, such as distortion, EQ, chorus, and compression. An example is the Line 6 Pod. You simply plug the electric guitar into the processor, adjust it for the desired sound, and record the signal direct. You wind up with a fully produced sound with a minimum of effort.

Reamping is a technique that lets you work on the amp's sound during mixdown rather than during recording. Record the guitar direct, then feed that track's signal into a guitar processor or miked guitar amp during mixdown. Use a low- to high-Z transformer between the track output and the processor or amp input. Record the processor or amp on an open track. In a digital audio workstation (DAW), you can start with a track of a direct-recorded guitar, then insert a guitar-amp modeling plug-in.

ELECTRIC BASS

BWAM, dik diddy bum. Do your bass tracks sound that clear? Or are they more muddy, like, "Bwuh, dip dubba duh"? Here's how to record the electric bass so it's clean and easy to hear in a mix.

As always, first you work on the sound of the instrument itself. Put on new strings if the old ones sound dull. Adjust the pickup screws (if any) for equal output from each string. Also adjust the intonation and tuning.

Usually, you record the electric bass direct for the cleanest possible sound. A direct pickup gives deeper lows than a miked amp, but the amp gives more midrange punch. You might want to mix the direct and miked sound. Use a condenser or dynamic mic with a good low-frequency response, placed 1 to 6 inches from the speaker.

When mixing a direct signal and a mic signal, make sure they are in phase with each other. To do this, set them to equal levels and reverse the polarity of the direct signal or the mic signal. The polarity that gives the most bass is correct.

Have the musician play some scales to see if any notes are louder than the rest. You might set a parametric equalizer to soften these notes, or use a compressor.

The bass guitar should be fairly constant in level (a dynamic range of about 6 dB) to be audible throughout the song, and to avoid clipping the recording on loud peaks. To do this, run the bass guitar through a compressor. Set the compression ratio to about 4:1, set the attack time fairly slow (8 to 20 msec) to preserve the attack transient, and set the release time fairly slow (1/2 second). If the release time is too fast, you get harmonic distortion.

EQ can make the bass guitar clearer. Try cutting around 60 to 80 Hz, or at 400 Hz. A boost at 2 to 2.5 kHz adds edge or slap, and a boost at 700 to 900 Hz adds "growl" and harmonic clarity. If you boost the lows around 100 Hz, try boosting at a lower frequency in the kick drum's EQ to keep their sounds distinct. A fretless bass will probably need different EQ or less EQ than a fretted bass.

Here are some ways to make the bass sound clean and well defined:

- Record the bass direct. Some direct boxes sound better than others, so experiment.
- Get a clear tone using the bass guitar's tone controls before applying any EQ.
- Use no reverb or echo on the bass.
- Have the bass player turn down the bass amp in the studio, just loud enough to play adequately. This reduces muddy-sounding bass leakage into other mics.
- Better yet, don't use the amp. Instead, have the musicians monitor the bass (and each other) with headphones.
- Have the bass player try new strings or a different bass. Some basses are better for recording than others. Use roundwound strings for a bright tone or flatwounds for a rounder tone.
- Ask the bass player to use the treble pickup near the bridge.
- Be sure to record the bass with enough edge or harmonics so the bass will be audible on small, cheap speakers.
- Try a bass-guitar signal processor such as the Zoom B1.
- Mix in a synth bass. For drum 'n' bass, dub-step, and electro-house genres, add some sub-bass at 30-60 Hz with space between notes, plus some mid-bass around 80-600 Hz. To create a pulsing effect, compress the synth bass with a kick drum as the sidechain input.

If the bass part is full and sustained, it's probably best to go for a mellow sound without much pluck. Let the kick drum define the rhythmic pattern. But if both the bass and kick are rhythmic and work independently, then you should hear the plucks. Listen to the song first, then get a bass sound appropriate for the music. A sharp, twangy timbre is seldom right for a ballad; a full, round tone will get lost in a fusion piece.

Often, a musician plays bass lines on a synth or sound module. The module is triggered from a keyboard, a sequencer, or a bass guitar plugged into a pitch-to-MIDI converter. Connect the module output to your mixer or audio interface line input.

Two effects boxes for the electric bass are the octave box and the bass chorus. The octave box takes the bass signal and drops it an octave in pitch. That is, it divides the bass signal's fundamental frequency in two. You put 82 Hz in; you get 41 Hz out. This gives an extra deep, growly sound. So does a 5-string bass.

A bass chorus gives a wavy, shimmering effect. Like a conventional chorus box, it detunes the signal and combines the detuned signal with the direct signal. Also, it removes the lowest frequencies from the detuned signal, so that the chorus effect doesn't thin out the sound.

SYNTHESIZER, DRUM MACHINE, AND ELECTRIC PIANO

For the most clarity, you usually DI a synth, MIDI sound module, drum machine, or electric piano. Set the volume on the instrument about three-quarters up to get a strong signal. Try to get the sound you want from patch settings rather than EQ.

Plug the instrument into a phone jack input on your mixer, or use a direct box. If you connect to a phone jack and hear hum, you probably have a ground loop. Here are some fixes:

- Power your mixer and the instrument from the same outlet strip. If necessary, use a thick extension cord between the outlet strip and the instrument.
- Use a direct box instead of a guitar cord, and set the ground-lift switch to the position where you monitor the least hum.
- To reduce hum from a low-cost synth, use battery power instead of an AC adapter.

A synth can sound dry and sterile. To get a livelier, funkier sound, you might run the synth signal into a power amp and speakers, and mike the speakers a few feet away.

If the keyboard player has several keyboards plugged into a keyboard mixer, you may want to record a premixed signal from that mixer's output. Record both outputs of stereo keyboards.

LESLIE ORGAN SPEAKER

This glorious device has a rotating dual-horn on top for highs and a woofer on the bottom for lows. Only one horn of the two makes sound; the other is for weight balance. The swirling, grungy sound comes from the phasiness and Doppler effect of the rotating horn, and from the distorted tube electronics that drive the speaker. Here are a few ways to record it (Figure 8.3):

- In mono: Mike the top and bottom separately, 3 inches to 1 foot away. Aim the mics into the louvers. In the top mic's signal, roll off the lows below 150 Hz.
- In stereo: Record the top horn with a stereo mic or a pair of mics out front. Put a mic with a good low end on the bottom speaker, and pan it to center.

FIGURE 8.3
Miking a Leslie organ speaker.

When you record the Leslie, watch out for wind noise from the rotating horn and buzz from the motor. Mike farther away if you monitor these noises.

Rather than recording an actual Hammond B3 organ and Leslie speaker, you might prefer to use a software emulation of those instruments: an organ soft synth or sample and a Leslie speaker plug-in. Trigger the synth or sample with a MIDI sequencer or MIDI controller (covered in Chapter 16 on MIDI). You can automate the horn rotation speed in the Leslie speaker plug-in.

DRUM SET

The first step is to make the drums sound good live in the studio. If the set sounds poor, you'll have a hard time making it sound great in the control room! You might put the drum set on a riser 1½ feet high to reduce bass leakage and to provide better eye contact between the drummer and the rest of the band. To reduce drum leakage into other mics, you could surround the set with goboes—padded thick-wood panels about 4 feet tall. For more isolation, place the set in a drum booth, a small padded room with windows. It's also common to overdub the set in a live room.

Tuning

One secret to creating a good drum sound lies in careful tuning. It's easier to record a killer sound if you tune the set to sound right in the studio before miking it.

First let's consider drum heads. Plain heads have the most ring or sustain, while heads with sound dots or hydraulic heads dampen the ring. Thin heads are best for recording because they have crisp attack and long sustain. Old heads become dull, so use new heads.

When you tune the toms, first take off the heads and remove the damping mechanism, which can rattle. Put just the top head on and hand-tighten the lugs. Then, using a drum key, tighten opposite pairs of lugs one at a time, one full turn. After you tighten all the lugs, repeat the process, tightening one-half turn. Then press on the head to stretch it. Continue tightening a half-turn at a time until you reach the pitch you want. You'll get the most pleasing tone when the heads are tuned within the range of the shell resonance. However, toms tuned very low can sound big and powerful.

To reduce ugly overtones, try to keep the tension the same around the head. While touching the center of the head, tap with a drumstick on the head near each lug. Adjust tension for equal pitch around the drum. Damping the head with a Dead Ringer or folded paper towel helps too. If you want a downward pitch bend after the head is struck, loosen one lug.

Keep the bottom head off the drum for the most projection and the broadest range of tuning. In this case, pack the bottom lugs with felt to prevent rattles. But you may want to add the bottom head for extra control of the sound. Projection is best if the bottom head is tighter than the top head—say, tuned a fourth above the top head. There will be a muted attack, an "open" tone, and some note bending. If you tune the bottom head looser than the top, the tone will be more "closed," with good attack.

With the kick drum (bass drum), a loose head gives lots of slap and attack, and almost no tone. The opposite is true for a tight head. Tune the head to complement the style of music. For more attack or click, use a hard beater.

Tune the snare drum with the snares off. A loose batter head or top head gives a deep, fat sound. A tight batter head sounds bright and crisp. With the snare head or bottom head loose, the tone is deep with little snare buzz, while a tight snare head yields a crisp snare response. Set the snare tension just to the point where the snare wires begin to "choke" the sound, then back off a little.

Damping and Noise Control

Usually the heads should ring without any damping. But if the toms or snare drum ring too much, put some plastic damping rings (Dead Ringers) on them. Or tape some gauze pads, folded paper towels, or folded handkerchiefs to the edge of the heads. Put masking tape on three sides of the pad so that the untaped edge is free to vibrate and dampen the head motion. Don't overdo the damping, or the drum set will sound like cardboard boxes.

Oil the kick drum pedal to prevent squeaks. Tape rattling hardware in place.

Sometimes a snare drum buzzes in sympathetic vibration with a bass-guitar passage or a tom-tom fill. Try to control the buzz by wedging a thick cotton wad between the snares and the drum stand. Or tune the snare to a different pitch than the toms.

Drum Miking

Now you're ready to mike the set. For a tight sound, place a mic near each drum head. For a more open, airy sound, use fewer mics or mix in some room mics placed several feet away. Typical room mics are omni condensers or boundary mics. Compressing the room mics can provide an explosive sound.

Figure 8.4 shows typical mic placements for a rock drum set. Let's look at each part of the kit.

SNARE

The most popular type of mic for the snare is a cardioid dynamic with a presence peak. The cardioid pattern reduces leakage; its proximity effect boosts the bass for a fatter sound. The presence peak adds attack. You might prefer a cardioid condenser for its sharp transient response.

Bring the mic in from the front of the set on a boom. Place the mic even with the rim, 2 inches above the head (Figure 8.5). Angle the mic down to aim where the drummer hits, or attach a mini condenser mic to the side of the snare drum so it "looks at" the top head over the rim.

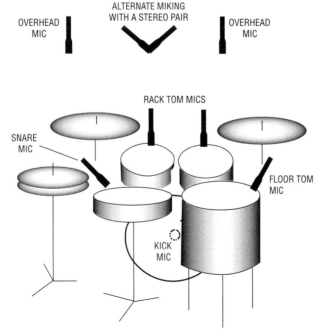

FIGURE 8.4
Typical mic placements for a rock drum set.

FIGURE 8.5
Snare-drum miking.

Some engineers mike both the top and bottom heads of the snare drum, with the microphones in opposite polarity. A mic under the snare drum gives a zippy sound; a mic over the snare drum gives a fuller sound. You might prefer to use just a top mic, and move it around until it picks up both the top head and snares. The sound is full with the mic near the top head, and thins out and becomes brighter as you move the mic toward the rim and down the side of the drum.

Whenever the hi-hat closes, it makes a puff of air that can "pop" the snare-drum mic. Place the snare mic so the air puff doesn't hit it. To prevent hi-hat leakage into the snare mic:

- Mike the snare closely.
- Bring the snare boom in under the hi-hat, and aim the snare mic away from the hi-hat.
- Use a piece of foam or pillow to block sound from the hi-hat.
- Use a de-esser on the snare.
- Don't play the hi-hat during tracking—overdub it later.

HI-HAT

Try a cardioid condenser mic about 6 inches over the cymbal edge that's farthest from the drummer (Figure 8.6). To avoid the air puff just mentioned, don't mike the hi-hat off its side; mike it from above aiming down. This also reduces snare leakage. Filter out the lows below about 500 Hz. You may not need a hi-hat microphone, especially if you use room mics. Usually the overhead mics pick up enough hi-hat.

TOM-TOMS

You can mike the toms individually, or put a mic between each pair of toms. The first option sounds more bassy. Place a cardioid dynamic about 2 inches over the drumhead and 1 inch in from the rim, angled down about 45 degrees toward

FIGURE 8.6
Hi-hat miking.

FIGURE 8.7
Tom-tom miking.

the head (Figure 8.7). Again, the cardioid's proximity effect gives a full sound. Another way is to clip mini condenser mics to the toms, peeking over the top rim of each drum.

FIGURE 8.8
Kick-drum miking.

If the tom mics pick up too much of the cymbals, aim the "dead" rear of the tom mics at the cymbals. If you use a supercardioid or hypercardioid mic, aim the null of best rejection at the cymbals (125° off axis for supercardioid; 110° for hypercadioid). It's common to gate the toms to reduce the low rumble of vibrating heads.

KICK DRUM

Place a blanket or folded towel inside the drum, pressing against the beater head to dampen the vibration and tighten the beat. The blanket shortens the decay portion of the kick-drum envelope. To emphasize the attack, use a wood or plastic beater—not felt—and tune the drum low.

A popular mic for a kick drum is a large-diameter, cardioid dynamic type with an extended low-frequency response. Some mics are designed specifically for the kick drum, such as the AKG D112, Audio-Technica AT AE2500, Electro-Voice N/D868, and Shure Beta 52A.

For starters, place the kick mic inside on a boom, a few inches from where the beater hits (Figure 8.8). Mic placement close to the beater picks up a hard beater sound; off-center placement picks up more skin tone, and farther away picks up a boomier shell sound.

How should the recorded kick drum sound? Well, they don't call it kick drum for nothing. THUNK! You should hear a powerful low-end thump plus an attack transient.

CYMBALS

To capture all the crisp "ping" of the cymbals, a good mic choice is a cardioid condenser with an extended high-frequency response, flat or rising at high frequencies. Place the overhead mics about 2 to 3 feet above the cymbal edges; closer miking picks up a low-frequency ring. The cymbal edges radiate the most highs. Place the cymbal mics to pick up all the cymbals equally. If your recording will be heard in mono, or for sharper imaging, you might want to mount the mic grilles together and angle the mics apart (Figure 8.4). This results in a narrow stereo spread. Another option is a stereo mic overhead. For wide stereo and sharp imaging, try a near-coincident pair aimed at the high-hat and floor tom.

Recorded cymbals should sound crisp and smooth, not muffled or harsh.

ROOM MICS

Besides the close-up drum mics, you might want to use a distant pair of room mics when you record drum overdubs. Place the mics about 10 or 20 feet from the set to pick up room reverb. When mixed with the close-up mics, the room mics give an open, airy sound to the drums. Popular room mics are omni condensers or boundary mics taped to the control-room window. You might compress the room mics for special effect. If you don't have enough tracks for room microphones, try raising the overhead mics.

BOUNDARY MIC TECHNIQUES

Boundary mics let you pick up the set in unusual ways. You can strap one on the drummer's chest to pick up the set as the drummer hears it. Tape them to hard-surfaced goboes surrounding the drummer. Put them on the floor under the toms and near the kick drum, or hang a pair over the cymbals. Try a supercardioid boundary mic in the kick drum.

RECORDING WITH TWO TO FOUR MICS

Sometimes you can mike the set simply. Place a single large-diaphragm cardioid condensor mic, two mics, or a stereo mic overhead and put another mic in the kick. If necessary, add a snare-drum mic (Figure 8.9). This method works well for acoustic jazz, and often for rock. If you want the toms to sound fuller, boost the lows with a narrow Q around 80 Hz in the overhead mics. Try reversing the polarity of the kick mic and see which position sounds best.

As an alternative try two SDC mics on a small kit (Figure 8.10). Place one mic over the snare about 8 inches up, 3 inches in from the rim, and midway between the hi-hat, rack tom, and snare drum. Place the other mic about 14 inches above the center of the

FIGURE 8.9
Miking a drum set with four mics.

FIGURE 8.10
Miking a drum set with a stereo pair of mics.

FIGURE 8.11
Possible EQ used on the overhead mics.

floor tom. Set the cymbals low. You might boost the lows around 60 Hz to add fullness to the toms and cut around 200 Hz to make the snare less tubby (Figure 8.11). Use a kick-drum mic too.

Audio clip 24 at www.elsevierdirect.com/companions/9780240811444 demonstrates several methods of miking a drum set.

Drum Recording Tips

After you set up all the mics, ask the drummer to play. Listen for rattles and leakage by soloing each microphone. Try not to spend much time getting a sound; otherwise you waste the other musicians' time and wear out the drummer.

To keep the drum sound tight during mixdown, mute or delete drum tracks that are not in use in a particular tune, use a noise gate on the kick and toms, or overdub the drums. One dated effect for the snare drum is gated reverb. It's a short splash of bright-sounding reverberation, which is rapidly cut off by a noise gate or expander. Many effects units have a gated-reverb program.

Another trick is recording "hot." Using an analog multitrack (or its plug-in equivalent), record the drums at a high level so they distort just a little. It's also common to compress the kick and snare.

A drummer might use drum pads, or drum triggers, fed into a sound module. Record directly off the module. You might want to mike the cymbals anyway for best sound. If you're recording a drum machine and it sounds too mechanical, add some real drums. The machine can play a steady background while the drummer plays fills.

When miking drums on stage for PA, you don't need a forest of unsightly mic stands and booms. Instead, you can use short mic holders that clip onto drum rims and cymbal stands (Mic-Eze at www.ac-cetera.com), or use mini condenser mics.

In a typical rock mix, the drums either are the loudest element, or are slightly quieter than the lead vocal. The kick drum is almost as loud as the snare. If you don't want a wimpy mix, keep those drums up front!

Try these EQ settings to enhance the recorded sound of the drums:

- Snare: Fat at 200 Hz, crack at 5 kHz, sizzle at 10 or 12 kHz. Some snare drums ring a lot at one note. To fix it, set up an equalizer on the snare-drum track (equalization is explained in Chapter 9). Set a narrow, high-Q boost around 500 Hz and sweep it in frequency until you amplify the frequency that is ringing. Apply a narrow, high-Q cut there to remove the ringing.
- Toms: Cut around 600 to 800 Hz to reduce the papery sound. Then if necessary, boost around 100-200 Hz for more fullness on rack toms or 80 to 100 Hz on floor toms. Boost around 5 kHz for more attack.
- Cymbals: Sizzle at 10 kHz or higher. If you are close-miking the toms, you might roll off the lows in the cymbal mics below 500 Hz to reduce low-frequency leakage.
- Kick drum: To remove the "cardboard" sound, cut at 300 to 600 Hz. Then if necessary, boost at 3 to 5 kHz for more click. Don't overdo the high-frequency boost; usually you don't want too much "point" on the kick sound. Boost 60 to 80 Hz if the kick sounds thin. Filter out highs above 9 kHz to reduce leakage from cymbals.

Try these tricks to come up with unusual drum sounds:

- Record with a cheap dynamic or crystal mic, maybe in a can.
- Run the drums through extreme processing: compression, gating, distortion, pitch shifting, tremolo, and so on.
- Substitute other objects for drums, cymbals, drumsticks, and brushes.
- Move a mic around a cymbal or drumhead while recording it.
- Put the drums in a reverberant room or hallway.
- Try the preverb effect described in Chapter 9.

Instead of recording an acoustic drum set, you might use an electronic drum set or CDs of drum samples. Copy the samples into a sampler or sampling software, then trigger them with a MIDI sequencer or MIDI controller with drum pads.

What if the drums sound too "live," "muddy," or reverberant? That means the drums suffer from too many early reflections because the drums are in a small, hard-surfaced room. Here are some solutions:

Add acoustical damping to the ceiling and walls. You can use acoustic foam, ATS Acoustics fiberglass panels, or Owens-Corning 703 pressed fiberglass insulation panels covered in muslin.

During mixdown, gate the snare and toms to remove early reflections. Here's how:

1. Solo the snare track.
2. Insert a gate plug-in in the snare track.
3. Solo the snare or tom track.
4. Set lookahead to 10 msec.
5. Gradually turn up the gate's input-level slider until the gate cuts off the reverb between snare or tom hits, but does not cut off the hits themselves.
6. Set the gate's hold time to 0-to-100 msec (experiment to see what sounds good).

There still will be some early snare reflections in the cymbal tracks. To reduce the snare sound in the cymbal tracks, roll off everything below about 1 kHz in the cymbal tracks. Here's how:

1. Solo the cymbal track(s).
2. Insert an equalizer plug-in in the cymbal tracks.
3. Solo the cymbal tracks.
4. Set band 1 to highpass.
5. Set band 1 Q to 1.7.
6. Set band 1 frequency to 1000 Hz. The cymbal tracks should sound thinner, with not as much snare leakage in them.

Now when you listen to the whole mix, the snare and tom sounds will come mostly from their tracks, and not from the cymbal tracks. It should sound a lot tighter and more professional.

Another way to tighten the sound is to signal-align the drum tracks. That way, the cymbal tracks and room-mic tracks play the drum hits at exactly the same time as the close-mic tracks. The result is a more focused sound. Zoom in on the track waveforms. Slide the distant-mic tracks to the left (earlier in time), or slide the other tracks right (later in time) until the waveforms align. It also helps to use a narrow stereo reverb on the snare rather than a wide, ambient reverb.

PERCUSSION

Let's move on to percussion, such as the cowbell, triangle, tambourine, or bell tree. A good mic for metal percussion is a condenser type because it has sharp transient response. Mike at least 1 foot away so the mic doesn't distort.

You can pick up congas, bongos, and timbales with a single mic between the pair, a few inches over the top rim, aimed at the heads. Or put a mic on each drum. It often helps to mike these drums top and bottom, with the bottom mic in opposite polarity. A cardioid dynamic with a presence peak gives a full sound with a clear attack.

For xylophones and vibraphones, place two cardioid mics 1½ feet above the instrument, aiming down. Cross the mics 135 degrees apart or place them about 2 feet apart. You'll get a balanced pickup of the whole instrument.

ACOUSTIC GUITAR

The acoustic guitar has a delicate timbre that you can capture through careful mic selection and placement. First prepare the acoustic guitar for recording. To reduce finger squeaks, try commercial string lubricant, a household cleaner/waxer, talcum powder on fingers, or smooth-wound strings. Ask the guitarist to play louder; this increases the "music-to-squeak" ratio!

Replace old strings with new ones a few days before the session. Experiment with different kinds of guitars, picks, and finger picking to get a sound that's right for the song.

For acoustic guitar, a popular mic is a pencil-type condenser with a smooth, extended frequency response from 80 Hz up. This kind of mic has a clear, detailed sound. You can hear each string being plucked in a strummed chord. Usually the sound picked up is as crisp as the real thing.

Now let's look at some mic positions. To record a classical guitar solo in a recital hall, mike about 3 to 6 feet away to pick up room reverb. Try a stereo pair (Figure 8.12A), such as XY, ORTF, MS, or a spaced pair (described in Chapter 7). If you record a classical guitar solo in a dead studio, mike about 1.5 to 2 feet away and add artificial reverb.

When you record pop, folk, or rock music, try a spot about 6 to 12 inches from where the fingerboard joins the guitar body (Figure 8.12B). That's a good starting point for capturing the acoustic guitar accurately. Still, you need to experiment and use your ears. Close to the bridge, the sound is woody and mellow.

In general, close miking gives more isolation, but tends to sound harsh and aggressive. Distant miking lets the instrument "breathe"; you hear a gentler, more open sound.

Another spot to try: tape a mini omni mic onto the body near the bottom of the sound hole, and roll off the excess bass. This spot gives good isolation (Figure 8.12C).

The guitar will sound more real if you record in stereo. Try one mic near the bridge, and another near the 12th fret (Figures 8.12D and E). Pan partway left and right. Another way to record stereo is with an XY pair of cardioid mics about 6 inches from the 12th or 16th fret, mixed with a 3-foot-spaced pair of omni mics about 3 feet away.

FIGURE 8.12
Some mic techniques for acoustic guitar.

Is feedback or leakage a problem? Mike close to the sound hole (Figure 8.12F). The tone there is very bassy, so turn down the low-frequency EQ on your mixer until the sound is natural. Also cut a few decibels around 3 kHz to reduce harshness.

You get the most isolation with a contact pickup. It attaches to the guitar, usually under the bridge. The sound of a pickup is something like an electric guitar. You can mix a mic with a pickup to add air and string noise to the sound of the pickup. That way, you get good isolation and good tone quality. You might record the acoustic guitar off its pickup while tracking to prevent leakage, then overdub the guitar later with a microphone for its better sound quality.

SINGER/GUITARIST

Normally you overdub the guitar and vocal separately. But if you have to record both at once, the vocal might sound filtered or hollow because of phase cancellations between the vocal mic and guitar mic. This can happen whenever two mics pick up the same source at approximately equal levels, at different distances, and both mixed to the same channel. Try one of these methods to solve the problem:

- Angle the vocal mic up and angle the guitar mic down to isolate the two sources. Follow the three-to-one rule described in Chapter 7.
- Mike the voice and guitar very close. Roll off the excess bass with your mixer's EQ.
- Use a pickup on the guitar instead of a mic.
- Place two bidirectional mics so the tops of their grilles touch. This gets rid of any delay between their signals. Aim the "dead" side of the vocal mic at the guitar; aim the dead side of the guitar mic at the mouth.

- Use just one mic, or a stereo mic, midway between the mouth and guitar about 1 foot out front. Adjust the balance between voice and guitar by changing the mic's height.
- Delay the vocal mic signal by about 1 msec. Then the signals of the two mics will be more in phase, preventing phase cancellations when they are mixed to the same channel. Some multitrack recorders have a track-delay feature for this purpose.
- Use Sound Radix Auto-Align ($149), an AU/RTAS plug-in. It automatically aligns the signals from two mics picking up the same source.

GRAND PIANO

This magnificent instrument is a challenge to record well. First have the piano tuned, and oil the pedals to reduce squeaks. You can prevent thumps by stuffing some foam or cloth under the pedal mechanism.

For a classical-music solo, record in a reverberant room such as a recital hall or concert hall. Reverb is part of the sound. Set the piano lid on the long stick. Use condenser mics with a flat response. Place a stereo mic, or a stereo pair of cardioid mics, about 7 feet away and 7 feet high, up to 9 feet away and 9 feet high (Figure 8.13). Move the mics closer to reduce reverb, farther to increase it. When using a pair of omni mics, place them 1.3 to 2 feet apart, 3 to 6.5 feet from the piano, and 4 to 5 feet high (Figure 8.14). You might need to mix in a pair of hall mics: try cardioids aiming away from the piano about 25 feet away.

FIGURE 8.13
Suggested grand-piano miking for classical music (using cardioid mics).

FIGURE 8.14
Suggested grand-piano miking for classical music (using omnidirectional mics).

When recording a piano concerto, give the piano a spot mic about 1 to 3 feet away. Put the mic in a shock mount.

Pop music demands close miking. Close mics pick up less room acoustics and leakage, and give a clear sound that cuts through the mix. Try not to mike the strings closer than 8 inches, or else you'll emphasize the strings closest to the mics. You want equal coverage of all the notes the pianist plays.

One popular method uses two spaced mics inside the piano. Use omni or cardioid condensers, ideally in shock mounts. Put the lid on the long stick. If you can, remove the lid to reduce boominess. Center one mic over the treble strings and one over the bass strings. Typically, both mics are 8 to 12 inches over the strings and 8 inches horizontally from the hammers (Figure 8.15A, bass and treble mics). Aim the mics straight down or angle them to aim at the hammers. Pan the mics partly left and right for stereo.

One alternative is to put the treble mic near the hammers, and put the bass mic about 2 feet toward the tail (Figure 8.15B). Another method uses two ear-spaced omni condensers or an ORTF pair about 12 to 18 inches above the strings. *Audio clip 25 at www. elsevierdirect.com/companions/9780240811444 demonstrates some mic techniques for grand piano.*

The spaced mics might have phase cancellations when mixed to mono, so you might want to try coincident miking (Figure 8.15A, stereo pair). Boom-mount a stereo mic, or an XY pair of cardioids crossed at 120 degrees. Miking close to the hammers sounds percussive; toward the tail has more tone.

For more clarity and attack, boost EQ around 10 kHz or use a mic with a rising high-frequency response.

Boundary mics work well, too. If you want to pick up the piano in mono, tape a boundary mic to the underside of the raised lid, in the center of the strings, near the hammers. Use two for stereo over the bass and treble strings. Put the bass mic near the

FIGURE 8.15
Suggested grand-piano miking for popular music.

tail of the piano to equalize the mic distances to the hammers (Figure 8.15C). If leakage is a problem, close the lid and cut EQ a little around 250 Hz to reduce boominess.

If your studio lacks a piano, consider using a software emulation of a piano. Some programs provide high-quality samples of piano notes that can be played with a sequencer or a MIDI controller. Examples are the Steinberg Grand VST 2.0 ($199 at www.steinberg.net) and Maxim Digital Audio Piano (freeware at http://mda.smartelectronix.com/ under VST synths).

UPRIGHT PIANO

Here are some ways to mike an upright piano (Figure 8.16):

- Remove the panel in front of the piano to expose the strings over the keyboard. Place one mic near the bass strings and one near the treble strings about 8 inches away. Record in stereo and pan the signals left and right for the desired piano width. If you can spare only one mic for the piano, just cover the treble strings.

FIGURE 8.16
Some mic techniques for upright piano.

- Remove the top lid and upper panel. Put a stereo pair of mics about 1 foot in front and 1 foot over the top. If the piano is against a wall, angle the piano about 17 degrees from the wall to reduce tubby resonances.
- Aim the soundboard into the room. Mike the bass and treble sides of the soundboard a few inches away. In this spot, the mics pick up fewer pedal thumps and other noises. Try cardioid dynamic mics with a presence peak.

ACOUSTIC BASS

The acoustic bass (string bass, double bass, upright bass) puts out frequencies as low as 41 Hz, so use a mic with an extended low-frequency response such as a large-diaphragm condenser mic or ribbon mic. As always, closer miking improves isolation, while distant miking tends to sound more natural but can pick up too much room sound. Try these techniques (Figure 8.17):

- 4 to 8 inches in front of the bridge, a few inches above the bridge.
- 4 to 6 inches under the bridge, a few inches from the strings. This mic will pick up a deep sound with good definition. You might mix in a second mic near the plucking fingers for clarity, a few inches from the side of the fingerboard.
- Mix a pickup with a mic, or use a pickup alone and use EQ to improve the sound.

With EQ you can emphasize slap at 5 kHz, or boxiness at 1.25 kHz. Try compressing the body mic 2:1 and the slap mic 6:1.

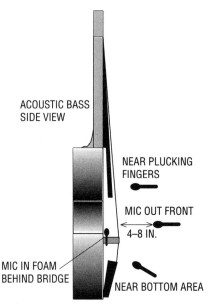

ACOUSTIC BASS
SIDE VIEW

NEAR PLUCKING
FINGERS

MIC OUT FRONT
←→
4–8 IN.

MIC IN FOAM
BEHIND BRIDGE

NEAR BOTTOM AREA

FIGURE 8.17
Some mic techniques for the acoustic bass.

Here are some methods that isolate the bass and let the player move around. They work well for PA:

- Wrap a mini omni condenser mic in foam rubber (or in a foam windscreen) and mount it in the bridge aiming up (Figure 8.17).
- Tape a mini omni mic to the bridge, or wedge it into a slot in the bridge.
- Wrap a regular cardioid mic in foam padding (except the front grille) and squeeze it behind the bridge (Figure 8.17) or tailpiece.
- For best isolation, try a direct feed from a pickup. This method adds clarity and deep bass, but probably will need some EQ. You might mix the pickup with a microphone.

BANJO

Try a flat-response condenser mic about 1 foot away from the center (Figure 8.18). Or try a cardioid dynamic mic 6 inches from the lower-right area or 6 inches from the lower rim, aiming at the bridge. If you need more isolation, mike closer and roll off some bass. The banjo sounds pleasantly mellow when miked toward the edge of the head, near the resonator holes (if the banjo has them). Cloth stuffed inside will tighten the sound and will reduce feedback in PA situations.

For the most isolation, tape a mini omni condenser mic to the head about halfway between the bottom edge and the bridge. You can wedge a pickup between the strings below the bridge and the banjo head. Put the pickup flat against the head surface.

BANJO

MINI OMNI
ON BRIDGE

MIC 1 FOOT
OUT FRONT

MINI OMNI
ON HEAD

MIC 6 INCHES
FROM LOWER-
RIGHT AREA

FIGURE 8.18
Four methods for miking a banjo.

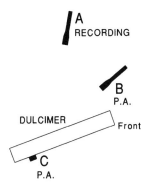

A
RECORDING

B
P.A.

DULCIMER

Front

C
P.A.

FIGURE 8.19
Some mic techniques for hammered dulcimer.

MANDOLIN, DOBRO, BOUZOUKI, AND LAP DULCIMER

Mike these about 6 to 8 inches away from the hole with a small-diaphragm condenser mic. Some engineers also mix in a mic near the neck or dobro resonator. If you need more lows and more isolation, tape a mini omni condenser mic near a hole and tweak EQ for the best sound.

HAMMERED DULCIMER

Place a flat-response condenser mic about 2 feet over the center of the soundboard (Figure 8.19A). On stage, place a cardioid dynamic or condenser 6 to 12 inches over the middle of the top end (Figure 8.19B). For the best gain-before-feedback in a PA system, mix in a mini omni condenser mic (or a cardioid with bass rolloff) very near the sound hole (Figure 8.19C).

OVERHEAD

FOR SINGING
FIDDLER

SIDE

FIGURE 8.20
Three fiddle-miking methods.

FIDDLE (VIOLIN)

Listen to the fiddle itself to make sure it sounds good. Correct any instrument problems before miking.

First try a flat-response condenser mic (omni or cardioid) about 2 feet over the bridge. This distant miking gives an airy, silky sound. Close miking (about 6 to 12 inches, see Figure 8.20) sounds more aggressive, which is desirable in old-time or bluegrass music. Aim the mic toward the f-holes for warmth or toward the fingerboard for clarity. A fiddle that sounds too bright and scratchy can be tamed using a ribbon mic, or by miking it from the side or even underneath. If the ceiling over the fiddle is low, nail a square yard of acoustic foam up there to prevent reflections.

If you have to mike close—say, for a singing fiddler—aim the mic at the player's chin from 6 to 12 inches above the end of the fingerboard. The mic will pick up both the singer and the fiddle.

If you need more isolation, try a mini omni mic or a Bartlett Fiddle Mic. Wrap its cable in foam rubber (or a windscreen) 1½ inches from the capsule. Wedge the foam under the tailpiece, and position the mic capsule halfway between the tailpiece and bridge, a half inch over the body (Figure 8.21). If necessary, cut a little at 10 kHz to reduce harshness and boost around 200 Hz for warmth. Another way to get a warm sound is to mount the mic near an f-hole.

A good spot for a pickup is on the left side of the top (player's view), on the player's side of the bridge.

To record a classical violin solo, try a stereo mic (or a stereo pair) 5 to 15 feet away in a reverberant room.

MINI OMNI
MIC PLACEMENTS
ON THE FIDDLE

OVER
F-HOLE,
CLIPPED TO
TAILPIECE

BETWEEN
TAILPIECE
AND BRIDGE

FIGURE 8.21
Two ways to close-mike a fiddle for isolation.

STRING SECTION

Place the strings in a large, live room and mike them at distance to pick up a natural acoustic sound. A common mic choice is a condenser with a flat response. First try a stereo mic or stereo pair of mics about 4 to 20 feet behind the conductor, raised about 13-15 feet.

If the room is noisy or too dead, or the balance is poor, you'll need to mike close and add digital reverb. Try one mic on every two to four violins, 6 feet off the floor, aiming down, with the same setup for the violas. Mike the cello about 1 to 2 feet from the bridge, to the right side between the bridge and f-hole. When you mix the strings to stereo, pan them evenly between the monitor speakers. Spread them left, center, and right to make a "curtain of sound." If you can spare only one track for the strings, use a stereoizer effect during mixdown.

STRING QUARTET

Record a quartet in stereo using a stereo mic or a pair of mics. Place them about 6 to 10 feet away to capture the room ambience. The monitored instruments should not spread all the way between speakers. If you want to narrow the stereo stage, angle or space the mics closer together.

BLUEGRASS BAND AND OLD-TIME STRING BAND

Suppose you're recording a group that has a good acoustic balance. Try a stereo mic or stereo pair of mics about 2 to 3 feet away and 5 feet high (lower if the group is seated). Move the players toward or away from the mics to adjust their balance. You'll have more control if you mike all the instruments up close and mix them. This also gives a more "commercial" sound. Arrange the players in a circle facing each other. The production style aims for a natural timbre on all the instruments, either with no effects or with slight reverb.

HARP

Use a condenser mic with a flat response. If the harp is playing with an orchestra, mike the harp about 18 inches from the front of the soundboard, or 18 inches from the player's left hand. With a Celtic group, try a mic 1/3 of the way up the sound board about 6 to 12 inches away.

Tape a mini omni condenser mic to the soundboard if you need isolation. A mic on the inside of the soundboard has more isolation; a mic on the outside sounds more natural. Also try a cardioid condenser wrapped in foam, stuck into the center hole from the rear.

HORNS

"Horns" in studio parlance refers to the brass instruments: trumpets, cornets, trombones, baritones, french horns, and tubas.

All the brass radiate strong highs straight out from the bell, but do not project them to the sides. A mic close to and in front of the bell picks up a bright, edgy tone. To mellow out the tone, mike the bell off-axis with a flat-response mic (Figure 8.22). The sound on-axis to the bell has a lot of spiky high harmonics that can overload a condenser mic, mixer input, or analog tape. That's another reason to mike off-axis.

Mike the trumpet with a dynamic or ribbon mic to take the edge off the sound. Use a condenser mic if you want a lot of sizzle. Mike about 1 foot away for a tight sound; mike several feet away for a fuller, more dramatic sound.

You can pick up two or more horns with one microphone. Several players can be grouped around a single omni mic, or around a stereo pair of mics. The musicians can play to a pair of boundary mics taped on the control-room window or on a large panel.

FIGURE 8.22
Miking for trumpet tone control.

Record a classical brass quartet in a reverberant room. Use a stereo mic, or a stereo pair of mics, about 6 to 12 feet away.

SAXOPHONE

A sax miked very near the bell sounds bright, breathy, and rather hard (Figure 8.23). Mike it there for best isolation. To get a warm, natural sound, mike the sax about 1½ feet away, halfway down the wind column (Figure 8.23). Don't mike too close, or else the level varies when the player moves. A compromise position for a close-up mic is just above the bell, aiming at the holes. You can group a sax section around one mic.

Figure 8.24 shows a typical miking setup for big-band jazz. It uses the techniques already described for the drums, bass, piano, electric guitar, trumpet, and sax.

FIGURE 8.23
Two ways to mike a saxophone.

FIGURE 8.24
Typical miking setup for big-band jazz.

WOODWINDS

With woodwinds, most of the sound radiates not from the bell, but from the holes. So aim a flat-response mic at the holes about 1 foot away (Figure 8.25).

When miking a woodwind section within an orchestra, you need to reject nearby leakage from other instruments. To do that, try aiming a bidirectional mic down over the woodwind section. The side nulls of the mic cut down on leakage.

To pick up a flute in a pop-music group, try miking a few inches from the area between the mouthpiece and the first set of finger holes (Figure 8.26). You may need a pop filter. If you want to reduce breath noise, roll off high frequencies or mike farther away. You also can attach a mini omni mic to the flute a few inches above the body, between the mouthpiece and finger holes.

For classical music solos, try a stereo pair 4 to 12 feet away.

FIGURE 8.25
Miking a clarinet from the side.

FIGURE 8.26
Two methods of miking a flute.

HARMONICA, ACCORDION, AND BAGPIPE

One way to mike a harmonica (harp) is to use a cardioid dynamic mic with a ball grille. Place the mic very close to the harmonica or have the player hold it. A condenser mic about 1 foot away gives a natural sound. To get a bluesy, dirty sound, use a "bullet"-type harmonica mic or play the harmonica through a miked guitar amp.

For an accordion, try a mic about 6 to 12 inches from the sound holes near the keyboard. Some accordions have sound holes on both sides, so you'll need two mics. Follow the three-to-one rule. The distance between mics should be at least three times the mic-to-source distance. One end of the accordion is in constant motion, so you might want to attach a mini omni mic to that end. A solo accordion or concertina could be miked with a stereo pair of flat-response cardioid condenser mics about 3 to 6 feet in front.

A bagpipe has two main sound sources: the chanter, which the musician plays with the fingers, and the drone pipes, which make a steady tone. Mike the chanter about a foot away from the side, and mike the drone pipes a foot from the end. Again, follow the three-to-one rule. You could also mike the bagpipe a few feet away with one mic.

LEAD VOCAL

The lead vocal is the most important part of a pop song, so it's critical to record it right. First set up a comfortable environment for the singer. Put down a rug, add some flowers or candles, dim the lights. Supply some water and let the singer stretch and warm up. Set up a good cue mix with effects to help the singer get into the mood of the song.

You might want to turn off the reverb in the singer's headphones; this makes it easier to hear pitch. If the vocalist is singing flat, reduce their headphone volume, and vice versa.

With any vocal recording, there are some problems to overcome, but we can deal with them. Among these are proximity effect, breath pops, wide dynamic range, sibilance, and sound reflections from the music stand. Let's look at these in detail.

Miking Distance

When you sing or talk close to most directional mics, the microphone boosts the bass in your voice. This is called the proximity effect. We've come to accept this bassy sound as normal in a PA system, but the effect just sounds boomy in a recording.

To prevent boomy bass, mike the singer at a distance, about 8 inches away (Figure 8.27). A popular mic choice is a flat-response condenser mic with a large diaphragm (1 inch or larger diameter). As always, you can use any mic that sounds good to you. If the mic has a bass rolloff switch, set it to "flat." Singers should maintain their distance to the

FIGURE 8.27
Typical miking technique for a lead vocal.

mic. I ask the singer to spread the fingers, touch lips with the thumb, and touch the mic with the pinky. The hand forms a spacer for keeping a constant distance.

If the singer's voice is too harsh or edgy, try miking partly to the side of the mouth, aiming at the mouth. Or try a ribbon mic, or a multiband compressor set to compress from 2000 Hz and up.

Some singers feel more comfortable singing into a handheld mic. You can give them a handheld mic but also mike them several inches away with a good condenser microphone. Record both mics on separate tracks and choose whatever sounds the best.

If you must record the singer and the band at the same time—as in a concert—you'll have to mike close to avoid picking up the instruments with the vocal mic. Try a cardioid or supercardioid mic with a bass rolloff and a foam pop filter. The sound will be bassy because of proximity effect, so roll off the excess lows at your mixer. For starters, try −6 dB at 100 Hz to 200 Hz. Some mics have a bass filter switch for this purpose. Aim the mic partly toward the singer's nose to prevent a nasal or closed-nose effect. This close-up method works well if you want an intimate, breathy sound.

When recording a classical-music singer who is accompanied by an orchestra, place the mic about 1 to 2 feet away. If the singer is a soloist (maybe accompanied by piano), use a stereo pair of boundary mics about 8 to 15 feet away to pick up room reverb.

Breath Pops

When you sing a word with "p" or "t" sounds, a turbulent puff of air shoots out of the mouth. The puff hits the mic and makes a thump or small explosion called a pop. To reduce it, put a foam-plastic pop filter on the mic. Some mics have a ball grille screen to cut pops, but foam works better. The pop filter should be made of special open-cell foam to pass high frequencies. For best pop rejection, allow a little air space between the foam and the front of the mic grille.

Foam pop filters reduce the highs a little. So they should be left off instrument mics, except for outdoor recording or dust protection. Pop filters do not reduce breathing sounds or lip noises. To get rid of these problems, mike farther away or roll off some highs.

The most effective pop filter is a hoop with a nylon stocking stretched over it (Figure 8.27) or a disk of perforated metal. You can buy those, or make one with a coat hanger and a crochet hoop. Place the filter a few inches from the mic.

Another way to get rid of pop is to put the mic at forehead height, aiming at the mouth. This way the puffs of air shoot under the mic and miss it. Make sure the vocalist sings straight ahead, not up at the mic, or the mic will pop.

Wide Dynamic Range

During a song, vocalists often sing too loud or too soft. They blast the listener or get buried in the mix. That is, many singers have a wider dynamic range than their instrumental backup. To even out these extreme level variations, ask the singer to use proper mic technique. Back away from the mic on loud notes; come in closer for soft ones. Or you can ride gain on singers: gently turn them down as they get louder, and vice versa. Usually this is done during mixdown with automation or volume envelopes.

Another solution is to pass the vocal signal through a compressor, which acts like an automatic volume control. Plug the compressor into the vocal channel's insert jacks or insert a compressor plug-in in the vocal track during mixdown. A typical compressor setting for vocals is a 2:1 ratio, −10 dB threshold, and about 3 to 6 dB of gain reduction. Of course, you should use whatever settings are needed for the particular singer. *Audio clip 26 at www.elsevierdirect.com/companions/9780240811444 demonstrates vocal compression, as well as breath pops and the effects of miking distance on recorded vocals.*

If the singer moves toward and away from the mic while singing, the average level will go up and down. Try to mike the singer at least 8 inches away, so that small movements of the singer won't affect the level.

An alternative to compressing the vocal is Waves Vocal Rider software (www.waves.com). It automatically adjusts the vocal volume in real time and provides a more natural sound than compression.

If you must mike close to prevent leakage or feedback, ask the vocalist to sing with lips touching the foam windscreen to keep the same distance to the mic. Turn down the excess bass using your mixer's low-frequency EQ (typically −6 dB at 100 Hz to 200 Hz).

Sibilance

Sibilance is the emphasis of "s" or "sh" sounds, which are strongest around 3 to 10 kHz. They help intelligibility. In fact, many producers like sizzly "s" sounds, which add a bright splash to the vocal reverb. But the sibilance should not be piercing or strident.

If you want to reduce sibilance, use a mic with a flat response—rather than one with a presence peak—or cut the highs little around 8 kHz on your mixer. Better yet, use a de-esser signal processor or plug-in, which cuts the highs only when the singer makes sibilant sounds. You can set a multiband compressor to act as a de-esser; see Chapter 9 under the heading "Suggested 'Ballpark' Compressor Settings."

Reflections from the Music Stand and Ceiling

Suppose that a lyric sheet or music stand is near the singer's mic. Some sound waves from the singer go directly into the mic. Other sound waves reflect off the lyric sheet or music stand into the mic (Figure 8.28, top). The delayed reflections will interfere with the direct sound, making a colored tone quality like mild flanging.

To prevent this, lower the music stand and tilt it almost vertically (Figure 8.28, bottom). This way, the sound reflections miss the mic.

If your studio has a low ceiling, the recorded vocal might have a colored tone quality due to phase cancellations from ceiling reflections. Try putting the mic lower and use a hoop-type pop filter. Also put a 3-foot square of acoustic foam or pressed-fiberglass panel on the ceiling over the singer and mic.

Vocal Effects

Some popular vocal effects are stereo reverb, echo, and doubling. You can record real room reverb by miking the singer at a distance in a hard-surfaced room. Slap echo provides a 1950s rock 'n' roll effect. Often a vocal is mixed dry, with no reverb. A little

FIGURE 8.28
Preventing reflections from a music stand.

distortion might even be effective on some songs. You might try a vocal processor, which offers a variety of effects. Try different EQ or different effects on each section of a song.

For the best sound, put vocal plug-ins in this order: (1) de-ess, (2) compressor, (3) EQ.

Doubling a vocal gives a fuller sound than a single vocal track. Overdub a second take of the vocal on an empty track, in sync with the original take. During mixdown, mix the second vocal take with the original, at a slightly lower level than the original. You can double a vocal track by running it through a digital delay set at 15 to 35 msec, or through a pitch shifter that is detuned 10 to 15 cents.

BACKGROUND VOCALS

When you overdub background vocals (harmony vocals), you can group two or three singers in front of a mic. The farther they are from the mic, the more distant they will sound in the recording. Pan the singers left and right for a stereo effect. Because massed harmonies can sound bassy, roll off some lows in the background vocals.

If you want independent control of each background singer, give each one a close-up mic and record them with separate mixer channels or separate tracks.

Barbershop or gospel quartets with a good natural blend can be recorded with a stereo mic or stereo pair of mics about 2 to 4 feet away. If their balance is poor, close-mike each singer about 8 inches away, and balance them with your mixer. This also gives a more "commercial" sound. If you close-mike, spread the singers at least 2 feet apart in a circle to prevent phase cancellations.

SPOKEN WORD

The tips given earlier for a lead vocalist also apply to recording the spoken word (including podcasts, audio books, etc.). Be sure to keep the miking distance constant and use a hoop-type pop filter. To prevent sound reflections into the mic, put the script on a padded music stand that is angled almost vertically, and put the mic in the plane of the stand near the top edge. Fold up a corner of each script page to form a handle for turning pages silently.

The engineer and announcer should both have the same script. Mark the beginning of each misread sentence. The announcer should reread each misread sentence from the beginning to make editing easier.

CHOIR AND ORCHESTRA

Figure 8.29 shows three ways of miking a choir. If the mics will also be used for PA, or if the venue is noisy or sounds bad, try miking close, pan the mic as desired, and add artificial reverb (Figure 8.29 A). Otherwise, try a near-coincident pair of cardioid mics

FIGURE 8.29
Choir miking suggestions. (A) Close-up panned mics. (B) Near-coincident stereo pair. (C) Spaced stereo pair.

(Figure 8.29B) or a pair of omni mics spaced about 2 feet apart (Figure 8.29C). Adjust the mic-to-choir distance until you hear the desired amount of hall acoustics in your monitors.

See Chapter 18 for suggestions on miking an orchestra.

SUMMARY

We can sum up mic placement like this: If leakage or feedback is a problem, place the mic near the loudest part of the instrument, and add EQ to get a natural sound. Otherwise, place the mic in various spots until you find a position that sounds good over your monitors. There is no single "correct" mic technique for any instrument. Just place the mic where you hear the desired tonal balance and amount of room reverb.

Try the techniques described here as a starting point, then explore your own ideas. Trust your ears! If you capture the power and excitement of electric guitars and drums, and the beautiful timbre of acoustic instruments and vocals, you've made a successful recording.

CHAPTER 9

Effects and Signal Processors

With effects, your mix sounds more like a real "production" and less like a bland home recording. You might simulate a concert hall with reverb. Put a guitar in space with stereo chorus. Make a kick drum punchy by adding compression. Used on all pop-music records, effects can enhance plain tracks by adding spaciousness and excitement. They are essential if you want to produce a commercial sound. But many jazz, folk, and classical groups sound fine without any effects.

This chapter describes the most popular signal processors and effects, and suggests how to use them.

Effects are available both as hardware and software (called plug-ins). To add a hardware effect to a track, you feed its signal from your mixer's aux send to an effects device, or signal processor (Figure 9.1). It modifies the signal in a controlled way. Then the modified signal returns to your mixer, where it blends with the dry, unprocessed signal.

SOFTWARE EFFECTS (PLUG-INS)

Most recording programs include plug-ins: software effects that you control on your computer screen. Each effect is a small program that runs either in your computer's CPU or in a DSP card. Some plug-ins come already installed with the recording software; others you can download or purchase on CD, then install on your hard drive. Each plug-in becomes part of your recording program (called the host), and can be called up from within the host.

You can use plug-ins made by your recording software company or by others. Some manufacturers make plug-in bundles, which are a variety of effects in a single package.

Plug-ins are the usual way to create effects in a DAW. Some DAWs let you configure your audio interface to produce an aux-send signal, which you feed to an external

FIGURE 9.1
Eventide H8000FW Multichannel Effects System, an example of a signal processor.

hardware processor. The processed signal returns to the interface and blends with the dry signal.

All the effects described below are available as plug-ins as well as hardware.

EQUALIZER

Recall from Chapter 2 that an equalizer (usually in the mixer) is a sophisticated tone control, something like the bass and treble controls in a stereo system. Equalization (EQ) lets you improve on reality: add crispness to dull cymbals or add bite to a wimpy electric guitar. EQ also can make a track sound more natural; for instance, it can remove tubbiness from a close-miked vocal.

To understand how EQ works, we need to know the meaning of a spectrum. Each instrument or voice produces a wide range of frequencies called its spectrum—the fundamentals and harmonics. The spectrum gives each instrument its distinctive tone quality or timbre.

If you boost or cut certain frequencies in the spectrum, you change the tone quality of the recorded instrument. EQ adjusts the bass, treble, and midrange of a sound by turning up or down certain frequency ranges. That is, it alters the frequency response. For example, a boost (a level increase) in the range centered at 10 kHz makes percussion sound bright and crisp. A cut at the same frequency dulls the sound.

Types of EQ

Equalizers range from simple to complex. The most basic type is a **bass and treble control** (labeled LF EQ and HF EQ). Figure 9.2 shows its effect on frequency response. Typically, this type has up to 15 dB of boost or cut at 100 Hz (for the low-frequency EQ knob) and at 10 kHz (for the high-frequency EQ knob).

With **3-band EQ** you can boost or cut the lows, mids, and highs at fixed frequencies (Figure 9.3). **Sweepable EQ** is more flexible because you can "tune in" the exact frequency range needing adjustment (Figure 9.4). If your mixer has sweepable EQ, one knob sets the center frequency while another sets the amount of boost or cut.

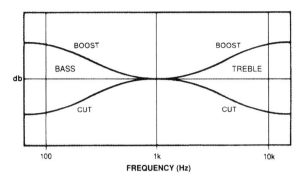

FIGURE 9.2
The effect of the bass and treble control.

FIGURE 9.3
The effect of 3-band equalization.

Parametric EQ lets you set the frequency, amount of boost/cut, and bandwidth—the range of frequencies affected. Figure 9.5 shows how a parametric equalizer varies the bandwidth of the boosted part of the spectrum. The "Q" or quality factor of an equalizer is the center frequency divided by the bandwidth. A boost or cut with a low-Q setting (such as 1.5) affects a wide range of frequencies; a high-Q setting (such as 10)

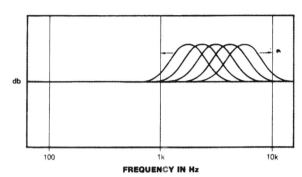

FIGURE 9.4
The effect of sweepable equalization.

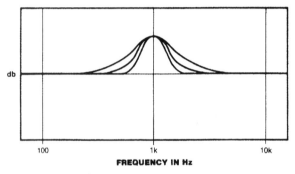

FIGURE 9.5
Curves that illustrate varying the bandwidth of a parametric equalizer.

FIGURE 9.6
Rane ME60S, an example of a graphic equalizer.

makes a narrow peak or dip. Generally, use narrow-band (high-Q) EQ to remove hum and resonances; use wide-band (low-Q) EQ for tonal changes.

A **graphic equalizer** (Figure 9.6) is usually outside the mixing console. This type has a row of slide pots that work on 5 to 31 frequency bands. When the controls are adjusted, their positions graphically show the resulting frequency response. Usually, a graphic equalizer is used for monitor-speaker EQ, or is patched into a channel for sophisticated tonal tweaking.

FIGURE 9.7
Peaking equalization at 7 kHz.

FIGURE 9.8
Shelving equalization at 7 kHz.

FIGURE 9.9
Lowpass, highpass, and bandpass filters.

Equalizers can also be classified by the shape of their frequency response. **Peaking EQ** shapes the response like a hill or peak when set for a boost (Figure 9.7). With **shelving EQ**, the shape of the frequency response resembles a shelf (Figure 9.8). The "corner frequency" of a shelf is midway between flat and full again. *Audio clip 27 at www.elsevierdirect.com/companions/9780240811444 demonstrates various types of EQ.*

A **filter** causes a rolloff at the frequency extremes. It sharply rejects (attenuates) frequencies above or below a certain frequency. Figure 9.9 shows three types of filters: lowpass, highpass, and bandpass. For example, a 10-kHz lowpass filter (high-cut filter) removes frequencies above 10 kHz. Its response is down 3 dB at 10 kHz and more above that. This reduces hiss-type noise without affecting tone quality as much as a gradual treble rolloff would. A 100-Hz highpass filter (low-cut filter) attenuates frequencies below 100 Hz. Its response is down 3 dB at 100 Hz and more below that. This removes low-pitched noises such as air handler rumble or breath pops. A 1-kHz bandpass filter cuts frequencies above and below a frequency band centered at 1 kHz.

The crossover filter in some monitor speakers consists of lowpass, highpass, and bandpass filters. They send the lows to the woofer, mids to the midrange, and highs to the tweeter.

A filter is named for the steepness of its rolloff: 6 dB per octave (first order), 12 dB per octave (second order), 18 dB per octave (third order), and so on.

How to Use EQ

If your mixer has bass and treble controls, their frequencies are preset (usually at 100 Hz and 10 kHz). Set the EQ knob at 0 to have no effect (flat setting). Turn it clockwise for a boost; turn it counterclockwise for a cut. If your mixer has multiple-frequency EQ or sweepable EQ, one knob sets the frequency range and another sets the amount of boost or cut.

Table 9.1 shows the fundamentals and harmonics of musical instruments and voices. The harmonics given represent an approximate range. Percussion, cymbals, and muted trumpet actually have some energy up to 80 to 100 kHz. For each instrument, turn up the lower end of the fundamentals to get warmth and fullness. Turn down the fundamentals if the tone is too bassy or tubby. Turn up the harmonics for presence and definition; turn down the harmonics if the tone is too harsh or sizzly.

Here are some suggested frequencies to adjust for specific instruments. If you want the effects described below, apply boost. If you don't, apply cut. Try these suggestions and accept only the sounds you like:

- Bass: Full and deep at 60 Hz to 100 Hz, growl at 600 Hz, presence at 2.5 kHz, string noise at 3 kHz and up. Cut around 200 to 500 Hz for clarity.
- Electric guitar: Thumpy at 60 Hz, full at 100 Hz, puffy at 500 Hz, presence or bite at 2 to 3 kHz, sizzly and raspy above 6 kHz.
- Drums: Full at 100 to 200 Hz, wooly at 250 to 800 Hz (try cutting in that range), trashy snare at 1 to 3 kHz, attack at 5 kHz. Try cutting toms around 600 Hz to reduce boxiness. Dull cymbals sound more sizzly and crisp with a boost at 10 to 12 kHz.
- Kick drum: Full and powerful below 60 Hz, papery at 300 to 800 Hz (cut at 400 to 600 Hz for better tone), click or attack at 2 to 6 kHz.
- Sax: Warm at 500 Hz, harsh at 3 kHz, key noise above 10 kHz.
- Acoustic guitar: Full or thumpy at 80 Hz, presence at 5 kHz, pick noise above 10 kHz.

Table 9.1	Frequency Ranges of Musical Instruments and Voices	
Instrument	**Fundamentals**	**Harmonics**
Flute	261–2349 Hz	3–8 kHz
Oboe	261–1568 Hz	2–12 kHz
Clarinet	165–1568 Hz	2–10 kHz
Bassoon	62–587 Hz	1–7 kHz
Trumpet	165–988 Hz	1–7.5 kHz
French horn	87–880 Hz	1–6 kHz
Trombone	73–587 Hz	1–7.5 kHz
Tuba	49–587 Hz	1–4 kHz
Snare drum	100–200 Hz	1–20 kHz
Kick drum	30–147 Hz	1–6 kHz
Cymbals	300–587 Hz	1–15 kHz
Violin	196–3136 Hz	4–15 kHz
Viola	131–1175 Hz	2–8.5 kHz
Cello	65–698 Hz	1–6.5 kHz
Acoustic bass	41–294 Hz	700 Hz–5 kHz
Electric bass	41–294 Hz	700 Hz–7 kHz
Acoustic guitar	82–988 Hz	1500 Hz–15 kHz
Electric guitar	82–1319 Hz	1–15 kHz (direct)
Elec. guitar amp	82–1319 Hz	1–4 kHz
Piano	28–4196 Hz	5–8 kHz
Bass (voice)	87–392 Hz	1–12 kHz
Tenor (voice)	131–494 Hz	1–12 kHz
Alto (voice)	175–698 Hz	2–12 kHz
Soprano (voice)	247–1175 Hz	2–12 kHz

- Acoustic guitar pickup: To make the guitar sound more "acoustic," try a narrow cut at 1.2 to 1.5 kHz and maybe some high-frequency cut.
- Voice: Full at 100 to 200 Hz (males), full at 200 to 400 Hz (females), honky or nasal at 500 Hz to 1 kHz, presence at 5 kHz, sibilance ("s" and "sh "sounds) around 3 to 10 kHz.

Example: Suppose a vocal track sounds too full or bassy. Reach for the LF EQ knob (say, 100 Hz) and turn it down until the voice sounds natural. Or suppose a snare drum sounds dull or muffled. Grab the mid-frequency EQ knob, set it to 5 kHz to 10 kHz, and turn it up until the snare sounds clear and crisp.

Set EQ to the approximate frequency range you need to work on. Then apply full boost or cut so the effect is easily audible. Finally, fine-tune the frequency and amount of boost or cut until the tonal balance is the way you like it.

What if an instrument sounds honky, tubby, or harsh, and you don't know what frequency to tweak? Set a sweepable equalizer for extreme boost. Then sweep the frequencies until you find the frequency range matching the coloration. Cut that range by the amount that sounds right. For example, a piano miked with the lid closed might have a tubby coloration—maybe too much output around 300 Hz. Set your low-frequency EQ for boost, and vary the center frequency until the tubbiness is exaggerated. Then cut at that frequency until the piano sounds natural.

In general, avoid excessive boost because it can distort the signal. Try cutting the lows instead of boosting the highs. To reduce muddiness or enhance clarity, cut 1 to 2 dB around 300 Hz—either on individual instruments or on the entire mix. Don't boost everything at the same frequency.

For the most effortless, natural sound and less phase shift, use minimum phase EQ, less gain, wider bandwidth (Q), and gentler slopes (6 dB/oct. instead of 12 or 18 dB/oct.).

Use a highpass filter (low-cut filter) on all tracks to remove low frequencies below each instrument's lowest fundamental frequency. This cleans up the sound and removes rumble from trucks and air handling.

When to Use EQ

Before using EQ, try to get the desired tone quality by changing the mic or its placement. This gives a more natural effect than EQ. Many purists shun the use of EQ, complaining of excessive phase shift or ringing caused by the equalizer—a "strained" sound. Instead, they use carefully placed, high-quality microphones to get a natural tonal balance without EQ.

The usual practice is to record flat (without EQ) and then equalize the track during mixdown.

Sometimes the instruments need a lot of EQ to sound good. If so, you might want to record with EQ so that the playback for the musicians will sound good. When you play the multitrack recording through your monitor mixer, the recording may not sound right unless the tracks are already equalized. (That's assuming the monitor mixer in your board has no EQ.)

Uses of EQ

Here are some applications for EQ:

Improve tone quality. The main use for EQ is to make an instrument sound better tonally. For example, you might use a high-frequency rolloff on a singer to reduce sibilance, or on a direct-recorded electric guitar to take the "edge" off the sound. You could boost 100 Hz on a floor tom to get a fuller sound, or cut around 250 Hz on a bass guitar for clarity. Cut around 100 Hz to reduce bass buildup on massed harmony vocals. The frequency response and placement of each mic affect tone quality as well.

Although you can set the EQ for each track when it is soloed, a better way is to set the equalizers when the entire mix is playing. That's because one instrument can mask or hide certain frequencies in another instrument. For example, the cymbals might mask the "s" sounds in the vocal, making the vocal sound dull—even though it might sound fine when soloed.

Create an effect. Extreme EQ reduces fidelity, but it also can make interesting sound effects. Sharply rolling off the lows and highs on a voice, for instance, gives it a "telephone" sound. A 1-kHz bandpass filter does the same thing. To make a mono keyboard track sound stereo, send it to two mixer channels. Boost lows and cut highs in one channel panned left; cut lows and boost highs in the other channel panned right.

Reduce noise and leakage. You can reduce low-frequency noises—bass leakage, air-conditioner rumble, mic-stand thumps—by turning down the lows below the fundamental-frequency range of the instrument you're recording. This information is shown in Table 9.1.

For example, a fiddle's lowest frequency is about 200 Hz, so you'd use a low-cut filter (highpass filter) set to 200 Hz (if possible). This low-cut filter won't change the fiddle's tone quality because the filtered-out frequencies are below the fiddle's lowest frequency. Similarly, a kick drum has little or no output above 9 kHz, so you can filter out highs above 9 kHz on the kick drum to reduce cymbal leakage. Filtering out frequencies below 100 Hz on most instruments reduces air-conditioning rumble and breath pops. Try rolling off the lows on audience mics to prevent muddy bass. To reduce hum, set a parametric EQ for a 24-dB cut, Q of 30, at these frequencies: 60, 120, and 180 Hz (in the United States) or 50, 100, and 150 Hz (in Europe).

Compensate for the Fletcher-Munson effect. As discovered by Fletcher and Munson, the ear is less sensitive to bass and treble at low volumes than at high volumes. So, when you record a very loud instrument and play it back at a lower level, it might lack bass and treble. To restore these, you may need to boost the lows (around 100 Hz) and the highs (around 4 kHz) when recording loud rock groups. The louder the group, the more boost you need. It also helps to use cardioid mics with proximity effect (for bass boost) and a presence peak (for treble boost).

Make a pleasing blend. If you mix two instruments that sound alike, such as lead guitar and rhythm guitar, they tend to mush together—it's hard to tell what each is playing. You can make them more distinct by equalizing them differently. For example, make the lead guitar edgy by boosting 3 kHz, and make the rhythm guitar mellow by cutting 3 kHz. Then you'll hear a more pleasing blend and a clearer mix. The same philosophy applies to bass guitar and kick drum. Because they occupy about the same low-frequency range, they tend to mask or cover each other. To make them distinct, either fatten the bass and thin out the kick a little, or vice versa. The idea is to give each instrument its own space in the frequency spectrum; for example, the bass fills in the lows, synth chords emphasize mid-bass, lead guitar adds edge in the upper mids, and cymbals add sparkle in the highs.

Compensate for mic placement. Sometimes you are forced to mike very close to reject background sounds and leakage. But a close mic emphasizes the part of the instrument that the mic is near. This gives a colored tone quality, but EQ can partly compensate for it. Suppose you had to record an acoustic guitar with a mic near the sound hole. The guitar track will sound bassy because the sound hole radiates strong low frequencies. But you can turn down the lows on your mixer to restore a natural tonal balance.

This use of EQ can save the day by fixing poorly recorded tracks in live concert recordings. During a concert, the stage monitors might be blaring into your recording/PA microphones, so you're forced to mike close in order to reject monitor leakage and feedback. This close placement, or the monitor leakage itself, can give the recording an unnatural tone quality. In this case, EQ is the only way to get usable tracks.

"Remix" a single track. If a track contains two different instruments, sometimes you can change the mix within that track by using EQ. Imagine a track that has both bass and synth. By using LF EQ, you can bring the bass up or down without affecting the synth very much. Mixing with EQ is more effective when the two instruments are far apart in their frequency ranges.

Improve the tonal balance of an entire mix. During mastering, you can EQ the stereo mix of each song to make it better, to make the songs on an album sound more similar, or to make the album sound more like commercial albums. An outstanding tool for this purpose is Harmonic Balancer (www.har-bal.com). It shows the long-term spectrum (level vs. frequency) of a mix, and you can shape the spectrum with your mouse, which affects the tonal balance of the song.

Whenever you record, the ideal situation is to use the right mic in the right position, and in a good-sounding room. Then you don't need or want equalization. Otherwise, though, your recordings will sound better with EQ than without it.

COMPRESSOR

A compressor acts like an automatic volume control, turning down the volume when the signal gets too loud. Here's why it's necessary.

Suppose you're recording a female vocalist. Sometimes she sings too softly and gets buried in the mix; other times she hits loud notes and blasts the listener. Or she may move toward and away from the mic while singing, so that her average recording level changes.

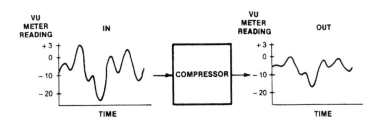

FIGURE 9.10
Compression.

To control this problem, you can ride gain—turn her down when she gets too loud; turn her up when she gets too quiet. But it's hard to anticipate these changes. You might prefer to use a compressor, which does the same thing automatically. It reduces the gain (amplification) when the input signal exceeds a preset level (called the threshold). The greater the input level, the less the gain. As a result, loud notes are made softer, so the dynamic range is reduced (Figure 9.10). *Play audio clip 28 at www.elsevier direct.com/companions/9780240811444.*

Compression keeps the level of vocals or instruments more constant, so they are easier to hear throughout the mix. Also, it can be used for special effects—to make drums sound fatter, or to increase the sustain on a bass guitar. In pro studios, compression is used almost always on vocals; often on bass guitar, kick drum, and acoustic guitar; and sometimes on other instruments.

Doesn't compression rob the music of its expressive dynamics? Yes, if overdone. But a vocal that gets too loud and soft is annoying. You need to tame it with a compressor. Even then, you can tell when the vocalist is singing loudly by the tone of the voice. It also helps to compress the bass and kick drum to ensure a uniform, driving beat.

You can avoid vocal compression if the singer uses proper mic technique. He or she should back away from the mic on loud notes, and come in close on soft notes. To tell whether you need a compressor, listen to your finished mix. If you can understand all the words, and no notes are too loud, omit the compressor.

Using a Compressor

Normally, you compress individual tracks or instruments, not the entire mix. You want to compress only the stuff that needs it. To compress a stereo mix, you need a 2-channel compressor with a stereo link, which keeps the left–right balance from changing during compression.

Multiband compression (covered later) is usually a better choice for compressing the stereo mix.

Let's describe the controls on the compressor. Some compressors have few controls; most of their settings are preset at the factory. Figure 9.11 shows a compressor plug-in.

FIGURE 9.11
Cakewalk Sonitus Compressor, an example of a compressor plug-in. (The graph labels are by the author.)

COMPRESSION RATIO OR SLOPE

This is the ratio of the change in input level to the change in output level. For example, a 2:1 ratio means that for every 2 dB change in input level, the output of the compressor changes 1 dB. A 20 dB change in input level results in a 10 dB change in the output, and so on. Common ratio settings are 2:1 to 4:1.

THRESHOLD

This is the input level above which compression occurs. Set the threshold high (about −5 dB) to compress only the loudest notes; set it low (−10 or −20 dB) to compress a broader range of notes. A setting of −10 is typical. Another method: while the music is playing, set the threshold high, then gradually bring it down until you get the amount of gain reduction you want on the gain-reduction meter (typically 6 dB). If the compressor has a fixed threshold, adjust the amount of compression with the input level control.

GAIN REDUCTION

This is the number of dB that the gain is reduced by the compressor. It varies with the input level. You set the ratio and threshold controls so that the gain is reduced on loud notes by an amount that sounds right. The amount of gain reduction shows up on a meter—3 to 10 dB is typical.

COMPRESSION GRAPH

Most compressor plug-ins display a compression graph of output level versus input level in dB (Figure 9.11). As the input level increases from zero up to the threshold level, the graph is an upward-sloping straight line in which the change in output level matches the change in input level. Where the input level is above the threshold, the graph transitions to a flatter slope where compression occurs: the change in output level is less than the change in input level.

The transition point is called the "knee" of the compression curve. The transition from no compression to compression can be abrupt (hard knee) or gradual (soft knee). A "soft knee" or "over easy" characteristic is a low compression ratio for low-level signals and a high ratio for high-level signals. Some manufacturers say that this characteristic sounds more natural than a fixed compression ratio. Typically, a soft-knee setting is 16 dB and hard-knee is 4 dB.

The slope of the compressed part of the curve (the part on the right) indicates the compression ratio. The more horizontal this part of the curve is, the greater the ratio. A horizontal line indicates limiting (explained later in this chapter). As you vary the compression ratio and threshold, the compression curve changes accordingly.

ATTACK TIME

This is how fast the compressor reduces the gain when it's hit by a musical attack. Typical attack times range from 0.25 to 50 msec. Some compressors adjust the attack time automatically to suit the music; others have a factory-set attack time. The longer the attack time, the larger the peaks that are passed before gain reduction occurs. So, a long attack time sounds punchy; a short attack time reduces punch by softening the attack. Bass instruments require a fairly long attack time (>50 msec) to prevent distortion.

RELEASE TIME

This is how fast the gain returns to normal after a loud passage ends. It's the time the compressor takes to reach 63% of its normal gain. You can set the release time from about 50 msec to several seconds; 0.2 second to 0.5 second is typical. For bass instruments, the release time must be longer than about 0.4 second to prevent harmonic distortion.

Short release times make the compressor follow rapid volume changes in the music, and keep the average level higher. But because the noise rises along with the gain, short release times can give a pumping or breathing sound. Long release times sound more natural. If the release time is too long, though, a loud passage will reduce the gain during a subsequent quiet passage. In some units, the release time varies automatically, or is factory-set to a useful value.

Some compressors disable the attack and release settings when the compressor is set to RMS or average mode. Those settings are adjusted automatically.

Figure 9.12 shows the effects of compressor attack time and release time on the envelope of a waveform. The grey block represents a musical note fed into the compressor. As Figure 9.12 shows, a long attack time tends to increase the attack portion of the envelope, giving a sound with a sharper attack or "edge." A short release time tends

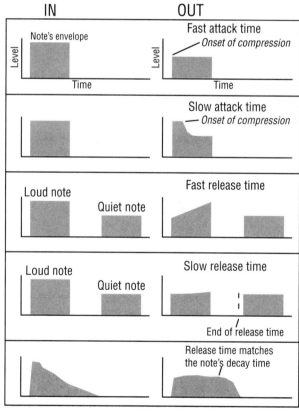

FIGURE 9.12
Effects of attack time and release time on the envelope of a note.

to bring up the level in between notes, giving a louder but more fatiguing sound. A release time similar to the note's decay time increases sustain.

OUTPUT-LEVEL CONTROL

Also called makeup gain, this control is used to increase the output level of the compressor by the amount of gain reduction. For example, if a compressor is causing 6 dB of level reduction, set the makeup gain to 6 dB to achieve unity gain. This also brings up the quiet parts of the track by 6 dB. Some compressors keep the output level constant when other controls are varied.

Spend some free time playing with all the settings so you learn how they affect the sound. Play various instruments and vocals through a compressor, vary the settings, and take notes on what you hear.

Some compressors have a side chain. This is a pair of in/out jacks for connecting an equalizer. To compress only the sibilant sounds on a vocal track, boost the side-chain EQ around 3 to 10 kHz. To compress only the breath pops on a vocal track, boost the side-chain EQ around 20 Hz.

A multiband or split-frequency compressor divides the audio band into three to five bands (bass, mids, treble) and compresses each band separately. That way, the compressor can squash a loud bass note, or soften "s" sounds, without bringing down the overall level. It works well when applied to the final mix of each song during mixdown or mastering.

If you recorded an acoustic guitar with finger squeaks, you can reduce the squeaks with a multiband compressor. Use a spectrum analyzer plugin to find the frequency band of the squeaks. Compress just that band.

Connecting a Compressor

To compress one track in a DAW, select a compressor plug-in for that track. Do not use an aux send for compression. Connect a hardware compressor in line with the signal you want to compress in one of the following ways:

- To compress one instrument or voice while recording: Locate the input module of the instrument you want to compress. On the back of that module, connect the insert send jack to compressor in; connect compressor out to the insert return jack. (Chapter 10 explains these terms.) Or, take a signal from the input module's direct out. Feed that into the compressor, and feed the compressor output to the recorder track input.
- To compress a group of instruments while recording: Locate the bus output of the instruments you want to compress. Go from bus out to compressor in, and go from compressor out to recorder-track in. If the bus has insert jacks, you could connect to them instead.

- To compress one track during mixdown: Go from track out to compressor in, and go from compressor out to mixer channel in. Or locate the mixer input module for that track, then find the insert send and return jacks in that module. (There might be a single insert jack with send and return terminals.) Connect the insert send to compressor in; connect compressor out to insert return.

Suggested "Ballpark" Compressor Settings

- Vocals: Ratio 2:1 to 3:1, soft knee, fast attack, 1/2 second release, set threshold for 3 to 6 dB of gain reduction. Singers with extreme dynamic range might need 12 dB of gain reduction and a ratio of 4:1.
- Bass and drums: Ratio 4:1, hard knee, set threshold for 3 to 6 dB of gain reduction on loud bass "pops." Adjust attack time depending on how much you want to soften the attack. Short attack time = soft attack, long attack time = loud attack. For starters, try 1.5 msec attack time and 150 msec release time. Bass needs a release time longer than about 400 msec to avoid distortion.
- Electric guitar: 4:1 to 8:1 ratio, 10 dB gain reduction, 400 msec release.
- To reduce breath pops: Use a multiband compressor. Enable only the lowest frequency band. Try these settings: ratio 30:1, upper frequency 100 Hz, makeup gain 0 dB, attack 1 msec, release 100 msec, threshold −18 dB. Experiment with the threshold setting.
- To reduce sibilance (de-ess mode): First, set an equalizer for a narrow boost around 6 kHz. While playing the vocal track, sweep the frequency up and down until you find the frequency area where sibilance ("s" and "sh" sounds) is exaggerated the most. Note that frequency. Remove the equalizer and insert a multiband compressor. Disable all bands except a band from about 3 kHz to 10 kHz. Try to set up a narrower band around the frequency you noted earlier. Suggested settings: ratio 10:1, attack 1 msec, release 100 msec, threshold −32 dB. Experiment with the threshold setting to get the desired amount of de-essing.
- To compress the stereo mix: Try these settings: 2:1, soft knee, attack 20 msec, release 200 msec. Set the threshold to get 5 to 10 dB of gain reduction. If you can hear the compressor squashing the sound, back off the gain reduction unless that's the sound you want.

Should you compress while tracking or mixing? If you compress while tracking, it will be difficult or impossible to change the amount of compression during mixdown. If you compress tracks during mixdown, you can change the settings at will.

When applying low-cut EQ to a track, insert the EQ before the compressor. Often there is too much bass on a track, and that extra bass will trigger the compressor unless you EQ it out first. When you apply a boost, put the EQ after the compressor so that the boost isn't compressed.

Parallel Compression

This effect blends compressed and uncompressed sounds of a track. It gives more punch but retains dynamics. Here's how to create it:

1. Clone a track.
2. Insert a compressor in the cloned track. Set it for extreme compression. Typical settings: attack time 1.5 msec, release 200 msec, threshold −50 dB, ratio 4:1 hard knee. Another setting might be ratio 10:1 and threshold −25 dB.
3. Mix the original track with the cloned (compressed) track. Gradually turn up the cloned track to get the effect you want.

Another method: Some compressors have a dry/wet mix control. Dry is uncompressed and wet is compressed. Adjust for the desired effect. You might parallel-compress an entire mix to "glue it together" without affecting dynamics.

Sidechain

A sidechain is an input to a compressor that controls its gain reduction. To make a kick drum control the level of a stereo mix, follow these steps:

1. Insert a compressor in the stereo mix bus. Be sure the two channels are linked.
2. In the kick-drum track, insert an aux send to the compressor's sidechain input.
3. Set the send to prefader.
4. Adjust the compression ratio, attack time, release time, and threshold so the kick drum modulates the mix. The mix will duck down every time the kick drum hits, and will come up between hits.

LIMITER

A limiter keeps signal peaks from exceeding a preset level. While a compressor reduces the overall dynamic range of the music, a limiter affects only the highest peaks (Figure 9.13). To act on these rapid peaks, limiters have a very fast attack time—1 microsecond to 1 millisecond. The compression ratio in a limiter is very high—10:1 or greater—and the threshold is set high, say at 0 dB. For input levels up to 0 dB, the output level

FIGURE 9.13
Limiting.

matches the input. For input levels above 0 dB, the output level stays at 0 dB. This prevents overload in the device following the limiter.

A compressor/limiter carries out both of the functions in its name. It compresses the average signal levels over a wide range, and limits peaks to prevent overload. It has two thresholds: one low for the compressor and one high for the limiter.

Limiters can be used to prevent recorder overload during field recording, or to prevent PA power amps from clipping. When you master a program of several mixed songs in a DAW, you might use limiting to reduce the level of signal peaks in the program. Set the threshold about 6 dB below the highest peak level. Then apply normalization, which raises the level of the entire program until the highest peak in the program reaches maximum level. Limiting and normalization create a louder program on your finished CD without compressing the music's dynamics.

A limiter plug-in with the "lookahead" feature checks the upcoming audio for peaks. Audio enters a lookahead buffer, and the limiter measures that audio signal and reacts quickly to reduce the incoming peaks.

NOISE GATE

A noise gate (expander) acts like an on-off switch that removes noises during pauses in an audio signal. It reduces the gain when the input level falls below a preset threshold. That is, when an instrument stops playing for a moment, the noise gate drops the volume, which removes any noise and leakage during the pause (Figure 9.14).

Note: The gate does not remove noise while the instrument is playing.

Where is it used? The noise gate helps to clean up drum tracks by removing leakage between beats. It can shorten the decay time of the drums, giving a very tight sound. If you're recording a noisy guitar amp, try a gate to cut out the buzz and hiss between phrases.

How do you use a noise gate? Patch it between a recorder-track output and a mixer line input, or use a gate plug-in in a DAW. Solo the track that you want to gate. Set the gate's threshold so that noise and leakage go away during pauses. If the gate chops off

FIGURE 9.14
Gating.

each note, the threshold is set too high—turn it down. Set the release time as short as possible, but long enough so that you hear the entire note (or tom hit) before the gate cuts off the sound. To fix a boomy kick drum, adjust the threshold until the kick sounds as "tight" as you want. That is, use the gate to shorten the decay portion of the kick-drum's envelope.

Excellent recordings can be made without gating. But if you want a tighter sound, gates come in handy. Some signal processors have compression, limiting, and noise gating in a single package.

Some gates have a sidechain input (also called a key input or trigger input). It's an input for an external signal that controls the gating action. For example,

1. Insert a gate in a bass track.
2. In the snare-drum track, insert an aux send to the gate's sidechain input.
3. Set the send to prefader.
4. Adjust the gate threshold so the snare-drum hits modulate the bass.
5. Change release time and hold time for best results. Remove high frequencies in the trigger (sidechain) input if necessary.

DELAY: ECHO, DOUBLING, CHORUS, AND FLANGING

A digital delay (or a delay plug-in) takes an input signal, holds it in memory, then plays it back after a short delay—about 1 msec to 1 second (Figure 9.15). Delay is the time interval between the input signal and its repetition at the output of the delay device.

If you listen to the delayed signal by itself, it sounds the same as the undelayed (dry) signal. But if you combine the delayed and dry signals, you may hear two distinct sounds: the signal and its repetition. By delaying a signal, a processor can create several effects such as echo, repeating echo, doubling, chorus, and flanging.

Echo

If the delay is about 50 msec to 1 second, the delayed repetition of a sound is called an echo. This is shown in Figure 9.16 by the two pulses. Echoes occur naturally when

FIGURE 9.15
Delaying the signal.

FIGURE 9.16
Echo.

sound waves travel to a distant room surface, bounce off, and return later to the listener—repeating the original sound. A delay unit can mimic this effect. Many people use the term delay to mean echo.

In setting up a mix with echo, you want to hear both the dry sound and its echo. You do this by creating an effects loop from the mixer, to the effects box, and back to the mixer. Here's how to set up echo with a hardware mixer and effects unit:

1. On the delay unit, set the dry/wet mix control all the way to "wet" or "100% mix." Then the output of the delay unit will be only the delayed signal.
2. Suppose you want to use aux 1 as the echo control. Connect aux 1 send to delay unit IN. Connect delay unit OUT to Bus 1 and 2 IN (or to the effects-return jacks).
3. Find the mixer module for the instrument you want to add echo to.
4. Assign the instrument to buses 1 and 2. Monitor buses 1 and 2.
5. Find the knobs labeled Bus 1 IN and Bus 2 IN. They might be called "Aux Return" or "Effects Return." Turn them up to 0, about three-fourths of the way up.
6. Turn up the aux 1 send knob, and there's your echo.

The delayed sound mixes with the dry sound in buses 1 and 2. You hear both sounds, which together make an echo. Each aux knob controls the amount of echo on each track, while the effect-return knobs control the overall amount of echo on all tracks that are feeding the echo unit.

To set up echo in a DAW, follow this procedure:

1. Create or use a stereo aux bus that has an echo or delay plug-in enabled.
2. In a track that you want to have echo, enable an aux send to that bus.
3. Open the echo plug-in. Set its dry/wet mix control all the way to wet or 100% mix.
4. In the track that you want to have echo, gradually turn up the aux send until you hear the desired amount of echo. In the echo plug-in, adjust the delay parameters for the desired effect.

FIGURE 9.17
Repeating echo.

SLAP ECHO

A delay from about 100 to 130 msec is called a slap echo or slapback echo. It was often used in 1950s rock 'n' roll and rockabilly tunes, and is still used today.

REPEATING ECHO

Most delay units can be made to feed the output signal back into the input, internally. Then the signal is redelayed many times. This creates a repeating echo—several echoes that are evenly spaced in time (Figure 9.17, and *audio clips 29 and 30 at www.elsevier direct.com/companions/9780240811444*). The regeneration (feedback) control sets the number of repeats.

Repeating echo is most musical if you set the delay time to create an echo rhythm that fits the tempo of the song. The formula is

$$\text{Delay in seconds} = 60/\text{tempo}$$

So if the tempo is 120 bpm, the delay is 0.5 sec (500 msec). That's one echo per quarter note. Use half that delay to get one echo per eighth note. Use one-third that delay for triplets. A slow repeating echo—0.5 second between repeats, for example—gives an outer-space or haunted-house effect. It often sounds good on lead vocal in a ballad.

Doubling

If you set the delay around 23 to 30 msec, the effect is called doubling or automatic double tracking (ADT). It gives an instrument or voice a fuller sound, especially if the dry and delayed signals are panned to opposite sides. The short delays used in doubling sound like early sound reflections in a studio, so they add some "air" or ambience. Try 40% pitch modulation with a modulation rate less than 1 Hz.

Doubling a vocal can be done without a delay unit. Record a vocal part, then overdub another performance of the same vocal part. Mix the parts, pan them both to center, or pan them left and right.

FIGURE 9.18
Flanging (or positive flanging).

Chorus

This is a wavy or shimmering effect. The delay is 15 to 35 msec, and the delay varies at a slow rate. Sweeping the delay time causes the delayed signal to bend up and down in pitch, or to detune. When you combine the detuned signal with the original signal, you get chorusing. Try about 10–30% pitch modulation at 1–10 Hz.

STEREO CHORUS

This is a beautiful effect. In one channel, the delayed signal is combined with the dry signal in the same polarity. In the other channel, the delayed signal is inverted in polarity, then combined with the dry signal. Thus, the right channel has a series of peaks in the frequency response where the left channel has dips, and vice versa. The delay is slowly varied or modulated. *Hear a demonstration with audio clip 33 at www.elsevier direct.com/companions/9780240811444.*

BASS CHORUS

This is chorus with a highpass filter so that low frequencies are not chorused, but higher harmonics are. It gives an ethereal quality to the bass guitar.

Flanging

If you set the delay around 0 to 20 msec, you usually can't resolve the direct and delayed signals into two separate sounds. Instead, you hear a single sound with a strange frequency response. The direct and delayed signals combine and have phase interference, which puts a series of peaks and dips in the frequency response. This is called a comb-filter effect (Figure 9.18). It gives a very colored, filtered tone quality. The shorter the delay, the farther apart the peaks and dips are spaced in frequency.

The flanging effect varies or sweeps the delay between about 0 and 20 msec. This makes the comb-filter nulls sweep up and down the spectrum. As a result, the sound is hollow, swishing, and ethereal, as if the music were playing through a pipe. Flanging is easiest to hear with broadband signals such as cymbals but can be used on any

FIGURE 9.19
Negative flanging.

instrument, even voices. *Hear a demonstration with audio clip 34 at www.elsevierdirect. com/companions/9780240811444.*

Some examples of flanging are on many Jimi Hendrix records, and on the oldies "Itchycoo Park" by the Small Faces and "Listen to the Music" by the Doobie Brothers. The first use of flanging was on "The Big Hurt" sung by Toni Fisher.

Positive flanging refers to flanging in which the delayed signal is the same polarity as the direct signal (Figure 9.18). With negative flanging, the delayed signal is opposite in polarity to the direct signal, which makes a stronger effect. The low frequencies are canceled (the bass rolls off), and the "knee" of the bass rolloff moves up and down the spectrum as the delay is varied. The high frequencies are still comb-filtered (Figure 9.19). Negative flanging makes the music sound like it's turning inside out.

When the flanger feeds some of the output signal back into the input, the peaks and dips get bigger. It's a powerful "science fiction" effect called resonant flanging.

To make a mono track sound stereo, you can pan the track to the right, and pan the flanged track to the left.

REVERBERATION

This effect adds a sense of room acoustics, ambience, or space to instruments and voices. To know how it works, we need to understand how reverb happens in a real room. Natural reverberation in a room is a series of multiple sound reflections that make the original sound persist and gradually die away or decay. These reflections tell the ear that you're listening in a large or hard-surfaced room. For example, reverberation is the sound you hear just after you shout in an empty gymnasium.

A reverb effect simulates the sound of a room—a club, auditorium, or concert hall— by generating random multiple echoes that are too numerous and rapid for the ear to resolve (Figure 9.20). Digital reverb is available either in a dedicated reverb unit, as part of a multieffects processor, or as a plug-in.

FIGURE 9.20
Reverberation.

The most natural sounding digital reverb is a sampling reverb or convolution reverb, which creates the reverb from impulse-response samples (wave files) of real acoustic spaces, rather than from algorithms. Convolution reverbs consume a lot of CPU processing power. One convolution reverb plug-in is SIR at http://www.knufinke. de/sir/. Free impulse-response samples are at www.noisevault.com and www.cksde. com/p_6_250.htm.

Reverb and echo are not the same thing. Echo is a repetition of a sound ("HELLO hello hello"); reverb is a smooth decay of sound ("HELLO-OO-oo-oo").

Some reverb plug-ins include Cakewalk Sonitus Reverb, Lexicon Pantheon, 2CAudio Aether, Overloud Breverb, IK Multimedia Classik Studio Reverb, Virsyn Reflect, Rob Papen RP-Verb, and D16 Group Toraverb.

Reverb Parameters

Here are some controls in a reverb unit or plug-in (Figure 9.21):

- Reverb Time (RT60): The time it takes for reverberation to decay 60 dB below its original level. Set it long (1 1/2 to 2 seconds) to simulate a large room; set it short (under 1 second) to simulate a small room. Generally you use short reverbs (or no reverb) for fast songs, and long reverbs for slow songs. Snare reverb should die out between hits to prevent masking.
- Predelay (pre-reverb delay): A short delay (23 to 100 msec) before the onset of reverb to simulate the delay that happens in real rooms before reverb starts. The longer the predelay, the bigger the room sounds. Using predelay on a track's reverb often helps to clarify the sound by removing the onset of reverb from the direct sound of the instrument or vocal. If your reverb unit does not have predelay, you can create it by connecting a delay unit between your mixer's aux send and the reverb input. If your reverb plug-in lacks predelay, insert a delay plug-in before the reverb plug-in. *Audio clip 31 at www.elsevierdirect.com/companions/9780240811444 demonstrates reverb: short reverb time, long reverb time, and predelay.*

FIGURE 9.21
Lexicon MPX Native Reverb plug-in.

- Density: A high density setting produces many echoes spaced close together. It gives a smooth decay but increases the load on the CPU. A low-density setting produces fewer echoes spaced farther apart, and may be adequate for vocals, synth pads, and organ. Use high density for percussive sounds to prevent grainy-sounding reverb.
- Diffusion: How spread out the reverb is.
- Early reflections: These are the first few sound reflections, usually within 100 msec of the direct sound. They contribute a sense of ambience and depth. Adding only early reflections to a dry signal can enhance a distorted guitar or enliven a dry vocal.
- Damping: Adjusts the reverb time or decay at high frequencies. Set the damping frequency high (say, 7 kHz) to simulate a hard-surfaced room; set it low (say, 2 kHz) to simulate a soft-surfaced room. The latter is also called a "warm room" reverb.
- EQ or rolloff: Some reverbs let you adjust the low- and high-frequency limit of the signal feeding the reverb. You might roll off the highs in the reverb to make it smoother and less obvious.
- Presets: Factory-supplied reverb settings of small rooms, auditoriums, halls, and so on. A plate reverb setting duplicates the bright sound of a metal-foil plate, which used to be the most popular type of reverb in pro studios for vocals and drums. Unnatural effects are available, such as nonlinear decay, reverse reverb that builds up before decaying, or gated reverb. With gated reverb, the reverb cuts off suddenly shortly after a note is hit. In the 1980s it was often used on a snare drum. A good example is the oldie "You Can Call Me Al" on Paul Simon's album Graceland.

Reverb Connections

To connect a reverb unit to a hardware mixer, connect a cable from the mixer aux send to the reverb input. Connect a cable (two for stereo) from the reverb outputs to the mixer aux returns (effects returns or bus inputs). Set the mix control on the reverb unit all the way to "wet" or "reverb." Turn the mixer's aux-return or bus-in knobs (if any) about two-thirds of the way up, and adjust the amount of reverb on each track with the aux-send knobs. Try to get an overall reverb-send level near 0 on the meter; then fine-tune the aux return level for the desired amount of reverb.

To enable reverb in a DAW, use the same procedure as for setting up echo, but choose a reverb plug-in.

PREVERB

Preverb is reverb that precedes a note rather than follows it. The reverb starts from silence and builds up to a note's attack (*audio clip 32 at www.elsevierdirect.com/ companions/9780240811444*). When used on a snare drum, preverb gives a whip-cracking kind of sound, like "shSHK!"

Here's how to add preverb to a snare drum track in a DAW:

1. Set up an aux 1 bus with reverb. Set the reverb all the way to wet or 100% mix.
2. Solo the drum track.
3. On the drum track, turn up the aux 1 send and set it to pre-fader. Find a good snare hit and play it.
4. Turn down the drum-track fader and play the snare hit. You should hear only the reverb from the aux 1 bus.
5. Highlight the snare hit up until the next hit. Export or save the mix (the snare hit) as "Drum reverb.wav".
6. Select a blank track and call it Drum Reverb. Import "Drum reverb.wav" into that track.
7. Select the drum-reverb clip, and then select reverse processing in your DAW. This reverses the drum-reverb clip.
8. On the original snare-drum track, reset the aux 1 send to post-fader and turn it back down. Turn up the track fader where it used to be.
9. Slide in time the reversed drum-reverb clip so that it ends just as a snare-drum hit starts (check the waveforms).
10. Play the reversed drum-reverb track along with the snare-drum track. You should hear preverb.

Some signal processors have a reverse reverb effect in which the reverb comes after the note that produced it, but builds up before it fades out. This is not quite the same as preverb. Reverse reverb can upset the musical timing; preverb doesn't.

ENHANCER

If a track or a mix sounds dull and muffled, you can run it through an enhancer to add brilliance and clarity. An enhancer works by adding slight distortion (as in the Aphex Aural Exciter).

OCTAVE DIVIDER

This unit takes a signal from a bass guitar and provides deep, growling bass notes one or two octaves below the pitch of the bass guitar. It does this by dividing the incoming frequency by 2 or 4: if you put 82 Hz in, you get 41 Hz out. Some MIDI sound modules have bass patches with extra-deep sound, and 5-string bass guitars have an extra string tuned especially low.

HARMONIZER

Basically a delay unit with delay modulation, a harmonizer makes a variety of pitch-shifting effects. It can create harmonies, change pitch without changing the duration of the program, change duration without changing pitch, and many other oddities. You've heard harmonizers on radio-station spots when the announcer's voice sounds like a Munchkin or Darth Vader. *Play audio clip 35 at www.elsevierdirect.com/ companions/9780240811444.*

VOCAL PROCESSOR

This device or plug-in can affect the vocal's inflection, add growls or whispers, correct the pitch, add vibrato, make the voice more or less nasal or chesty, and so on. The latest vocal formant-corrected pitch-shifters maintain the voice formant structure when they shift pitch; this prevents the "chipmunk" effect. An example is the TC-Helicon VoiceModeler plug-in.

Another type of vocal processor is called a channel strip. It includes a high-quality mic preamp or two, plus EQ, compression, gating, de-esser, and perhaps some tube saturation distortion. Examples are the Focusrite ISA One and Sonivox Vocalizer.

PITCH CORRECTION

This plug-in provides automatic or manual pitch correction of a single track (but not of chords). In automatic mode, it corrects flat or sharp notes by changing their pitch to match a musical scale of your choice. In manual mode (if available), you see a graph of the notes' pitches on your monitor screen, and slide certain notes up or down to correct their pitch. Manual mode is less obvious than automatic; it's better to correct only a few notes. Rerecording an out-of-tune phrase sounds more natural than pitch correction.

Some pitch-correction plug-ins are Antares Auto-Tune Evo, GSnap (free), Celemony Melodyne, TC Helicon Intonator HS, and Roland's V-Vocal VS plug-in for Cakewalk Sonar Producer.

You also can use pitch correction as a "robotic" effect where the sung notes change pitch in a stepwise, yodeling, jerky way rather than smoothly. It's also called the Cher effect because she first used it in her song "Believe." Although this effect has already become a cliché, you should know how to do it. Set Vibrato to minimum, Note to maximum, Sense to maximum, and Attack and Release to 0. "Note" is amount of pitch correction. "Sense" is the range of pitches that are pitch-corrected. "Attack and Release" are the speed of reaction of the pitch correction. Your pitch-correction software might have different names for those functions.

Melodyne Direct Note Access lets you edit individual notes within chords. See http://www.celemony.com.

TUBE PROCESSOR

This device or plug-in uses a vacuum tube or a simulation of one. Tubes have euphonic even-order harmonic distortion, which is claimed to add "richness" or "warmth" when the tube distorts (*audio clip 36 at www.elsevierdirect.com/companions/9780240811444*). There are tube mics, tube mic preamps, tube compressors, and standalone tube processors.

ROTARY SPEAKER SIMULATOR

This effect simulates the sound of a Leslie organ speaker, which plays music through rotating horns. It's a complex sound effect of pitch shifting, tremolo, and phase shifting. The speed and depth of the effect are adjustable.

ANALOG TAPE SIMULATOR

Analog tape saturation is mainly third-harmonic distortion and compression. An analog tape simulator adds this distortion to digital recordings in an attempt to smear or warm up the sound in a pleasant way (*audio clip 37 at www.elsevierdirect.com/companions/9780240811444*).

Some tube or tape saturation plug-ins are Avid Heat, Crane Song Phoenix II, McDSP Analog Channel, PSP Audio PSP Mixsaturator2 and VintageWarmer2, SoundToys

Decapitator, Universal Audio Fatso, Antares Warm, Massey TapeHead, Metric Halo Character, Nomad Factory Magnetic, and Virsyn VTPAE Suite.

SPATIAL PROCESSOR

Spatial processors enhance the stereo imaging or spatial aspects of a mix heard over two speakers. Some units have joystick-type pan pots, which move the image of each track anywhere around the listener. Other units make the stereo stage wider, so that images can be placed to the left of the left speaker, and to the right of the right speaker. The listener might hear images toward the sides of the listening room. In 5.1 surround systems, this spatial processing is done by surround panning and surround reverbs.

MICROPHONE MODELER

Antares and Roland offer a microphone modeler or simulator. You tell it which mic you are using and which mic you want it to sound like. A wide variety of vintage and current mic simulations are available. Mic modeling comes in three forms: hardware device, plug-in, and firmware (programmed into a chip) in a recorder-mixer.

GUITAR AMPLIFIER MODELER

Another simulator takes the sound of a direct-recorded guitar, and makes the guitar sound like it is played through a guitar amp (*audio clip 38 at www.elsevierdirect.com/companions/9780240811444*). Several amp models can be simulated, as well as effects, tone, drive, the mic used to pick up the amp, and the mic's position.

What are some hardware and software amp modelers? Two hardware examples are the Line 6 Pod and the Johnson J Station. Amp Farm is a guitar modeling plug-in for Pro Tools. Other notable plug-ins are Peavey ReValver MK III.V, Avid Eleven Rack Expansion Pack, Scuffham S-Gear, Line 6 POD Farm 2.5, IK Multimedia AmpliTube Custom Shop, Softube Metal Amp Room, and Ableton Amp. Roland's digital workstations offer COSM mic modeling and guitar-amp modeling.

Guitar processors or guitar stomp boxes can be used on any instrument or vocal to add distortion.

DISTORTION

Some plug-ins add intentional distortion of various types to generally "shred" the sound for a low-fi effect. Two examples are iZotope Trash and Camel Audio CamelCrusher (Figure 9.22). Bit-depth reduction changes a 16-bit or 24-bit signal to 4 or 8 bits. Parallel distortion is a clean track mixed with a distorted clone of the same track.

FIGURE 9.22
CamelCrusher distortion plug-in.

DECLICK AND DENOISE

Also called "audio restoration programs," these are plug-ins—or standalone programs—that can remove the clicks and pops from LP records, or remove hiss and hum from noisy recordings. Some examples are Dart Pro 24, iZotope RX, Bias SoundSoap, and Waves Native Restoration Bundle.

SURROUND SOUND

Recent plug-ins for surround sound are surround panning, surround reverb, and surround encoding/decoding.

MULTIEFFECTS PROCESSOR

This provides several effects in a single device or plug-in. Some units let you combine up to four effects in any order. Others have several channels, so you can put a different effect on different instruments. With most processors, you can edit the sounds and save them in memory as new programs. *At www.elsevierdirect.com/ companions/9780240811444, audio clips 29–38 demonstrate various effects on voice.*

Some processors offer 100 or more programmable presets with MIDI control over any parameter. For example, with some units you can place an instrument in a simulated room, and use a MIDI controller to continuously change the size of that room. Many signal processors can be controlled by MIDI program-change commands. You can quickly change the type of effect, or effect parameters, by entering certain program changes into a sequencer.

Suppose you want each tom-tom hit in a drum fill to have a different size room added to it. For example, put the high-rack tom in a small room, put the low-rack tom in a concert hall, and put the floor tom in a cave. To do this, first assign a different program number (patch or preset number) to each effects parameter. You do this with the effects device. Then, using the sequencer, punch in the appropriate program number for each note.

A MIDI program-change footswitch lets guitarists call up different effects on MIDI signal processors. By tapping a footswitch, they can get fuzz, flanging, wah-wah, spring reverb, and so on.

A MIDI mapper lets you control some effects parameters with any controller. For example, vary reverb decay time with a pitch wheel, or vary a filter with key velocity.

CHANNEL STRIP PLUG-IN

A channel strip in a mixing console is a vertical row of controls for one channel: preamp trim, aux sends, pan pot, fader, EQ, in-line compressor, and so on. Channel strip plug-ins emulate those functions. Some examples are Eventide E-Channel, Focusrite Forte Suite, McDSP Channel G, Metric Halo Channel Strip, TC Electronic VoiceStrip, Universal Audio Channel Strips, URS Classic Console Strip Pro, Wave Arts Track Plug 5, and Waves AudioTrack, Renaissance Channel, and SSL E-Channel.

LOOKING BACK

We've come a long way with effects. Looking back over the past few decades, each era had its own "sound" related to the effects used at the time. The '50s had tube distortion and slap echo; the '60s used fuzz, wah-wah, and flanging. Much of the early '70s sounded dry, and the early '90s emphasized synthesizers, drum machines, and gated reverb. Now vacuum tubes and acoustic instruments are back, along with occasional low-fi (tinny, distorted, or noisy) sounds and dry vocals. Whatever effects you choose, they can enhance your music if used with taste.

SOUND-QUALITY GLOSSARY

The sound of mic techniques, effects, and EQ in a recording can be hard to translate into engineering terms. For example, what EQ should you use to get a "fat" sound or a "thin" sound? The glossary below may help. It's based on conversations with producers, musicians, and reviewers over 30 years. Not everyone agrees on these definitions, but they are common. This glossary does not suggest the cause of the sound quality or how to change it; that's up to you to determine.

AIRY Spacious. The instruments sound like they are surrounded by a large reflective space full of air. A pleasant amount of reverb or early reflections. High-frequency response that extends to 15 or 20 kHz.

BALLSY OR BASSY Emphasized low frequencies below about 200 Hz.

BLOATED Excessive mid-bass around 250 Hz. Poorly damped low frequencies, low-frequency resonances.

BLOOM Adequate low frequencies. Spacious. Good reproduction of dynamics and reverberation. Early reflections or a sense of "air" around each instrument in an orchestra.

BOOMY Excessive bass around 125 Hz. Poorly damped low frequencies or low-frequency resonances.

BOXY Having resonances as if the music were enclosed in a box. Speaker cabinet diffraction or vibration. An emphasis around 250 Hz to 600 Hz.

BREATHY Audible breath sounds in vocals, flute, or sax. Good high-frequency response.

BRIGHT High-frequency emphasis. Harmonics are strong relative to fundamentals.

BRITTLE High-frequency peaks, or weak fundamentals. Slightly distorted or harsh highs. Opposite of round or mellow. See THIN. Objects that are physically thin and brittle emphasize highs over lows when you crack them.

CHESTY A vocal signal with a bump in the low-frequency response around 125 to 250 Hz.

CLEAN Free of noise, distortion, and leakage.

CLEAR See TRANSPARENT.

CLINICAL Too clean or analytical. Emphasized high-frequency response, sharp transient response. Not warm.

COLORED Having timbres that are not true to life. Nonflat response, peaks, or dips.

CLOUDY See WOOLY.

CONSTRICTED Poor reproduction of dynamics. Dynamic compression. Distortion at high levels. Also see PINCHED.

CRISP Extended high-frequency response. Like a crispy potato chip, or crisp bacon frying. Often referring to cymbals.

CRUNCH Pleasant guitar-amp distortion.

DARK Opposite of bright. Weak high frequencies.

DELICATE High frequencies extending to 15 or 20 kHz without peaks. A sweet, airy, open sound with strings or acoustic guitar.

DEPTH A sense of closeness and farness of instruments, caused by miking them at different distances. Good transient response that reveals the direct/reflected sound ratio in the recording.

DETAILED Easy to hear tiny details in the music; articulate. Adequate high-frequency response, sharp transient response.

DISTANT Too much leakage. Low direct-to-reverb ratio.

DRY Without effects. Not spacious. Reverb tends towards mono instead of spreading out. Overdamped transient response.

DULL See DARK.

ECHOEY Having audible echoes or reverberation.

EDGY Too much high frequencies. Trebly. Harmonics are too strong relative to the fundamentals. When you view the waveform on an oscilloscope, it even looks edgy or jagged, because of excessive high frequencies. Distorted, having unwanted harmonics that add an edge or raspiness to the sound.

EFFORTLESS Low distortion, usually coupled with flat response.

ETCHED Clear but verging on edgy. Emphasis around 10 kHz or higher.

FAT See FULL and WARM. Also, a diffuse spatial effect, or smeared out in time, with some reverberant decay, or the sound of a snare drum tuned low.

FOCUSED Referring to the image of a musical instrument which is easy to localize, pinpointed, having a small spatial spread.

FORWARD Sounding close to the listener. Emphasis around 2 kHz to 5 kHz.

FULL Opposite of thin. Strong fundamentals relative to harmonics. Good low-frequency response, not necessarily extended, but with adequate level around 100 to 300 Hz.

GENTLE Opposite of edgy. The harmonics—highs and upper mids—are not exaggerated, or may even be weak.

GLARE, GLASSY A little less extreme than edgy. A little too bright or trebly.

GRAINY The music sounds like it's segmented into little grains, rather than flowing in one continuous piece. Not liquid or fluid. Suffering from harmonic or IM (intermodulation) distortion. Some early A/D converters sounded grainy, as do current ones of inferior design. "Powdery" is finer than "grainy"!

GRUNGY Lots of harmonic or IM (intermodulation) distortion.

HARD Too much upper midrange, usually around 3 kHz. Or, good transient response, as if the sound is hitting you hard.

HARSH Too much upper midrange. Peaks in the frequency response from 2 kHz to 6 kHz. Or, excessive phase shift.

HEAVY Good low-frequency response below about 50 Hz. Suggesting great weight or power, like a diesel locomotive or thunder.

HOLLOW See HONKY. Also, too much reverberation, or a mid-frequency dip, or comb filtering.

HONKY The music sounds the way your voice sounds when you cup your hands around your mouth. A bump in the response around 500 to 700 Hz.

IN YOUR FACE Dry (without effects, without reverb), possibly with compression.

LIQUID Opposite of grainy. A sense of seamless flowing of the music. Flat response and low distortion. High frequencies are flat or reduced relative to mids and lows.

LOW-FI (low fidelity) "Trashy" sounding. Tinny, distorted, noisy, or muddy.

MELLOW Reduced high frequencies, not edgy.

MUDDY Not clear. Weak harmonics, smeared time response, IM distortion. Too much reverb at low frequencies. Too much emphasis around 200 to 350 Hz. Too much leakage.

MUFFLED The music sounds covered up. Weak highs or weak upper mids.

MUSICAL Conveying emotion. Flat response, low distortion, no edginess.

NASAL The vocalist sounds as if he or she is singing with the nose closed. Also applies to strings. Bump in the response around 500 to 1000 Hz. See HONKY.

NEUTRAL Accurate tonal reproduction. No obvious colorations. No serious peaks or dips in the frequency response.

PAPERY Referring to a kick drum that has too much output around 400 to 600 Hz.

PHASEY Having phase interference (comb filtering). The sound of a direct signal and its delayed repetition mixed to the same channel (delay usually under 20 msec). Might be due to multiple mics picking up the same source, or one mic picking up direct sound and delayed reflected sound. Also, a delayed signal mixed with itself undelayed, or some opposite-polarity crosstalk between stereo channels, or one monitor speaker is reversed in polarity.

PINCHED Narrowband. Midrange or upper-midrange peak in the frequency response. Pinched dynamics are overly compressed.

PIERCING Strident, hard on the ears, screechy. Having sharp, narrow peaks in the response around 3 kHz to 10 kHz.

PRESENT, PRESENCE Adequate or emphasized response around 5 kHz for most instruments, or around 2 to 5 kHz for kick drum and bass. Having some edge, punch, detail, closeness, and clarity.

PUFFY Bump in the response around 400 to 700 Hz.

PUNCHY Good reproduction of dynamics. Good transient response. Or conversely, referring to highly compressed transients (especially snare drum and kick drum) that sound like hitting a punching bag. Sometimes a bump around 5 kHz or 200 Hz.

RASPY Harsh, like a rasp. Peaks in the response around 6 kHz, which make vocals sound too sibilant or piercing.

RICH See FULL. Also, having euphonic distortion made of even-order harmonics.

ROUND High-frequency rolloff or dip. Not edgy.

SHARP See CRISP, STRIDENT, and TIGHT.

SIBILANT, ESSY Exaggerated "s" and "sh" sounds in vocals, too much output around 5 to 10 kHz.

SIZZLY See SIBILANT. Also, too much highs on cymbals.

SMEARED Lacking detail. Poor transient response. This may be a desirable effect in large-diameter mics. Also, poorly focused images.

SMOOTH Easy on the ears, not harsh. Flat frequency response, especially in the midrange. Lack of peaks and dips in the response. Low distortion.

SPACIOUS Conveying a sense of space, ambience, or room around the instruments. To get this effect, mike farther back, mix in an ambience mic, add reverb, or record in stereo. Components that have opposite-polarity or out-of-phase crosstalk between channels may add false spaciousness.

SQUASHED Overly compressed.

STEELY Emphasized upper mids around 3 to 6 kHz. Peaky, nonflat high-frequency response. See GLASSY, HARSH, and EDGY.

STRAINED The component sounds like it's working too hard. Distorted. Inadequate headroom or insufficient power. Opposite of effortless.

STRIDENT See HARSH and EDGY.

SWEET Not strident or piercing. Flat high-frequency response, low distortion. Lack of peaks in the response. Highs are extended to 15 or 20 kHz, but they are not bumped up. Often used when referring to cymbals, percussion, strings, and sibilant sounds.

THIN Fundamentals are weak relative to harmonics. Note that the fundamental frequencies of many instruments are not very low. For example, violin fundamentals are around 200 to 1000 Hz. So if the 300 Hz area is weak, the violin may sound thin—even if the mic's response goes down to 40 Hz.

TIGHT Good low-frequency transient response. Absence of ringing or resonance when reproducing the kick drum or bass. Good low-frequency detail. Absence of leakage.

TINNY, TELEPHONE-LIKE Narrowband, weak lows, peaky mids. The music sounds like it's coming through a telephone or tin can.

TRANSPARENT Easy to hear into the music, detailed, clear, not muddy. Wide, flat frequency response, sharp time response, very low distortion and noise.

TUBBY See BLOATED. Having low-frequency resonances as if you're singing in a bathtub.

VEILED The music sounds like you put a silk veil over the speakers. Slight noise or distortion, or slightly weak high frequencies.

WARM Good bass, adequate low frequencies, adequate fundamentals relative to harmonics. Not thin. Also, excessive bass or midbass, or pleasantly spacious, with adequate reverberation at low frequencies, or gentle highs, like from a tube amplifier. See RICH.

WASHED OUT Phase interference from multiple mics picking up the same source. Too much leakage or reverberation.

WOOLY, BLANKETED The music sounds like there's a wool blanket over the speakers. Weak high frequencies or boomy low frequencies. Sometimes, an emphasis around 250 to 600 Hz.

CHAPTER 10

Mixers and Mixing Consoles

The heart of your recording studio is the mixer. It's a control center where you plug in all sorts of signals; mix or blend them; add effects, EQ, and stereo positioning; and route the signals to recorders and monitor speakers. A mixing console (also called a board or desk) is a large mixer with many controls.

A recorder-mixer combines a mixer and a multitrack recorder in a single portable chassis. This convenient unit is also called a mini studio, portable studio, or digital multitracker. Low-end recorder-mixers record 4 tracks, and high-end units record 8 to 32 tracks.

We'll look at analog mixers, digital mixers, and software mixers. The first two types are hardware, but their concepts apply to a software mixer in a DAW as well.

STAGES OF RECORDING

First let's review the three stages of using a mixer during a session: recording, overdubbing, and mixdown.

1. **Recording (tracking):** The mixer accepts mic-level signals and amplifies them up to line level. You send the line-level signal from each mic to a separate track in a multitrack recorder. The multitrack recorder records several tracks on hard disk. One track might be a lead vocal, another track might be a saxophone, and so on.
2. **Overdubbing:** While listening to recorded tracks over headphones, the musician records new parts on open (unused) tracks. You set up the mixer to monitor both the recorded tracks and the new live part.
3. **Mixdown:** After all the tracks are recorded, you use the mixer to combine or mix the tracks into 2-channel stereo with panning, EQ, and effects.

MIXER FUNCTIONS AND FORMATS

Although the knobs, buttons, and meters on a mixer may appear intimidating, you can understand them if you read the manual and practice with the equipment. A mixer is complicated because it lets you control many aspects of sound:

- The loudness of each instrument (the balance among instruments in the mix)
- The tone quality of each instrument (bass, treble, midrange)
- The room that the instruments are in (reverberation)
- The left-to-right position of each instrument (panning)
- Effects (flanging, echo, reverb, chorus, etc.)
- Track assignments (which instrument goes on which track)
- Recording level (the signal voltage going to the recorder tracks)
- Monitor selection (what you want to listen to)

Mixers come in many formats:

- An **analog mixer** is a control device that works on analog signals, and sends them to an external multitrack recorder and 2-track recorder.
- A **digital mixer** is the same, but works internally with digital signals. It accepts analog or digital signals.
- A **software mixer** exists only in your computer as part of digital recording software. You control it with a mouse or a control surface (described next). The recording is done on your hard drive. Whether you use a hardware or software mixer, be sure to read the section in this chapter on analog mixers because many of the same principles apply.
- A **control surface** is a device that looks like a mixer with faders and knobs. It plugs into your computer's USB or FireWire port and controls the software mixer.

A mixer can be specified by the number of inputs and outputs it has. For example, an 8-in, 2-out mixer (8×2 mixer) has 8 signal inputs that can be mixed into 2 output channels (buses) for stereo recording. Similarly, a 16-in, 8-out (16×8) mixing board has 16 signal inputs and 8 output channels for multitrack recording. A 16×4×2 mixing board has 16 inputs, 4 submixes or groups (explained under the heading "Output Section" later in this chapter), and 2 master outputs. There also are connectors for external equipment, such as effects devices and a monitor power amplifier. The more inputs your mixer has, the more instruments you can record at the same time. If you're recording only yourself, you may need only two inputs.

Let's look at the analog mixer in more detail. Knowing how it works will help you understand the other types.

ANALOG MIXER

A mixer can be divided into three sections: input, output, and monitor. Here are the main parts of each section and what they do:

Input section

- Inputs connect to your mics, electric instruments, and recorder outputs.
- Faders are sliding volume controls that affect the loudness of each instrument. This lets you control the balance among instruments in the mix.
- Equalization (EQ) knobs adjust the tone quality of each instrument (bass, treble, midrange).
- Aux knobs set the amount of reverb or other effects, and also can be used to set up a monitor mix or headphone mix.
- Pan pots place the monitored sound of each track where desired between your stereo speakers—left, center, right, or anywhere between. In some mixers, the pan pot is also used with the assign switch during recording to send signals to the desired tracks.
- Channel assign buttons route each input signal to the desired recorder track.

Output section

- Master faders set the overall level of the entire stereo mix.
- Group faders set the overall level of each group or submix.
- Outputs connect to your recorder inputs and the monitor power amplifier.
- Meters help you set the correct recording level (to prevent distortion and noise).

Monitor section

- Monitor controls select what you want to listen to.
- Aux knobs or channel faders set up the monitor mix.

Let's look at each part in more detail.

Input Section

A mixer is made of groups of controls called modules. An input module (Figure 10.1) or channel strip affects a single input signal—from a microphone, for instance. The module is a narrow vertical strip, one per input. Several modules are lined up side-by-side. Each input module is the same, so if you know one, you know them all.

Let's follow the signal flow from input to output through a typical input module (Figure 10.2). Every mixer is a little different, but you are likely to find features like those described here.

FIGURE 10.1
A typical input module.

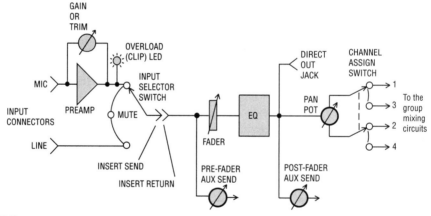

FIGURE 10.2
Signal flow through a typical input module.

INPUT CONNECTORS

On the back of each input module are input connectors with these labels:

MIC: Accepts signals from a microphone or direct box
LINE: Accepts an electric musical instrument, or a track output of a multitrack recorder
TRACK: Accepts a track output of a multitrack recorder; not included in all mixers

Some units use a single jack for both mic and line inputs; others have separate jacks for each. The mic input is either an unbalanced 1/4-inch phone jack (a 1/4-inch hole) or an XLR-type connector (with three small holes). The line input is either a 1/4-inch phone jack, an RCA phono jack (like you see on a stereo system), or an XLR-type connector.

You can plug a synthesizer directly into a phone-jack line input without using a direct box if the cable is less than 10 feet; a longer cable may pick up hum. In that case, use a direct box plugged into a mic input.

In some mixers, a phone-jack input is switchable between low impedance (for micro-phones) and high impedance (for guitar pickups). Other mixers might have a separate low-impedance mic input and a high-impedance instrument input.

PHANTOM POWER (P48, +48)

This switch (not shown) turns on phantom power (48V DC) for condenser micro-phones. In the mic input connector, the 48V appears on pins 2 and 3 relative to pin 1. The microphone receives phantom power and sends audio along the same two cable leads.

MIC PREAMP

After entering the mic connector, the microphone signal goes to a mic preamplifier inside the mixer. This preamp boosts or amplifies the weak microphone signal up to a higher voltage, making it a line-level signal.

TRIM (GAIN)

The TRIM or GAIN control adjusts the amount of amplification in the mic preamp. If the GAIN control is turned up fully, and the incoming mic signal is very strong due to a loud instrument or vocal, this signal can overload the mic preamp. This causes dis-tortion—a gritty sound. To get a good signal-to-noise ratio (S/N) in your mixer, set the GAIN control as high as possible, but not so high that the preamp distorts.

Here's how: Start with the TRIM turned all the way down (counterclockwise). In most mixers, each input module has a tiny light (LED) labeled "clip," "peak," or "OL" (over-load). It flashes when the mic preamp is distorting. When an instrument is playing its loudest signal through an input module, gradually turn up the TRIM control until the clip LED starts flashing. Then turn down the TRIM control just to the point where the light stays off, and turn it down another 6 dB for extra headroom.

Some low-cost mixers do not have a TRIM control. The input fader serves this function.

INPUT SELECTOR SWITCH

This switch lets you select the type of signal you want to work with. Some common switch labels are below:

MIC: Mic or direct box
LINE: For line-level signals: a synthesizer, drum machine, electric guitar, or multi-track recorder's track output
INPUT: Mic or line
TRACK: Multitrack recorder's track output

Using the input selector is simple. If you plugged in a microphone or direct box to record its signal, set the input selector to MIC or INPUT. If you plugged in a synth, drum machine, or electric guitar, set the selector to LINE or INPUT. When you're ready to mix, select TRACK (if available) or LINE (if the recorder track outputs are plugged into mixer line inputs).

Some mixers have no input selector. The mixer processes whatever signal is plugged in.

In some recorder-mixers, a single 1/4-inch phone jack is used both for mic-level and line-level signals. The two levels are handled either by a MIC/LINE switch or a TRIM control.

INSERT JACKS

Following the input selector switch are the insert-send jack and insert-return jack (on the back of each input module). Or the mixer might have a single insert jack, a TRS type that has the send on the tip terminal and the return on the ring terminal. Inside the mixer, the send is connected to the return so that the signal passes through to the rest of the mixer.

If you insert a plug into the insert jack(s), you break the signal path so you can insert an external device there in series with the input module's signal. You might insert a compressor into the signal path of one module for automatic volume control, or insert any other signal processor (reverb/delay, for instance). That way, if all your aux sends are tied up, you can add another signal processor. On the reverb/delay unit, set the dry/wet mix control for the desired amount of effect.

Another use for the insert jack is to send the mic-preamp output signal to a multi-track recorder track. The output of each track returns to the insert jack and continues through your mixer. In this case, you use the TRIM controls to set recording levels, and use the faders, EQ, and aux sends to set up a monitor mix.

Low-cost mixers omit the insert jacks. Some units have insert jacks on only two inputs.

INPUT FADER (CHANNEL FADER)

Next, the input signal that you selected goes to a fader. This is a sliding volume control for each input signal. During recording, you can use the fader in two ways. If you are recording one instrument per track, record off the insert send and use the fader to adjust the instrument's level in the monitor mix. If you are recording two or more instruments on one track, record off a group output, and use each instrument's fader to set up a mix within that group. For example, if you are recording several drum mics onto one track, set up a drum mix with the drum-mic faders.

During mixdown, you use the faders to set the loudness balance among instruments.

EQUALIZATION (EQ)

The signal from the input fader goes to an equalizer, which is a tone control. With EQ you can make an instrument sound more or less bassy, and more or less trebly, by boosting or cutting certain frequencies. Details are in Chapter 9 under the heading "Equalizer"

DIRECT OUT

The direct out is an output connector following each input fader and equalizer. The signal at the direct-out jack is an amplified, equalized version of the input signal. The fader controls the level at the direct-output jack. You can use the direct-out jack when you want to record one instrument per track (with EQ) on an external multitrack recorder. Connect the direct-out jack to a multitrack recorder track input. Because the direct output bypasses the mixing circuits farther down the chain, the result is a cleaner signal.

Suppose your mixer has 8 inputs and 2 outputs. You can use this mixer with an 8-track recorder. Just connect the direct-out jack in each input module to a separate track input. Or do the same with the insert-send jacks.

CHANNEL ASSIGN SWITCH

The equalized signal also goes through a pan pot (explained in the section below) to the channel assign switch or track selector switch. It lets you send the signal of each instrument to the recorder track you want to record that instrument on.

A mixer with four groups or buses would have an assign switch labeled 1, 2, 3, 4. If you want to record bass on track 1, for instance, find the assign switch for the input module the bass is plugged into, and push assign switch 1. If you want to record four drum mics on track 2, push assign switch 2 for all those input modules.

Some mixers assign tracks with two controls: a selector switch and a pan pot.

PAN POT

This knob sends a signal to two channels in adjustable amounts. By rotating the pan-pot knob, you control how much signal goes to each channel. Set the knob all the way left and the signal goes to one channel. Set it all the way right and the signal goes to the other channel. Set it in the middle and the signal goes to both channels equally.

Here's how you might use a pan pot to assign an instrument to a track. The channel-assign switch might have two positions labeled 1–2 and 3–4. If you turn the pan pot left, the signal goes to odd-numbered tracks (either 1 or 3, depending on how you set the assign switch). If you turn the pan pot right, the signal goes to even-numbered tracks (2 or 4). Suppose you want to assign the bass to track 1. Set the assign switch to 1–2, and turn the pan pot far left to choose the odd-numbered track (track 1).

During mixdown, the pan pot has a different function: it places images between your speakers. An image is an apparent source of sound, a point between your speakers where you hear each instrument or vocal. Set the pan pot to locate each instrument at the left speaker, right speaker, or anywhere in between. If you set the pan pot to center, the signal goes equally to both channels, and you hear an image in the center.

A stereo track panned to the center still ends up in stereo. A stereo source on two tracks can be panned hard left and right. Or, suppose you have two stereo sources—drums and piano—and you want one to be toward the left and one toward the right. Pan the two drum tracks hard left and center, and pan the two piano tracks hard right and center.

AUX

The aux or aux-send function (Figures 10.2 and 10.3) sends some of the input module's signal to equipment outside the mixer. A pre-fader aux send is before the fader, and usually goes to a power amp that feeds monitor speakers or headphones. You can use the pre-fader aux knobs to set up a monitor mix—a balanced blend of input

FIGURE 10.3
Aux sends.

signals you hear over speakers or headphones. A post-fader aux send is after the fader and EQ, and usually goes to an effects unit. You use the post-fader aux knobs to set the amount of effects (reverb, echo) heard on each instrument in a mix.

Some mixers have no aux sends, some have one aux-send control per module, and some have two or more (labeled aux 1, aux 2, etc.). The more aux sends you have, the more you can play with effects, but the greater the cost and complexity. The aux number (1 or 2) is not necessarily assigned a specific function; you decide what you want aux 1 and aux 2 to do.

During recording and overdubbing, the aux knobs of all the input modules can be used to create a monitor mix. The monitor mix that you create with the aux knobs is independent of the levels going to the multitrack recorder. You use the gain-trims during recording to set recording levels, and you use the aux knobs to create an independent mix that is heard over your monitor system.

In Figure 10.3, the aux 2 send control is just before the fader. In this mixer, the signals from all the aux 2 knobs in the mixer combine at a connector jack labeled "aux 2 send." You can connect that jack to your power amplifier, which drives monitor speakers and headphones.

In Figure 10.3, the aux 1 send control is just after the fader. During mixdown, each aux 1 knob controls how strong the effects (reverb, echo) are on each track. In each input module, the aux knob adjusts how much of that input signal is sent to an external effects unit. The effected signal returns to the mixer's aux-return or bus-in jacks, where it blends with the original signal.

For example, suppose the aux 1 send is connected to a digital reverb. The more you turn up the aux 1 send knob, the more signal goes to the reverb. The output of the reverb unit returns to the mixer's bus-in jack, and blends with the original dry signal, adding a spacious effect.

A few mixers have an aux-return control (also called effects-return or bus-in) that sets the overall effects level returning to the mixer.

There might be a pre/post switch next to the aux-send knob. When an aux knob is set to pre (pre-fader), its level is not affected by the fader setting. You use the pre setting for a headphone mix during recording or overdubbing because you don't want the fader settings to affect the monitor mix.

The post setting (post-fader) is used for effects during mixdown. In this case, the aux level follows the setting of the fader. The higher you set the track volume with the fader, the higher the effects level. But the dry/wet mix stays the same.

Output Section

The output section is the final part of the signal path; the section that feeds mixed signals to the recorder tracks. It includes mixing circuits, submaster or group faders (sometimes), master faders, and meters (Figure 10.4).

FIGURE 10.4
Input module and output section of a mixer.

MIXING CIRCUITS (ACTIVE COMBINING NETWORKS), GROUP FADERS, AND BUS OUTPUT CONNECTORS

The group mixing circuits are in the center of Figure 10.4. Recall that you use the assign switches to send each input signal to the desired channel or bus, and each bus feeds a different recorder track. A bus is a channel in a mixer containing an independent mix of signals. The bus 1 or group 1 mixing circuit accepts the signals from all the inputs you assigned to bus 1 and mixes them together to feed track 1 of the recorder. The bus 2 mixing circuit mixes all the bus 2 assignments, and so on.

Following each group mixing circuit is a group fader. If you assigned all the drum mics to group 1, the group 1 fader adjusts the overall level of the drum mix. The signal from each group fader goes to a group or bus output connector in your mixer. You can connect each bus output to a recorder track input.

STEREO MIX BUS, MASTER FADERS, MAIN OUTPUT CONNECTORS

The stereo mix bus includes two group mixing circuits: one for channel 1 and one for channel 2. Three types of signals feed into the stereo mix bus:

1. The group output signals.
2. Signals from input modules. You can assign an input module's signal directly to the stereo mix bus, bypassing the groups. This results in lower noise.
3. Effects-return signals, such as the reverberated signal from an external digital reverb.

Located on the right side of your mixer, the stereo master faders are one or two sliding volume controls that affect the overall level of the stereo mix bus. Usually, you set the master fader(s) within design center, the shaded area about three-quarters of the way up on the scale. This setting minimizes mixer noise and distortion. You can fade out the end of a mix by turning down the master faders gradually.

After the master faders, the signal goes to a pair of main output connectors, which feed a 2-track recorder of your choice.

You feed the multitrack recorder either from group outputs, direct outs, or insert sends. If you're mixing several instruments to track 5, for example, assign those instruments' signals to group 5. Connect the group 5 output to recorder track 5 in. If you're recording one instrument on track 5, however, connect that instrument's direct out or insert send to track 5 in. The signal is cleaner at the direct out or insert send than at the group output.

Some mixing consoles have voltage controlled amplifier (VCA) group faders. A VCA group fader acts like a remote control for groups of channel faders. For example, you could assign each drum mic's fader to a group, which is an audio path, and also assign each drum mic's fader to a VCA group, which is a fader that controls all the drum mic channels at once.

METERS

Meters are an important part of the output section. They measure the voltage level of various signals. Usually, each group or bus output has a meter to measure its signal level. If these buses feed the recorder tracks, you use the meters to set the recording level for each track.

A mixer has either VU meters or LED bar graph meters.

A VU meter (now rarely used) is a voltmeter that shows approximately the relative loudness of various audio signals. Set the record level so that the meter needle reaches +3 VU maximum for most signals, and about −6 VU maximum for drums, percussion, and piano. That's necessary because the VU meter responds too slowly to show the true level of percussive sounds.

An LED bar graph meter has a column of lights (LEDs) that shows peak recording level. Usually you set the recording level to peak near −6 dB maximum.

If your console has VU meters, and it is feeding a digital recorder with LED peak meters, set a 0 VU tone to read −20 dB on the recorder meters. This allows some headroom for peaks.

Monitor Section

The monitor section is used to control what you're listening to. It lets you select what you want to hear, and lets you create a mix over headphones or speakers to approximate the final product. The monitor mix has no effect on the levels going to your recorder.

During recording, you want to monitor a mix of the input signals. During playback or mixdown, you want to hear a mix of the recorded tracks. During overdubs, you want to hear a mix of the recorded tracks and the instrument that you're overdubbing. The monitor section lets you do this.

MONITOR SELECT BUTTONS

These buttons let you choose what signal you want to monitor or listen to. Because the configuration of these buttons varies widely among different mixers, they are not shown in Figure 10.4.

If you want to use aux 2 as the monitor-mix bus, select the aux 2 bus as the monitor source. That lets you listen just to the aux 2 mix. Some mixers have no monitor-select switches. Instead, you always monitor the stereo mix bus.

MONITOR MIX CONTROLS AND CONNECTORS

One way to set up a monitor mix is with the aux knobs. Suppose you want aux 2 to be the monitor mix. Connect the aux 2 send jack to your power amp and speakers. Or if you're using headphones, find the monitor-select switch for the headphones, and set it to aux 2. Turn up all the aux 2 knobs about halfway, then turn each knob up or down to set a good loudness balance. You do this during recording or overdubbing.

Here's another way to set up a monitor mix using the faders. Connect your multitrack recorder ins and outs to the insert jacks (Figure 10.5). Connect the insert-jack 1 tip

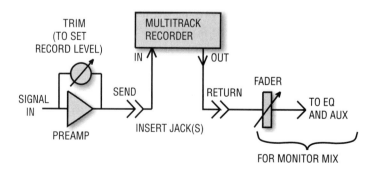

FIGURE 10.5
Using insert jacks to send each input signal to a recorder track. The track signal returns to the mixer, where you adjust level, panning, EQ, and effects.

(send) to track 1 in; connect track 1 out to the insert-jack ring (return). Make similar connections for the other tracks. Also connect the mixer's monitor-out jacks to your power amp and speakers. Monitor the stereo mix bus. With this setup, use the TRIM controls to set recording levels. Use the faders to set up a monitor mix, cue mix, or mixdown with EQ and effects. It's a convenient way to work.

During mixdown, monitor the stereo mix bus. This is automatic in a DAW.

SOLO

The SOLO button in an input module is actually a monitor function. It lets you monitor one instrument or vocal at a time so you can hear it better or so you can edit one track. By pressing two or more SOLO buttons, you can monitor more than one input signal at a time.

Suppose you hear a buzz in the audio and suspect it may be in the bass guitar signal. If you push the SOLO button in the bass guitar's input module, you'll monitor only the bass guitar. Then you can easily hear whether the buzz is in that input.

On British consoles, the SOLO function is called pre-fader listen (PFL) or after-fader listen (AFL).

MUTE

When you press the MUTE button in a channel, the signal in that channel is turned off. During mixdown, it's a good idea to mute tracks that have nothing playing at the moment to reduce noise and leakage. During overdubs, you might want to mute previous takes or certain instruments to make a cleaner headphone mix for the musician.

Additional Features in Large Mixing Consoles

Large mixing consoles have more features than small mixers do. If you're working only with a small mixer or a recorder-mixer, you might want to skip this section.

FOLDBACK (FB): Another name for cue, or headphone mix.

PHASE (POLARITY INVERT): Used only with balanced lines, this switch inverts the polarity of the input signal. That is, it switches XLR pins 2 and 3 to flip the phase 180 degrees at all frequencies. You might use it to correct a miswired mic cable whose polarity is reversed. If you mic a snare drum top and bottom, you need to invert the polarity of the bottom mic.

AUTOMATED MIXING CONTROLS: These controls (Read, Write, Update, Record Automation, Play Automation) set up the console for various automation functions. With automation, a memory circuit in the mixer remembers your console settings and mix moves. After leaving a mix for several days, you can recall the settings with the push of a button, then continue working on the mix. More details on automated mixing can be found at the end of Chapter 11.

EFFECTS PANNING: This feature places the images of the effects signals wherever desired between the monitor speakers. Some consoles let you pan effects in the monitor mix as well as in the final program mix.

EFFECTS RETURN TO CUE: This is an effects-return level control that adjusts the amount of effects heard in the studio headphone mix. These monitored effects are independent of any effects being recorded.

EFFECTS RETURN TO MONITOR: This effects-return control adjusts the amount of effects heard in the monitor mix. These monitored effects are independent of any effects being recorded.

BUS/MONITOR/CUE switch for effects return: A switch that feeds the effects-return signal to your choice of three destinations: program bus (for mixdown), monitor mix (control-room speakers), cue mix (headphones), or any combination of the three.

METER SWITCHES: In many consoles, the meters can measure signal levels other than console output levels. Switches near the meters can be set so that the meters indicate bus level, aux-send level, aux-return level, monitor-mix level, etc.

Those readings help you set optimum levels for the outboard devices receiving those signals. Too low a level results in noise; too high a level causes distortion in the outboard unit. For example, if the aux-return signal sounds garbled or distorted, the cause may be an excessive aux-send level. Verify that condition by checking the meters switched to read the aux or effects bus.

DIM: A switch that reduces the monitor level by a preset amount so you can talk (as in "dim the lights").

TALKBACK: An intercom between the control room and studio. A mic built into the console lets you talk to the musicians in the studio when you push the talkback button.

SLATE: This function routes the control-room microphone signal to all the buses so you can record the name of the tune and take number.

OSCILLATOR or TONE GENERATOR: This is used to put a level-calibration tone on a tape, or to reference the recorder's meters to those on the console. You also can use it to check signal path, levels, and channel balance.

Some mixers include a FireWire or USB port. It sends the output signal of each mic preamp to a separate track in your recording software, and returns the stereo mix back to the mixer for monitoring. See Chapter 4 under the heading "Computer Digital Audio Workstation."

DIGITAL MIXER

So far we've looked at the analog mixing console, which works entirely with analog signals. A digital mixing console accepts analog or digital signals. It converts the analog input signals to digital, and processes all signals internally in digital format. The signal stays in the digital domain for all mixer processing. Level changes, EQ, and so on are done by digital signal processing (DSP or computer calculations) rather than by analog circuits.

Some of the digital output signals from the mixer go through digital-to-analog (D/A) converters. The resulting analog signals feed a power amp, effects unit, and so on.

Analog and digital consoles operate differently. To use EQ in an analog console, you find the channel you want to EQ, and adjust its EQ knobs. To use EQ in a digital console, you press a button to select the channel you want to equalize, and press an EQ button. Then the EQ settings for that channel show up on an LCD screen. You press buttons and turn a knob to adjust the EQ frequency and boost/cut for that channel.

Because one knob controls the EQ for all the channels, digital consoles have fewer controls than analog consoles. One knob or switch can have several functions. This makes digital consoles a little harder to operate because you can't just reach for an EQ knob for a particular channel. You have to press a few buttons to set the EQ parameters.

Some models have an LCD touch screen that shows a bank of faders with small EQ curves, compression curves, and virtual aux knobs. To tweak EQ for one channel, find the EQ curve and tap it. The screen zooms to a larger view of the EQ settings for that channel. Rotary encoders (knobs) under the touch screen, or built into it, let you adjust parameters.

Digital consoles vary a lot in how they implement functions, so you need to read the manual.

Some digital mixers have a bank switch or layer switch. It lets you select the channels that the faders control. For example, suppose a mixer has 16 faders. Press the layer switch once to make the faders access channels 1-16. Press the switch again to make the faders access channels 17-32, and so on.

Several digital consoles have built-in effects, multitrack recording to hard disk, and a USB port for a flash drive that saves presets and mixes.

Many digital mixers offer automated mixing. One type of automated mixing is called scene or snapshot automation. When you press the snapshot button, a memory circuit in the mixer takes a "snapshot" of all the mixer settings for later recall. Another type of automation is dynamic. Your mix moves are stored in memory for later recall.

When you recall a mix, some mixers make the faders move into the positions you set up. This feature is called "flying faders" or "motorized faders." A few mixers do not move the faders when you recall a mix. You have to set them manually by looking at a display, which is a disadvantage. Motorized faders are more expensive but easier to use.

Digital Mixer Features

Look for the following features in digital mixers when making a buying decision:

- Number of mic inputs and number of output buses
- Number and type of digital inputs and outputs: S/PDIF, AES3, MADI, TDIF, Lightpipe, USB, and FireWire. Chapter 13 describes those standards under the heading "Digital Audio Signal Formats."
- Word-clock input and output
- Routing section that lets you route any input signal to any channel, or route any output signal to any output bus
- Number and type of option card slots: extra I/O, DSP, sync, effects, plug-ins
- Number of effects processors
- PC USB interface for control and/or backup
- Ease of use
- Snapshot or dynamic automation, or no automation
- Motorized or nonmotorized faders
- Varispeed clock for pitch shift
- Surround-sound monitoring
- Touch screen with virtual controls

SOFTWARE MIXER

Here's another type of mixer. A software mixer (virtual mixer) is a simulated mixer you see on a computer screen (Figure 10.6). It exists only in your computer as part of digital recording software or a DAW. You control it with a mouse or with a controller surface. Recordings made with this mixer, and all the mixer settings, are stored on your hard drive.

In a software mixer, the input and output connectors are in an audio interface (such as a sound card) that connects to the computer. This interface is 2-channel or multichannel. The software mixer works with effects that are either plug-ins (software) or external (hardware). Automation is a standard feature. For more on software mixers, see Chapter 14.

CONTROL SURFACE

A control surface or controller resembles a hardware mixer with faders and knobs (Figure 14.10). It plugs into your computer's USB or FireWire port and controls the software mixer. Some people prefer a controller because it is easier to use than a

FIGURE 10.6
Cakewalk Sonar Producer mix window, an example of a software mixer.

mouse. This device is discussed in more detail in Chapter 14 under the heading "Control Surface."

Now that you understand the typical features of mixers and mixing consoles, you are ready to learn how to use them.

CHAPTER 11

Mixer Operation

Get your hands on those knobs. You're going to operate a mixer as part of a recording session. This will be a basic run-through. Detailed session procedures are described in Chapter 15. Most of these procedures also apply to operating a virtual mixer in a digital audio workstation (DAW).

First recall the stages in making a recording:

1. Session preparation
2. Recording
3. Punching in
4. Overdubbing
5. Mixdown

This chapter considers each stage in turn.

SESSION PREPARATION

If you're recording on a hard drive, make sure you have enough drive space for the project. Chapter 14 has simple equations and a table that show how much space you need.

Plan your track assignments. Write a track sheet that tells what instrument goes on which track. Note: If you assign multiple instruments to the same track, you can't separate their images in the stereo stage. That is, you can't pan them to different positions; all the instruments on one track sound as if they're occupying the same point in space. If you're recording 4-track, you may want to do a stereo mix of the rhythm section on tracks 1 and 2; then overdub vocals and solos on tracks 3 and 4.

Studio setup for the musicians is covered in Chapter 15.

Set Up the Mixer and Recorder

To start the process, first zero or neutralize the mixer by setting all the controls to "off," "flat," or "zero." This establishes a point of reference and avoids surprises later on. Set all faders down.

If you have a separate mixer and multitrack recorder, you need to make their meter readings match. Follow one of these procedures:

If your mixer and recorder have VU meters, play a steady tone into the mixer to get a 0 reading on the meters for all channels. Then set the multitrack recorder's record level (if any) to get 0 VU readings on all the tracks.

If your mixer has VU meters and your recorder has LED peak-reading meters, set 0 VU on your mixer to equal −20 dB on the recorder's peak meters.

If your mixer and recorder have LED peak-reading meters, set 0 dB on the mixer to equal 0 dB on the recorder's peak meters.

If you are feeding the multitrack recorder from your mixer's insert jacks or direct-out jacks, you'll need to watch the recorder meters.

Suppose you're ready to record a vocal or an acoustic instrument. Place the microphone and plug it into a mic input. If you want to record the audio output of a hardware synth or drum machine, connect a cable between the instrument's output and a line input on the mixer.

Attach a "scribble strip" of masking tape or white removable tape along the bottom of the faders, and label each fader according to what instrument you plugged into that input. Some consoles and software mixers have scribble strips that you can type on. In Figure 11.1, such a strip is on the mixer at the right.

Set the input selector (if any) to Mic or Line depending on what is plugged into each input.

Plug in headphones to hear what you're recording. Turn up the headphone volume control. Or, if you're in a control room and the musicians are in a studio, turn up the monitor level to listen over the monitor loudspeakers. On a hardware mixer, set the MONITOR SELECT switch to hear the signal you're recording, and turn up the musicians' cue mix.

Set the master faders about three-quarters of the way up, at 0, or within the shaded portion of fader travel. This is called design center.

If you are recording a soft synth—a software synthesizer in a DAW—there will be a short delay in the monitored signal because of the latency (signal processing delay) in the software. Set the latency or buffer setting as low as possible without getting dropouts. A delay under 5 msec or so should be acceptable.

FIGURE 11.1
A studio control room with a DAW, mixer, and monitor speakers.

If you are recording a vocal or real instrument with a DAW, do not enable "Echo input monitor" because this causes a latency delay in the monitored signal. Instead, monitor the mic or input signal directly from your mixer, USB mic, or audio interface.

Assign Inputs to Tracks

If your equipment wiring is mics > mixer > insert sends > track inputs, your inputs are already assigned. Input 1 goes to track 1, input 2 goes to track 2, and so on.

If you're using a computer DAW, select the input signal source for each track. For example, you might set track 5's input to channel 5 in your audio interface. Then, whatever is plugged into interface channel 5 will go to track 5. If your sound card has just 2 channels, set track 5's input to channel 1 or 2 in your audio interface, and likewise for the other tracks. Some DAWs record stereo tracks, so you'd assign each track's input to stereo channels 1 and 2 in the audio interface.

If your equipment wiring is mics > mixer > buses > track inputs, assign each input signal to the desired output channel (bus) as specified on your track sheet. Each bus is connected to the corresponding numbered track on the multitrack recorder.

Note: If only one instrument is assigned to a track, you can eliminate the noise of the console's combining amplifier by patching the instrument's signal directly to the recorder track. To do that, locate the direct output jack (or the insert jack) of the input module for that instrument and patch it to the desired track. Some mixing boards also require pressing a Direct button on the input module.

Set Recording Levels

Now you're ready to "get a level." Have each instrument play the loudest part of the music, one at a time or all at once. For each input signal, set the mixer's TRIM or GAIN control so the recording level is as high as possible without causing distortion. While setting levels for a digital multitracker, peak each track around −6 dBFS (−6 decibels full scale) maximum in peak meter mode, not RMS mode. That allows some headroom for surprises. Also, musicians generally play louder during a performance than during a level check. If you exceed 0 dBFS you'll hear digital clipping, which makes a loud click.

Most audio interfaces have gain knobs to set recording levels. A few interfaces use on-screen volume controls for that purpose. Some interfaces come with software that lets you set a monitor mix on-screen, independent of the recording levels.

If you are mixing several instruments to one or two groups (as in a drum submix), follow this procedure:

1. Monitor the group(s).
2. Set its submaster fader (group fader) to design center.
3. Set the submix balances, panning, and recording level with the input faders.
4. Fine-tune each submix level with the submaster fader.

Set a Monitor Mix

Once the recording levels are good, it's time to set up a monitor mix for the engineer and for the musicians' headphones. In other words, create a good balance among the vocals and instruments for the musicians to play with.

Here are two ways to set a monitor mix with a hardware mixer:

- Set all the aux 1 knobs to pre-fader. Monitor (listen to) the aux 1 bus. Create a mix with the aux 1 knobs.
- If you are recording from your mixer's inserts to a multitrack recorder, set up a monitor mix with the faders, EQ knobs, and effects sends.

Here are two ways to set a monitor mix with DAW recording software:

- Use the monitor mixer software that came with the audio interface (if any).
- Create an aux bus. In each track, insert an aux send to that bus. Monitor the aux bus. Set monitor levels with all the aux sends.

RECORDING

Next, set the track(s) you want to record to "record ready" mode. Now start recording. Write down the counter time for this take. "Slate" the recording: record the name of the tune and the take number. Then record a two-measure count-off. This is done to set the tempo for overdubs. For example, if the time signature is 4/4, you say, "1, 2, 3, 4, 1, 2, (rest) (rest)." The rests are silent beats. You need some silence before the song starts to make editing easier later on. A DAW has a built-in metronome (click track) that can be used for count-offs.

PLAYBACK

After recording the track, go to the beginning of the song using the return-to-zero or locate function. If necessary, set the monitor selector to "track," or "mix," and play back the recording to check the performance and sound quality. You can set a rough mix with the monitor mix (aux) knobs. If you are recording on a computer, or if your multitrack is patched to the insert jacks, use the faders, pan pots, EQ, and aux knobs to set a rough mix.

Get or write a sheet of the song's arrangement. It shows the lyrics, verses, choruses, bridge, and so on. Play the song and set a locate point (marker) at the start of each verse and chorus. That way, when the musician says, "Let's fix that flat note at the end of the second chorus," you can go instantly to that part of the song.

OVERDUBBING

After your first track is recorded with a good performance, you might want to add more musical parts. This procedure is called overdubbing. When you overdub, the musician listens to tracks already recorded, and records a new part on an unused track.

Ready to overdub? Here is one way:

1. Turn off the control room speakers and listen on headphones.
2. Set up a headphone monitor mix of the recorded tracks. Some mixers use the aux knobs for this. If you are recording on a computer, or if your multitrack recorder is connected to the mixer's insert jacks, monitor the stereo mix, and set up a mix with the channel faders.
3. Plug in the mic or direct box for the instrument or vocal you want to record. Assign it to an unused track. Turn up its fader to design center (the 0 point about three-quarters of the way up).
4. Set up your multitrack recorder, mixer, or audio interface to monitor the playback of recorded tracks and the input signal of the instrument or vocal you want to record.

5. Have the musician play or sing. Can you hear the signal in the phones? Set the recording level using the TRIM (GAIN) control for that mic's channel.

6. Play the recording. As the musician plays or sings along, set up a good mix of the recorded tracks and the live instrument you're going to record. Ask the musician if he is hearing what he needs to hear. Change the mix if needed. Some musicians want to hear effects when they overdub; some want it dry. If you're overdubbing backup vocals one at a time, often it helps to remove (mute) certain other vocals from the headphone mix.

7. When you're ready to record the new part, go to a point about 10 seconds before the part of the song where the musician plays. Set the recorded tracks to SAFE and set the track you're recording on to RECORD READY.

8. Before you hit that RECORD button, stop! Are you recording on the correct track(s)? Are you going to accidentally erase any tracks? Double-check your track sheet, and make sure the record-enable buttons are on only for the tracks that are safe to record over.

9. Start recording and have the musician play along with the tracks. If the musician makes a mistake, you can rerecord or punch-in the new part without affecting the parts already recorded.

In a DAW, if you accidentally record over something, you can simply undo the recording (Ctrl-Z or Command-Z) and start over.

PUNCHING IN

Punching in is used to fix mistakes in a recorded performance, or to record a musical part in segments. A punch is also called an insert. You enable record mode on a track, play the multitrack recording, then "punch" or press the RECORD button at the right spot, record a new part, then punch out of record mode.

Some musicians record the same musical part over and over until the part is perfect. This process is tedious, but you have to pay attention. You need to be aware of where you are in the song at all times, and not erase anything you want to keep.

Some musicians like to record a performance a phrase at a time, perfecting each phrase as they go. Others record a complete take, then go back and fix the weak parts.

To do a punch-in, grab your song-arrangement sheet and follow these steps:

1. Go to a point about 10 seconds before the point where you want to start recording. Note the counter time and write it down. Also enter a memory location point there if your recorder can do that.

2. Play the song to the musician over headphones. The musician plays along to practice the part. Write down the counter times where you want to punch in and punch out. Otherwise you might erase a good take.

3. Finally you're ready to record. Before you hit that RECORD button, stop! Tell the musician what you're going to do so there's no chance of a mistake. Be very clear. For example, "I'm recording your keyboard part on a new track." Or, "I'm punching in over your old performance—is that what you wanted?"

4. Okay, ready to go. Play the recording. During a rest or pause in the music just before the part needing correction, punch in the RECORD button (or use a footswitch). Have the musician record the new part, and punch out right away. You don't want to erase the rest of the track! If you are using a DAW, though, you can undo the recording.

5. Press LOCATE or GO TO MARKER to go to the location point you set, about 10 seconds before the punch. Play the recording to see if the punch was okay. If necessary, you can rerecord the punch. After you go to the locate point, notice whether the musician wants to practice the part. Don't redo the punch until he or she is ready.

Some multitrack recorders have an autopunch function. You set the punch-in and punch-out points into the machine's memory. As the recording plays, the recorder automatically goes into and out of record mode at those points. Some recorders can loop repeatedly between those two points.

COMPOSITE TRACKS

If several open tracks are available, you can record a solo performance in several takes, each on a separate track or virtual track. Then combine the best parts of each track into a single track. Use only that track in the final mix, and you'll hear the best parts of all the takes in succession. This is called "recording composite tracks" or "comping." Here's an example of comping vocal tracks with a multitrack recorder and mixer:

1. Suppose you have four takes of a vocal recorded on tracks 10, 11, 12, and 13. Solo each track, and mark on the lyric sheet which tracks have the best performance of each section. For example, you might prefer track 13 on Verse 1, track 10 on Chorus 2, and so on.

2. Assign all the vocal tracks to an open track, which we'll call the comp track. Match their levels.

3. Start recording on the comp track.

4. As the song plays, mute and unmute the tracks to copy the best performances to the comp track. To do that, you can stop the recorder between sections, change the mute settings, and record a section at a time.

5. Once you're happy with the comp track, you can erase or archive the original tracks.

Comping with a DAW is even easier. Create a blank comp track, then copy and paste good sections from other tracks into the comp track. Or divide the tracks into short sections, delete sections you don't like, and play the mix of all the tracks. Some DAWs let you simply click on an audio clip in a take to add it to a comp track.

Cakewalk Sonar Producer software lets you record several takes on one track, view all the take waveforms as "lanes" in that track, select the best parts of each take, and create a composite track of those parts.

Some recorders have virtual tracks—explained in Chapter 13 under the heading "Recorder-Mixer (Digital Personal Studio)." They let you comp a performance with virtual tracks rather than real tracks.

GETTING MORE TRACKS

Most DAWs can record nearly an unlimited number of tracks. But what if you want to overdub more parts on a recorder-mixer, and all the tracks are full? With care, you can punch in more instruments by recording them in the pauses on recorded tracks. For example, suppose all the tracks are full but you want to add a cymbal crash at the beginning of the chorus. Find a track that has a pause at that moment, and punch in the cymbal crash there.

As an alternative, you can bounce tracks—mix several tracks to one or two open tracks, and record the mix on that track. Then you can erase the original tracks, freeing them for more overdubs. Bouncing procedures are in the instruction manual for your recorder or software.

DRUM REPLACEMENT

Suppose you've recorded the drum tracks, but you don't like the sound of them. Maybe the kick drum is too flabby, and no amount of EQ or gating seems to help. You might try a technique called drum replacement. In a MIDI/audio recording program, you can replace an audio drum track with a MIDI track that plays drum samples. First, set up a MIDI track with a drum sample ready to play (for details, see Chapter 16). Then use one of these methods:

- Start recording on the MIDI drum track, and tap your MIDI controller key in sync with the playback of the audio drum track.
- Select the audio drum track, then enable the "Extract Rhythm" or "Extract Beat" feature, if any. Copy and paste the extracted MIDI beats to the MIDI drum-sample track. Mute the original audio drum track and press PLAY. You should hear the replacement drum sample playing. To get a more natural sound, you might mix in some of the original acoustic drum track.
- In the audio drum track, insert a drum-replacement VST plug-in like Boxsounds Replacer. It makes the audio signal trigger wave files of drum hits.
- Purchase drum replacement software such as Drumagog or Steven Slate Digital Trigger Advanced Drum Replacer.

FIGURE 11.2
One input channel in a mixer, showing the fader, pan pot, EQ controls, and aux knobs.

MIXDOWN

After all your tracks are recorded (maybe with some bouncing), it's time to mix or combine them to 2-track stereo. You will use the mixer faders to control the relative volumes of the instruments, use panning to set their stereo position, use EQ to adjust their tone quality, and use the aux knobs to control effects. Figure 11.2 shows the controls you will be using.

Set Up the Mixer and Recorders

Set all the mixer controls to "off," "zero," or "flat." You should start from ground zero in building a mix. Some engineers like to set all the faders to −12 dB and go from there.

Tape a strip of paper or masking tape along the front of the mixer to write which instrument(s) each fader affects. Keep this strip with the multitrack tape for use each time you play it. Or type in this information in your DAW or mixing console.

If necessary, set the input-selector switches on the mixer to "track" because you'll be mixing down the multitrack recording. Monitor the 2-track stereo mix bus. These steps are unnecessary if you are using a DAW.

For starters, put the master faders at design center (about three-quarters of the way up, at the shaded portion of fader travel). This sets the mixer gain structure for the best compromise between noise and headroom.

Erase or Delete Unwanted Material

Mixing will be a lot easier if you first erase noises and leakage before and after each song, and within each track.

Play the multitrack recording and listen to each track alone. Erase or delete unwanted sounds, silent (unplayed) sections, outtakes, and entire segments that don't add to the song. In a DAW you can do this visually by looking for parts of each track where there is no waveform. Zoom into a single track, highlight a silent portion, and select Edit > Cut (or a similar command). To avoid mistakes, it's best to do this while the musicians are around. If you aren't sure whether it's okay to delete part of a track, set that part's volume envelope to zero or mute the track during that part.

You can use editing to improve a song's arrangement. Typically, you'd put instrumental solos and fills only in the "holes" where the vocal is silent. Add instruments to the mix gradually so that it builds in excitement. Add vocal harmonies only in the choruses. Reserve "all-out improvs" (where everybody plays) for the very end.

Panning

You need to pan the tracks before doing the mix, because the loudness of a track depends on where it's panned. Assign each track to buses 1 and 2 (or the stereo mix bus), and use the pan pots to place each track where desired between your stereo speakers. Typically the bass, snare, kick drum, and vocals go to center; guitars can be panned left or right, and stereo keyboards and drum overheads go partly left and right (see Figure 12.2). Sometimes a mono drum mix can be punchier than a stereo one.

You can give an acoustic guitar a spacious effect by recording it twice. Pan the first take hard left and pan the second hard right. The second take could be capoed higher to add interest.

Pan tracks to many points between the monitors: left, half-left, center, half-right, right. Try to achieve a stereo stage that is well balanced either side of center. For clarity, pan to opposite sides any instruments that cover the same frequency range.

You may want some tracks to be unlocalized. Harmony singers and strings should be spread out rather than appearing as point sources. Stereo keyboard sounds can wander between speakers. You could fatten a lead-guitar solo by panning it left, and panning the solo delayed to the right. (This delay might come from a distant room mic you used while recording, or from copying and sliding a track a few milliseconds in a DAW.) Pan doubled vocals left and right for a spacious effect.

Consider creating some front-to-back depth. Leave some instruments dry so they sound close; add reverb to others so they sound farther away.

If you want the stereo imaging to be realistic (say, for a jazz combo), then pan the instruments to simulate a band as viewed from the audience. If you're sitting in an audience listening to a jazz quartet, you might hear drums on the left, piano on the right, bass in the middle, and sax slightly right. The drums and piano are not point sources, but are somewhat spread out. If spatial realism is the goal, you should hear the same ensemble layout between your speakers. In some rock recordings, the piano and drums are spread all the way between speakers—interesting but unrealistic.

Pan-potted mono tracks often sound artificial; each instrument sounds isolated in its own little space. It helps to add some stereo reverb. It surrounds the instruments and "glues" them together.

When you monitor the mix in mono, you'll likely hear center channel buildup. Instruments in the center of the stereo stage will sound louder in mono than they did in stereo, so the mix balance will change in mono. To prevent this, note which tracks are panned hard left or right, and bring them a little toward the center: 9 and 3 o'clock on the pan knobs.

Use a Highpass Filter on Each Track

To reduce muddy buildup, highpass all tracks by ear. Set up a highpass (low-cut) filter in each track. Adjust the filter frequency: start at a low frequency, then gradually raise it until the sound starts to thin out. Then back off a little (lower the filter frequency slightly).

Compression

Sometimes the lead vocal track might be too loud or too quiet relative to the instruments because vocals have a wider dynamic range than instruments. You can control this by running the vocal track through a compressor. It will keep the loudness of the vocal more constant, making it easier to hear throughout the mix. If you're using a hardware mixer, patch the compressor's inputs and outputs into the insert jack(s) of the vocal input module. Or connect the compressor between the recorder's vocal-track output and the mixer input. Set the desired amount of compression ratio, attack time, release time, and gain reduction. It's also common to compress the kick drum and bass. (For more information, see Chapter 9 under the heading "Compressor")

Set a Balance

Now comes the fun part. The mixdown is one of the most creative parts of recording. Here are some tips to help your mixes sound terrific.

Before doing a mix, tune up your ears. Play over your monitors some CDs whose sound you admire. This helps you get used to a commercial balance of the highs, mids, and lows.

Choose a CD with tunes like those you're recording. Check out the production. How is the balance set? How about EQ, effects, and sonic surprises? Try to figure out what techniques were used to create those sounds, and duplicate them. Of course, you might prefer to break new ground.

Using the input faders, adjust the volume of each track for a pleasing balance among instruments and vocals. You should be able to hear each instrument clearly. Some mixing consoles have trim knobs that set the playback gain of the multitrack recorder tracks. In that case, set all faders in use to design center, and adjust the trims to get a rough mix.

Recording engineer Roger Nichols offered this suggestion: Change the level of one track by +1 dB, then by −1 dB. Is it better? Is it worse? If you can hear less than 1 dB change, the fader is close to where it should be.

Here's one way to build the mix. Make all the instruments and vocals equally loud. Then turn up the most important tracks and turn down background instruments. Or, bring up one track at a time and balance it with the other tracks. For example, first bring up the kick drum to about −10 dB, then add bass and balance the two together. Next add drums and set a balance. Then add guitars, keyboards, and finally vocals.

In a ballad, the lead vocal is usually on top. You might set the soloed lead-vocal level to peak at −6 dB. Bring up the monitor level so that the vocal is as loud as you like to hear it, then leave the monitor level alone. Bring in the other tracks one at a time and mix them relative to the vocal track.

When the mix is right, everything can be heard clearly, yet nothing sticks out too much. The most important instruments or voices are loudest; less important parts are in the background. In a typical rock mix, the snare is loudest, and the kick is nearly as loud. The lead vocal is next in level. Note that there's wide latitude for musical interpretation and personal taste in making a mix.

Sometimes you don't want everything to be clearly heard. Once in a while, you might mix in certain tracks very subtly for a subconscious effect.

Recording engineer Michael Cooper offers this advice: To check the vocal level, turn the monitors way down. The lead vocal should disappear just after the instruments do. To avoid washy sound, mike acoustic guitars in mono, keep most tracks dry, use delay instead of reverb, or use predelay with short reverbs.

Engineer Michael Stavrou (who wrote *Mixing with Your Mind*) suggests: Turn up the most inspiring player. Mix other tracks with it, one at a time, to see which tracks fight it and which don't. Mute or delete the tracks that fight it.

If you recorded some MIDI tracks, consider replacing a MIDI instrument with another one that sounds better. You might replace a grand piano with a honky-tonk piano, replace a fretted bass with a fretless bass, and so on. Swap out the snare sound with a new one during choruses.

If you hear a poorly played note or transition that you can't fix, you might cover it up with a cymbal roll or other sound.

It's a good idea to monitor around 85 dBSPL. If you monitor louder, the bass and treble will be weak when the mix is played softly.

To test your mix, occasionally play the monitors very quietly and see if you can hear everything. Switch from large monitors to small, and make sure nothing is missing.

Set EQ

Set EQ for the tonal balance you want on each track. If a track sounds too dull, turn up the highs or add an enhancer. If a track sounds too bassy, turn down the lows, and so on. Cymbals should sound crisp and distinct, but not sizzly or harsh; kick drum and bass should sound deep, but not overwhelming or muddy. Be sure the bass is recorded with enough edge or harmonics to be audible on small speakers.

For suggestions on EQ settings, see Chapter 9 under the heading "How to Use EQ."

You'll need to readjust the mix balances after adding EQ. The EQ that sounds right on a soloed track seldom sounds right when all the tracks are mixed together. So make EQ decisions when you have the complete mix happening. Switch EQ in and out to make sure it's an improvement.

In pop-music recordings, the tone quality or timbre of instruments does not have to be natural. Still, many listeners want to hear a realistic timbre from acoustic instruments, such as the guitar, flute, sax, or piano.

The overall tonal balance of the mix shouldn't be bassy or trebly. That is, the perceived spectrum should not emphasize lows or highs. You should hear the low bass, mid-bass, midrange, upper midrange, and highs roughly in equal proportions. Frequency bands that are too loud can tire your ears.

When your mix is almost done, switch between your mix and a commercial CD to see whether you're competitive. If the tonal balance of your mix matches a commercial CD, you know your mix will translate to the real world. This works regardless of what monitors you use. An effective tool for this purpose is Harmonic Balancer (www.harbal.com). It displays the spectrum (level vs. frequency) of a mix, and shows where certain frequencies may be too strong or too weak. You can adjust the frequency balance with a mouse to make the spectrum (and tone quality) more like that of a commercial reference mix.

Add Effects

With the balances and EQ roughed in, it's time to add effects. If you're using a DAW, use the plug-ins. If you're using a hardware mixer, you might want to plug in a reverb or delay to add spaciousness to the sound (see Chapter 9). This device connects between your mixer's aux-send and aux-return jacks (or aux-send and bus-in jacks).

1. On the effects unit, set the dry/wet mix control all the way to "wet" or "effect."
2. Find the AUX RETURN or BUS IN controls (if any), set them half of the way up, and pan them hard left and right.

3. Turn up the aux-send knob for each input, according to how much effect you want to hear on that input signal. Suppose you're using reverb as an effect. You might turn it up by different amounts for the vocals, drums, and lead guitar, and leave it turned down for the bass and kick drum.

4. As you're setting the aux levels, check the overload indicator on the effects unit. If it's flashing, turn down the input level on the effects unit just to the point where the overload light stops flashing. Then turn up the output level on the effects unit (or turn up the aux return on the mixer) to achieve the same amount of effect you heard before.

Too much effects and reverb can muddy the mix. You might turn up the reverb only on a few instruments or vocals. Once you have the reverb set, try turning it down gradually and see how little you can get by with. Typically the bass gets no reverb so it retains its clarity.

Usually, short reverb decay times (under about 1 second) are best for up-tempo songs. Longer reverb times are better for slow ballads.

You might use 45–120 msec of predelay on vocals or certain instruments to separate the dry sound from the reverb. This makes the voice sound up front even with reverb added.

The producer of a recording is the musical director and decides how the mix should sound. The producer might be the band members or yourself. Ask to hear recordings having the kind of sounds the producer desires. Try to figure out what techniques were used to create those sounds.

Also try to translate the producer's sound-quality descriptions into control settings. If the producer asks for a "warmer" sound on a particular instrument, turn up the low frequencies. If the lead guitar needs to be "fatter," try a stereo chorus on the guitar track. If the producer wants the vocal to be more "spacious," try adding reverb, and so on. *Audio clip 39 at www.elsevierdirect.com/companions/9780240811444 demonstrates a mixdown.* Other examples are at www.prosoundweb.com (search for Bruce Bartlett).

Set Levels

Set the overall recording level as you're mixing. To maintain the correct gain staging, keep the master faders at design center. Then adjust all the track faders by the same amount so your stereo output level peaks around −3 dB maximum. You can touch up the master faders a few decibels if necessary. Don't exceed a 0 dBFS recording level if you're recording to a digital medium.

You might prefer to start with the master faders at +5 dB. When the mix is nearly finished, bring the master faders down to 0 dB. Check that the stereo output level peaks around −3 dB maximum (in peak meter mode). That way, you may not need to bring down all the track faders by the same amount.

Judging the Mix

When you mix, your attention scans the inputs. Listen briefly to each instrument in turn and to the mix as a whole. If you hear something you don't like, fix it. Is the vocal too tubby? Roll off the bass on the vocal track. Is the kick drum too quiet? Turn it up. Is the lead-guitar solo too dead? Turn up its effects send.

Check the mix while listening from another room, where the lows and highs are weakened. Is the balance still good?

The mix must be appropriate for the style of music. For example, a mix that's right for rock music usually won't work for folk music or acoustic jazz. Rock mixes typically have lots of production EQ, compression, and effects; and the drums are way up front. In contrast, folk or acoustic jazz is usually mixed with no effects other than slight reverb, and the instruments and vocals sound natural. A rock guitar typically sounds bright and distorted; a straight-ahead jazz guitar usually sounds mellow and clean.

Suppose you are mixing a pop song, and you're aiming for a realistic, natural sound. Listen to the reproduced instruments and try to make them sound as if they're really playing in front of you. That is, instead of trying to make a pleasant mix or a sonically interesting recording, try to control the sound you hear to simulate real instruments—to make them believable. To do this you must be familiar with the sound of real instruments.

It's like an artist trying to draw a still-life as realistically as possible. The artist compares the drawing to the real object, notes the difference, and then modifies the drawing to reduce the difference.

When you're striving for a natural sound—as your ears hear an instrument live—compare the recorded instrument with your memory of the real thing. How does it sound different? Turn the appropriate knob on the console that reduces the difference. You're creating an illusion of accuracy. To make a recorded instrument sound real—like it's really playing in front of you—often you need to exaggerate things and add some processing.

Alternatively, when you're mixing, imagine that you're creating a sonic experience between the monitor speakers, rather than just reproducing instruments. Sometimes you don't want a recording to sound too realistic. If a recording is very accurate, it sounds like musical instruments, rather than just music itself.

This approach contradicts the basic edict of high fidelity—to reproduce the original performance as it sounded in the original environment. Some songs seem to require unreal sounds. That way, you don't connect the sounds you hear with physical instruments, but with the music behind the instruments—the composer's dream or vision.

Here's one way to reproduce pure music rather than reproducing instruments playing in a room: Mike closely or record direct to avoid picking up studio ambience. Then add reverb. Also add EQ, double-tracking, and effects to make the instrument or voice slightly unreal. The idea is to make a production, rather than a documentation—a record, rather than a recording.

Try to convey the musician's intentions through the recorded sound quality. If the musician has a loving, soft message, translate that into a warm, smooth tone quality. Add a little mid-bass or slightly reduce the highs. If the musical composition suggests grandeur or space, add reverberation with a long decay time. Ask the musicians what they are trying to express through the music, and try to express that through the sound production as well. Listen to the lyrics to see if the song is intimate (suggesting a warm, dry sound) or a proclamation (maybe with lots of reverb).

Try to keep the mix clean and clear. A clean mix is uncluttered; not too many parts play at once. It helps to arrange the music so that similar parts don't overlap. Usually, the fewer the instruments, the clearer the sound. Mix selectively, so that not too many instruments are heard at the same time. Have guitar licks fill in the holes between vocal phrases, rather than playing on top of the vocals. Sustained parts (like synth pads) mask other tracks more than transients, so keep pads low in level.

In a clear-sounding recording, instruments do not "crowd" or mask each other's sound. They are separate and distinct. Clarity arises when instruments occupy different areas of the frequency spectrum. For example, the bass provides lows; keyboards might emphasize mid-bass; lead guitar may provide upper mids, and cymbals fill in the highs.

Often the rhythm guitar occupies the same frequency range as the piano, so they tend to mask each other's sound. You can aid clarity by equalizing them differently. Boost the guitar at, say, 3 kHz, and boost the piano around 10 kHz. Or pan them to opposite sides. Distorted electric guitars can mask the vocal, so try cutting the guitars' EQ at 1 kHz.

More on judging sound quality is in Chapter 12.

Changes during the Mix

It's rare to do a mix in which you set the faders and leave them there. Often you need to change fader levels, EQ, or effects-send levels during a mix.

Level changes during the mix should be subtle, or else instruments will "jump out" for a solo and "fall back in" afterwards. Set faders to preset positions during pauses in the music. Nothing sounds more amateurish than a solo that starts too quietly then comes up as it plays. You can hear the engineer working the fader. If you need to reduce the level of a loud passage, do so at the end of the preceding soft passage before the loud one begins.

If you need to change fader levels during the mix, and you don't have automation, you might mark these levels next to each fader on a thin piece of tape. Make a cue sheet that notes the mixer changes at various tape-counter times. For example:

    0:15 Unmute vocal
    1:10 Lead solo -5

1:49 Lead -10
2:42 Synth EQ +6 at 12K
3:05 Start fade, out by 3:15

Most recording software lets you automate the mix. Details are in this chapter under the heading "Recording Software with Automated Mixing."

Record or Export the Mix

When you're happy with the mix and recording levels, record the mix on your hard drive with recording software. If you're mixing with a DAW, export or save the mix as a 24-bit WAV file (with dither off) for future mastering or 16-bit (with dither on) to burn a CD track.

If you don't have automation, and the mix is very difficult, you can record it a section at a time and then edit the sections together.

To fade out the end of the tune with a hardware mixer, pull down the mixer master faders slowly and smoothly. In a DAW, either (1) select all the tracks and insert a fade-out at the end of all the tracks or (2) export the mix, import the mix, and insert a fade-out at the end of the mix.

Use a fade-out that starts quickly and ends slowly. Try to have the music faded out by the end of a musical phrase. The slower the song, the slower the fade should be. The musical meaning of a fade is something like, "This song is continuing to groove, but the band is leaving on a slow train." You might want to postpone fades until mastering several songs to make an album.

The mixdown is complete. You might record several different mixes of one song, then choose the best mix. It's common to record a mix with the lead vocal up 1 dB, and another with the lead vocal down 1 dB.

Repeat these mixdown procedures for the rest of the good takes. Give your ears a rest every few hours! Otherwise, your hearing loses highs and you can't make correct judgments.

After a few days, listen to the mix on a variety of systems—car speakers, a boom box, headphones, a home system. The time lapse between mixdown and listening will allow you to hear with fresh ears. Do you want to change anything? If so, make it right. You'll end up with a mix to be proud of.

If you lack good multitrack recordings with which to practice mixing, go to www.raw-tracks.com. There you can download individual tracks in WAV or MP3 format, or purchase a CD of raw tracks.

20 QUESTIONS TO ASK YOURSELF WHILE MIXING

When mixing a song, I've found it helpful to ask some questions about what I'm hearing. The questions keep me focused, and they make sure that I'm not overlooking anything. Here are a few things to ask about your own mixes.

1. **Can I hear everything?** This is the most important question. Seems obvious, but sometimes a musical part becomes hidden. While the mix is playing, listen just for the sound of each track and make sure it's there. No matter the genre of music, the minimum requirement for a good mix is that you can hear all the instruments and vocals. Nothing is missing and nothing sticks out. Sometimes you need to mute or turn down some tracks to make a hidden track come out.

2. **Can I understand the lyrics?** Another important query. If you can't tell what the words are in certain spots, raise the vocal level there with a volume envelope (automation). Also, you might compress the vocals, make sure they have enough clarity around 5 to 10 kHz, and maybe reduce 3–6 kHz in instruments that compete in the same range as the vocals. Don't overdo the vocal effects. Some engineers use this guideline: the lead vocal in rock music should be just loud enough so you can understand the lyrics without straining. In ballads, traditional country, or folk music, the lead vocal can be a few dB louder than that.

3. **Is there too much reverb or other effects?** A little goes a long way. If the mix seems to be distant rather than present and engaging, try adding about 25 msec of predelay. Also try reducing the reverb sends a dB at a time, and see how little you can get away with. Some engineers ask, "Can I notice the reverb only when it's turned off?"

4. **Is there enough stereo spread?** If you pan vocals and most instruments to the middle, you have essentially a mono recording! Try panning two similar guitar parts hard left and right, or pan guitars left and keys right. If too many parts are panned to center, that center area can sound congested. Spread things out a little.

5. **Is each instrument's sound appropriate for the song?** For example, a twangy bass or an edgy kick seldom work in a ballad. Turn down the upper mids if those sounds are too bright and distracting.

6. **Similarly, is the mix appropriate for the genre?** For example, if you're mixing punk rock, a clean, tight sound probably won't work. If you're mixing a folk song done by an acoustic group, you probably don't want to hype the highs and lows—leave it natural.

7. **Is each instrument in its own spectral space?** If multiple instruments play in the same range of frequencies, they can mask or cover up each other's sound. Then they blur together and sound indistinct. You might roll off the lows in the guitars so they don't compete for space with the bass guitar. Thin out the kick and keep the bass full, or vice versa.

8. **Is the mix competitive with commercial CDs?** Plug a CD player into your monitoring system. Put in a CD (or several) of the same genre that you are mixing.

Switch back-and-forth between your mix and the CD playback. You'll quickly hear if your mix has enough bass, midrange, and treble compared to the commercial CD. Compare your mix balances to those on the CD. This can be very enlightening.

9. **Are the vocals too sibilant?** Are the "s" and "sh" sounds too piercing and annoying? Some singers are very sibilant, or the mic used on the singer is too bright. Solutions: Use a de-esser, which is a multiband compressor set to compress only the range from about 3 to 20 kHz. Or use a flatter sounding mic like a ribbon microphone, or options like a Neumann U87, ElectroVoice RE20, Shure SM7, or Kel Audio HM-2D. A high-frequency cut around 7–10 kHz helps too.

10. **Are the vocals too loud or too quiet sometimes?** Either apply compression or adjust the vocal levels with a volume envelope (automation). The latter sounds more natural.

11. **Do the vocals sound too "small" or "squashed"?** Usually that means you are applying too much compression. You might reduce the compression ratio to 3:1 or less, and/or raise the threshold so that the gain reduction is 6 dB or less.

12. **Is the overall sound harsh or is it warm and pleasant?** If it's harsh, maybe there is too much 2–4 kHz in the mix. Or maybe there's some distortion caused by excessive track levels or clipping plug-ins. Try reducing the amount or type of compression, too. If the mix sounds "digital" and edgy, reduce the highs a little, or use a tube or tape plug-in.

13. **Is the overall sound muffled?** If the mix seems lackluster or weak in the treble, maybe you need to boost the upper mids or highs a little. Try boosting electric guitars around 2–4 kHz, vocals around 5–10 kHz, toms around 5 kHz, kick around 4 kHz, cymbals around 12 kHz. Or cut a little around 250–600 Hz instead.

14. **Is the mix dynamic?** Do the choruses get more sonically exciting than the verses? If not, you might need to bring up the overall level a dB or two in the choruses, switch to a different guitar timbre, add a doubled vocal, add harmonies, increase the panning width, and so on.

15. **Is the mix creative and exciting?** Are you employing unusual effects or instrument sounds? Or are they like everybody else's record? Try to do something different but tasteful.

16. **Are solos at the right level?** Generally, a guitar solo should be just as loud as the lead vocal. Guitar licks in the "holes" (vocal pauses) should be quieter than that so they are not too distracting.

17. **Does the mix seem to have a focal point?** At any part in the song, is there something that grabs your attention, or is everything equally loud? You know the vocal is too quiet when it doesn't stand out from the background a bit.

18. **Are vocal harmonies at the right level?** Generally, a harmony vocal's level should be below the lead vocal just enough so that the melody of the lead vocal is clear. If a harmony line is too loud, the listener isn't quite sure who's singing the melody line.

19. Is the arrangement too busy? If too many instruments play at the same time, a mix can turn to mush. Consider having guitar licks just in the holes, not playing continuously. Think call-and-response. Start the mix with fewer instruments and gradually bring them in so that the mix builds.

20. Is the mix "musical"? This one is hard to define in tech terms. Can you feel the emotion expressed in the lyrics? Does the song make you want to move or dance? That depends on the song and its performance, but it also depends on the mix.

When you no longer hear anything you want to change, the mix is almost done. A day later come back with fresh ears, and see if anything needs tweaking. If not, congratulations on crafting a great mix!

SUMMARY

The following are summaries of the procedures for recording, overdubbing, and mixdown. Use these steps for easy reference.

RECORDING

1. Turn up the headphone or monitor volume control. Monitor the aux bus that you're using for the monitor mix. If your multitrack is wired to the insert jacks, monitor the stereo bus instead. If you're using a DAW, monitoring is automatic.
2. Assign instruments to tracks. To record one instrument per track, connect its direct out (or insert send) to a track input. If you're using a DAW, select each track and assign it an input signal from your audio interface.
3. Turn up the input faders, submaster, and master faders to design center (the shaded portion of fader travel, about three-quarters of the way up).
4. Adjust the input attenuators (trim) to set submixes and recording levels. If you're recording with a DAW, use the volume controls of the audio interface (they might be in a software application).
5. Set the monitor mix.
6. Record onto the multitrack recorder.

OVERDUBBING

1. Assign the instruments or vocals to be recorded to open tracks. An open track is blank or has already been bounced.
2. Turn up the monitor/cue system.
3. Turn up the submasters and master to design center.
4. Play the multitrack recording and set up a cue mix of the already recorded tracks.
5. While a musician is playing, adjust the input attenuation and recording level.

6. Set the monitor/cue mix to include the sound of the instrument or vocal being added.

7. Record the new parts on open tracks.

8. Punch in and comp tracks as needed.

MIXDOWN

1. If necessary, set the input selectors to accept the multitrack recorder output signals.

2. Monitor buses 1 and 2 (or the stereo mix bus). Monitoring is automatic with a DAW.

3. Assign tracks to buses 1 and 2 (or the stereo mix bus, or the stereo output of your audio interface).

4. Turn up the master fader to design center. In some mixers, the submasters should also be up.

5. Set preliminary panning.

6. Set a rough mix with the input faders. Maybe start with all faders at −12 dB, then adjust from there.

7. Set equalization and effects.

8. Perfect the mix and set recording levels. Set up automation if your system has it (described next).

9. Record the mix onto a hard drive using recording software. If you have a DAW, export or save the mix as a WAV file.

AUTOMATED MIXING

A multitrack mixdown is often a complicated procedure. It can be difficult to change the mixer settings correctly at all the right times. So you might want to use automated mixing—have computer memory store and reset the changes for you.

As you mix a song, you might adjust the mixer controls several times. For example, raise the piano's volume during a solo, then drop it back down; or mute a track to reduce noise during pauses in the performance. An automated mixing system can remember your mix moves, and later recall and reset them accordingly each time you play back the mix. You can even overdub mix moves; for example, do the vocal-fader moves on the first pass, drum moves on the second pass, and so on. You also can punch in fader moves to correct them. Effects changes can be automated as well in some units. Automated mixing is a feature in digital mixers, software DAWs, and many recorder-mixers.

Automated mixing has many advantages. With it you can:

- Perform complicated mixes without errors
- Fine-tune the mix moves

- Recall mixes weeks or months after storing them, without having to reset the mixer manually each time
- Listen to the mix without the distraction of having to adjust faders

Types of Automation Systems

Three types of automation systems are:

1. Automated mixer with nonmotorized faders
2. Automated mixer with motorized faders (Flying Faders)
3. Recording software with automated mixing

Each of these is worth a closer look.

AUTOMATED MIXER WITH NONMOTORIZED FADERS

Suppose you have mixer with nonmotorized faders. The mix is playing and you are adjusting the faders. As you set the position of each fader, this action produces a MIDI signal that is recorded by a sequencer. This sequencer is either built into the mixer or is external.

When the mix plays back, so does the sequence of mix moves. MIDI signals from the sequencer set the volume level for each channel. This is done by varying the gain of a voltage controlled amplifier (VCA) or digitally controlled amplifier (DCA) in each channel. Because the faders do not move as the mix plays back, the fader positions do not represent the mix you are hearing.

AUTOMATED MIXER WITH MOTORIZED FADERS

This system works as follows. First, you adjust the faders to set up a mix. This generates MIDI signals that are recorded by a sequencer. When you play back the mix, the sequencer makes the motorized faders move up and down as if controlled by a ghost, matching your mix moves. The positions of the mixer faders show the mix levels. VCAs are not needed, which is a benefit because VCAs can degrade the audio slightly.

RECORDING SOFTWARE WITH AUTOMATED MIXING

Suppose you use a computer digital audio workstation to record tracks and edit them. The monitor screen shows virtual faders that you adjust with a mouse or control surface. As you set up a mix, the automation feature in the DAW remembers your fader moves, EQ, panning, and plug-in settings. All those moves and settings are reproduced each time you open a project and play it back.

Automation is easy in DAWs that let you insert a volume envelope on a track. This envelope is a graph of a track's fader setting versus time (Figure 11.3). You might prefer

FIGURE 11.3
In a track, a volume envelope controls the track's volume in the mix over time.

to automate the mix by tweaking a track's volume envelope on-screen instead of moving the fader. As the song plays, the fader for that track will move up and down, following the envelope. Many DAWs let you automate other parameters as well, such as reverb-send level, panning, and so on.

What if a track needs a major change in EQ or reverb at a certain point? Find the section in the track that needs different EQ or reverb, create a clip of that section, and move it to another track with those plug-in settings. Or simply insert a new plug-in into the clip (if your DAW has that feature).

As we said earlier, a MIDI/audio recording program includes both digital audio tracks and sequencer tracks. You automate the audio track levels by adjusting their gain (fader levels or volume envelopes). You automate the MIDI tracks by adjusting their MIDI volume or key-velocity scaling. Individual MIDI notes can be tweaked, too.

Snapshot versus Continuous Automation

Two types of mixer automation are snapshot and continuous (dynamic). With snapshot automation, you push a button to take a "snapshot" or "scene" of the mixer settings. The sequencer stores the snapshot as MIDI "program changes." To reset the mixer to any of these stored settings, you punch up the appropriate number (MIDI program change) as the song plays. Recording software also can record and play snapshots of the mixer settings.

With continuous or dynamic automation, the sequencer records the motion of the mixer controls—not just their position. Continuous automation costs more than snapshot and consumes more memory, but permits finer resolution of mix moves. Some digital mixers do both snapshot and continuous automation with their internal sequencer. Other mixers require an external MIDI sequencer to record continuous automation moves.

Some automated mixers let you specify a fade time between snapshots. When the sequencer changes from one snapshot to the next, the sequencer fades or adjusts the control settings gradually between the two snapshots. This acts like continuous or dynamic automation.

Automated Mixing Procedure

Here is a suggested order of steps in doing an automated mix:

1. Set up a rough mix manually. You might prefer to start with a song section where all the instruments are playing.
2. Record the mutes. Mute tracks that aren't being used, or that don't add to the arrangement. For example, a chorus might sound better without the horns. Unmute each muted track just before you want it to be heard. If you're using a DAW, you have already deleted unwanted portions of each track.
3. Record the fader moves or, in a DAW, draw a volume envelope for each track. For example, turn up the guitar during the solo, turn up the toms during the fills, and turn down the background vocals when they get too loud. Save your changes often.
4. Record the effects changes. You might bring up the chorusing on the acoustic guitar during the bridge, and so on.
5. Sit back and listen to the mix. Is there anything you want to change?

Another way to do an automated mix is with snapshots. At several points in the song where you need mutes or fader changes, set the mute or fader level on the appropriate track and take a snapshot.

LO-FI RECORDING: HOW TO TRASH YOUR TRACKS

It's an exciting time for home recording. We have low-cost digital recorders and mixers, soft synths, and good-sounding cheap mics. All these tools make it affordable to record with sonic purity and accuracy. That's a wonderful development.

But now that anyone can record high-quality sounds, it's not such a big deal anymore. Sure, there's a place for clean, accurate recordings. But in many of today's records, you'll hear instances of lo-fi sound: fuzzy vocals, tinny drums, and humming guitar amps.

Lo-fi is the opposite of hi-fi. High fidelity recordings sound like real musical instruments. Technically, hi-fi sound implies a flat frequency response, no noise or distortion, and low wow and flutter. In contrast, lo-fi sounds might have a narrow frequency response (a thin, cheap sound), and might include noise such as hiss or record scratches. They could be distorted or wobbly in pitch.

Also, the lo-fi attitude snubs its nose at "clean" recordings. "Clean" can mean a lot of things: free of noise, free of leakage, free of excess room reverb. So a lo-fi sound might have tape hiss, some leakage, and some room acoustics as part of the mix.

Lo-fi really took off with rap music, in which the drum sound was the opposite of the usual polished studio sound. Instead of a tight kick, we heard a boomy kick. Instead of a wide-range snare with a full thump and crisp attack, rap gave us a tinny, trashy snare that was all midrange.

Want to put some lo-fi sonic interest in your mixes? Let's look at a few ways to trash those clean sounds that are so easy to record.

Lo-Fi Frequency Response

A flat frequency response in a mic or mixer tends to give an accurate, natural sound. If the response is flat, all the musical fundamentals and harmonics are reproduced in their original proportions. So the recording sounds accurate or hi-fi.

You can easily make a lo-fi effect by messing up the response. Cut the highs and lows, boost the mids. Or create a raggedy response with lots of bumps and dips. Some ways to do this are with EQ, mic choice, and mic placement.

Play a snare-drum track through your mixer, and turn down the LF EQ and HF EQ. Boost around 1 kHz or nearby frequencies. Your snare sound will change from high-budget to bargain-basement. Beck's "Soul-Suckin' Jerk" from the EP "Loser" is a good example of a lo-fi drum set.

Or try recording a child's drum set with its small heavy cymbals and boomy kick drum. You might loop a hi-hat beat made from this set, and mix it with a full-range recording of a quality drum set.

Find a toilet-paper tube, or a flexible plastic tube that extends gutter downspouts. Put the tube in front of a mic and sing through the tube. The resonances in the tube will color the sound in a wild way.

You might track down some cheap old mics at a garage sale, on eBay, or from vintage mic collectors. Record a few tracks using those mics. Their frequency response tends to be a complex series of peaks and valleys that you can't duplicate with EQ.

FIGURE 11.4
The old mic-in-a-wastebasket trick.

Unusual mic placements are fun. Record a guitar amp or vocal with the mic placed in a wastebasket (Figure 11.4). Hit a cymbal with a cheap mic while recording its signal. Mike a snare drum from underneath for a thin, zippy effect. If you pick up a crash cymbal off its edge, the sound will waver as the cymbal tilts when it is struck.

Distortion

Distortion is the addition of harmonics that did not exist in the original sound. Here's an obvious way to create distortion: drive a piece of recording gear at very high levels—beyond what it can handle. For example, record drums on a cassette recorder with the meters pinning. Or yell into a "bullet" type harmonica mic so that the mic distorts. In a DAW, use a distortion plug-in such as iZotope Trash (www.izotope.com) (Figure 11.5).

Another method to add distortion to a track: feed the signal through an effects box with a distortion setting. Run a drum track through a guitar stomp box, or through a broken vintage compressor. Feed a vocal through an Amp Farm or Line 6 Pod processor (Figure 11.6). You might record some instruments on a cheap cassette recorder (Figure 11.7). The Rolling Stones did that to create the beginning of "Jumping Jack Flash."

FIGURE 11.5
iZotope Trash.

FIGURE 11.6
Line 6 Pod.

FIGURE 11.7
Recording with a cheap cassette recorder.

Noise

Try iZotope's Vinyl simulation plug-in to add record scratches and other noises. Another way to have noises in your mix is to record noisy instruments! You might find an old, creaky upright piano and do the keyboard tracks with it. Or use an old guitar amp with blown tubes to record a sound full of hum and hiss.

Leakage

If your mixes are too sterile or studio-clean, you might want to record some leakage. Leakage (or bleed or spill) is the pickup of an instrument by another instrument's microphone. For example, if a guitar mic picks up the drums from across the room, that pickup of the drums is called leakage. Since the guitar mic picks up the drums at a distance, the leakage changes the recorded sound of the drums from tight to muddy.

Normally we try to isolate instruments or mike close to prevent leakage. But if you want to add some creative leakage to your mixes, mike a little farther away than you normally do, and record all the instruments at once.

Room Sound

In the quest for clean, tight recordings, it is standard practice to partly cover the walls of the studio with sound-dampening material. This reduces early reflections, which are less than about 20 msec after the direct sound from the instrument being recorded. Those early reflections tell the ear that the instrument was recorded in a small room. Normally we get rid of those early reflections and replace them with artificial reverb. But a lo-fi recording often includes the sound of the room as part of the sound of the recorded instrument.

To pick up room reflections, mike farther away than usual and leave the walls uncovered. Use the room for its coloration, rather than rejecting the room.

Want to make a synth track more organic and spacious? Run the synth track through a guitar amp, and mike the amp in stereo from several feet away. The guitar amp rolls off the highs and adds distortion as well. You could even record a vocal or harmonica through the amp this way.

For a spacious effect, consider recording several instruments in stereo with two mics. Pick up instruments or vocals in a hallway, a bathroom, a box, or even outdoors.

Lo-Fi Aesthetics

It's common to include hi-fi sounds along with lo-fi sounds in the same mix. This makes a statement to your listeners: "I can record hi-fi sounds, but I'm choosing not to. The trashy sounds are due to a conscious choice rather than a lack of recording chops." If you have nothing but lo-fi sounds in your mixes, it might sound like you don't know what you're doing—you have no command over the recording process.

So, your mix might be mostly pristine, full and sparkling ... but tainted with a vocal that sounds like it came through a telephone. The contrast of clean and dirty sounds, modern and vintage, can add a lot of sonic interest.

The ear delights in complexity. We can add some of that to our recordings with intriguing, colorful lo-fi sounds.

I recommend the album *Mule Variations* by Tom Waits. It is a brilliantly creative lo-fi masterpiece. So is Beck's *Timebomb*. Others are "digital hardcore" genre albums by Ronin, Technology Scum, and Cheap Czad.

CHAPTER 12

Judging Sound Quality

Seat an engineer behind a mixing console and ask him or her to do a mix. It sounds great. Then seat another engineer behind the same console and again ask for a mix. It sounds terrible. What happened?

The difference lies mainly in their ears—their critical listening ability. Some engineers have a clear idea of what they want to hear and how to get it. Others haven't acquired the essential ability to recognize good sound. By knowing what to listen for, you can improve your artistic judgments during recording and mixdown. You are able to hear errors in microphone placement, equalization, and so on, and correct them.

To train your hearing, try to analyze recorded sound into its components—such as frequency response, noise, reverberation—and concentrate on each one in turn. It's easier to hear sonic flaws if you focus on a single aspect of sound reproduction at a time. This chapter is a guide to help you do this.

CLASSICAL VERSUS POPULAR RECORDING

Classical and popular music have different standards of "good sound." One goal in recording classical music (and often folk music or jazz) is to accurately reproduce the live performance. This is a worthy aim because the sound of an orchestra in a good hall can be quite beautiful. The music was composed and the instruments were designed to sound best when heard live in a concert hall. The recording engineer, out of respect for the music, should always try to translate that sound to disc with as little technical intrusion as possible.

By contrast, the accurate translation of sound to disc is not always the goal in recording popular music. Although the aim may be to reproduce the original sound, the producer or engineer may also want to play with that sound to create a new sonic experience, or to do some of both.

In fact, the artistic manipulation of sounds through studio techniques has become an end in itself. Creating an interesting new sound is as valid a goal as re-creating the original sound. There are two games to play, each with its own measures of success.

If the aim of a recording is realism or accurate reproduction, the recording is successful when it matches the live performance heard in the best seat in the concert hall. The sound of musical instruments is the standard by which such recordings are judged.

When the goal is to enhance the sound or produce special effects (as in most pop-music recordings), the desired sonic effect is less defined. The live sound of a pop group could be a reference, but pop-music recordings generally sound better than live performances—recorded vocals are clearer and less harsh, the bass is cleaner and tighter, and so on. The sound of pop music reproduced over speakers has developed its own standards of quality apart from accurate reproduction.

GOOD SOUND IN A POP-MUSIC RECORDING

Currently, a good-sounding pop recording might be described as follows (there are always exceptions):

- Well mixed
- Wide range
- Tonally balanced
- Clean
- Clear
- Smooth
- Spacious

It also has:

- Presence
- Sharp transients
- Tight bass and drums
- Wide and detailed stereo imaging
- Wide but controlled dynamic range
- Interesting sounds
- Suitable production

The next sections explore each one of these qualities in detail so that you know what to listen for. Assume that the monitor system is accurate, so that any colorations heard are in the recording and not in the monitors.

A Good Mix

In a good mix, the loudness of instruments and vocals is in a pleasing balance. Everything can be clearly heard, yet nothing is obtrusive. The most important instruments or voices are loudest; less important parts are in the background.

A successful mix goes unnoticed. When all the tracks are balanced correctly, nothing sticks out and nothing is hidden. Of course, there's wide latitude for musical interpretation and personal taste in making a mix. Dance mixes, for example, can be very severe sonically.

Sometimes you don't want everything to be clearly heard. On rare occasions you may want to mix in certain tracks very subtly for a subconscious effect.

The mix must be appropriate for the style of music. For example, a mix that's right for loud rock music usually won't work for a pop ballad. A rock mix typically has the drums way up front and the vocals only slightly louder than the accompaniment. In contrast, a pop ballad has the vocals loudest, with the drums used just as "seasoning" in the background.

Level changes during the mix should be subtle, or should make sense. Otherwise, instruments jump out for a solo and fall back in afterwards. Move faders slowly, set them to preset positions during pauses in the music, or use automation. Nothing sounds more amateurish than a solo that starts too quietly and then comes up as it plays—you can hear the engineer working the fader.

Wide Range

Wide range means extended low- and high-frequency response. Cymbals should sound crisp and distinct, but not sizzly or harsh; kick drum and bass should sound deep, but not overwhelming or muddy. Wide-range sound results from using high-quality microphones, DIs, virtual instruments, and adequate EQ.

You might want to combine "hi-fi" and "lo-fi" sounds in a single mix. The lo-fi sounds generally cover a narrow frequency range and might be distorted.

Good Tonal Balance

The overall tonal balance of a recording should be neither bassy nor trebly. That is, the perceived spectrum should not emphasize low or high frequencies. Low bass, mid-bass, midrange, upper midrange, and highs should be heard in equal proportions (Figure 12.1). Emphasis of any one frequency band over the other eventually causes listening fatigue. Dance club mixes, however, are heavy on the bass end to get the crowd moving.

Recorded tonal balance is inversely related to the frequency response of the studio's monitor system. If the monitors have a high-frequency rolloff, the engineer will compensate by boosting highs in the recording to make the monitors sound correct. The result is a bright recording.

Before doing a mix, play over the monitors some commercial recordings whose sound you admire, to become accustomed to a commercial spectral balance. After your mix is recorded, play it back and alternately switch between your mix and a commercial recording. This comparison indicates how well you matched a commercial spectral

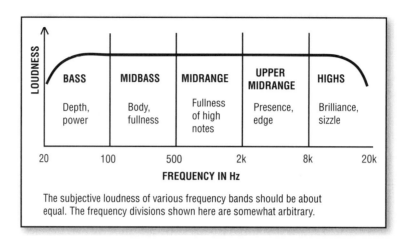

The subjective loudness of various frequency bands should be about equal. The frequency divisions shown here are somewhat arbitrary.

FIGURE 12.1
Loudness versus frequency of a pop recording with good sound

balance. An effective tool for this purpose is Harmonic Balancer (www.har-bal.com). Of course, you may not care to duplicate what others are doing.

In pop-music recordings, the tonal balance or timbre of instruments does not necessarily have to be natural. Still, many listeners want to hear a realistic timbre from acoustic instruments, such as the guitar, flute, sax, or piano. The reproduced timbre depends on microphone frequency response, microphone placement, the musical instruments themselves, and equalization.

Clean Sound

Clean means free of noise and distortion. Hiss, hum, and distortion are inaudible in a good recording. Distortion in this case means distortion added by the recording process, not distortion already present in the sound of electric-guitar amps or Leslie speakers. There are exceptions to this guideline; some popular recordings have noise or distortion added intentionally.

Clean also means "not muddy" or free of low-frequency ringing and leakage. A clean mix is one that is uncluttered or free of excess instrumentation. This is achieved by arranging the music so that similar parts don't overlap, and not too many instruments play at once in the same frequency range. Usually, the fewer the instruments, the cleaner the sound. Too many overdubs can muddy the mix.

Clarity

In a clear-sounding recording, instruments do not crowd or mask each other. They are separate and distinct. As with a clean sound, clarity arises when instrumentation is sparse, or when instruments occupy different areas of the frequency spectrum. For example, the bass provides low frequencies, keyboards might emphasize mid-bass, lead guitar provides upper midrange, and cymbals fill in the highs.

In addition, a clear recording has adequate reproduction of each instrument's harmonics. That is, the high-frequency response is not rolled off.

Smoothness

Smooth means easy on the ears, not harsh, uncolored. Sibilant sounds are clear but not piercing. A smooth, effortless sound allows relaxation; a strained or irritating sound causes muscle tension in the ears or body. Smoothness is a lack of sharp peaks or dips in the frequency response, as well as a lack of excessive boost in the midrange or upper midrange. It is also low distortion, such as provided by a 24-bit recording.

Presence

Presence is the apparent sense of closeness of the instruments—a feeling that they are present in the listening room. Synonyms are clarity, detail, and punch.

Presence is achieved by close miking, overdubbing, and using microphones with a presence peak or emphasis around 5 kHz. Using less reverb and effects can help. Upper-midrange boost helps, too. Most instruments have a frequency range that, if boosted, makes the instrument stand out more clearly or become better defined. Presence sometimes conflicts with smoothness because presence often involves an upper-midrange boost, while a smooth sound is free of such emphasis. You have to find a tasteful compromise between the two.

Spaciousness

When the sound is spacious or airy, there is a sense of air around the instruments. Without air or ambience, instruments sound as if they are isolated in stuffed closets. (Sometimes, though, this is the desired effect.) You achieve spaciousness by adding reverb, recording instruments in stereo, using room mics, or miking farther away.

Sharp Transients

The attack of cymbals and drums generally should be sharp and clear. A bass guitar and piano may or may not require sharp attacks, depending on the song.

Tight Bass and Drums

The kick drum and bass guitar should "lock" together so that they sound like a single instrument—a bass with a percussive attack. The drummer and bassist should work out their parts together so they hit accents simultaneously, if this is desired.

To further tighten the sound, damp the kick drum and record the bass direct. Rap music, however, has its own sound—the kick drum usually is undamped and boomy, sometimes with short reverb added. Equalize the kick and bass in complementary ways so that they don't mask each other; for example:

> Kick: Boost 60 to 80 Hz, cut 600 Hz, boost 3 kHz.
> Bass: Cut 60 to 80 Hz, boost 600 Hz, cut above 3 kHz.

Wide and Detailed Stereo Imaging

Stereo means more than just left and right. Usually, tracks should be panned to many points across the stereo stage between the monitor speakers. Some instruments should be hard-left or hard-right, some should be in the center, and others should be half-left or half-right (panned to 9 o'clock and 3 o'clock). Try to achieve a stereo stage that is well balanced between left and right (Figure 12.2). Instruments that occupy the same frequency range can be made more distinct by panning them to opposite sides of center.

You may want some tracks to be unlocalized. Backup choruses and strings should be spread out rather than appearing as point sources. Stereo keyboard sounds can wander between speakers. A lead-guitar solo can have a fat, spacious sound.

There should also be some front-to-back depth. Some instruments should sound close or up front; others should sound farther away. Use different miking distances or different amounts of reverb on various tracks.

If you want the stereo imaging to be realistic (for a jazz combo, for example), the reproduced ensemble should simulate the spatial layout of the live ensemble. If you're sitting in an audience listening to a jazz quartet, you might hear drums on the left, piano on the right, bass in the middle, and sax slightly right. The drums and piano are not point sources, but are somewhat spread out. If spatial realism is the goal, you should hear the same ensemble layout between your speakers. On some commercial

FIGURE 12.2
An example of image placement between speakers.

CDs, the piano and drums are spread all the way between speakers—an interesting effect, but unrealistic.

Pan-potted mono tracks often sound artificial in that each instrument sounds isolated in its own little space. It helps to add some stereo reverberation around the instruments to "glue" them together.

Often, TV mixes are heard in mono. Hard-panned signals sound weak in mono relative to center-panned signals. So pan sound sources to 3 and 9 o'clock, not hard-right and hard-left.

Wide but Controlled Dynamic Range

Dynamic range is the range of volume levels from softest to loudest. A recording with a wide dynamic range becomes noticeably louder and softer, adding excitement to the music. To achieve this, don't add too much compression (automatic volume control). An overly compressed recording sounds squashed—crescendos and quiet interludes lose their impact, and the sound becomes fatiguing.

Vocals often need some compression or gain-riding because they have more dynamic range than the instrumental backup. A vocalist may sing too loudly and blast the listener, or sing too softly and become buried in the mix. A compressor can even out these extreme level variations, keeping the vocals at a constant loudness. Bass guitar also can benefit from compression.

Interesting Sounds

The recorded sound may be too flat or neutral, lacking character or color. In contrast, a recording with creative production has unique musical instrument sounds, and typically uses effects. Some of these are equalization, echo, reverberation, doubling, chorus, flanging, compression, distortion, and stereo effects.

Making sounds interesting or colorful can conflict with accuracy or fidelity. That's okay, but you should know the trade-off.

Suitable Production

The way a recording sounds should imply the same message as the musical style or lyrics. In other words, the sound should be appropriate for the particular tune being recorded.

For example, some rock music is rough and raw. The sound should be, too. A clean, polished production doesn't always work for high-energy rock 'n' roll. There might even be a lot of leakage or ambience to suggest a garage studio or nightclub environment. The role of the drums is important, so they should be loud in the mix. The toms should ring, if that is desired.

New Age, disco, rhythm and blues, contemporary Christian, and pop music are typically slickly produced. The sound is usually tight, smooth, and spacious. Folk music and acoustic jazz typically sound natural and have little or no reverb. Music in the digital hardcore genre has lots of distortion.

Actually, each style of music is not locked into a particular style of production. You tailor the sound to complement the music of each individual tune. Doing this may break some of the guidelines of good sound, but that's usually okay as long as the song is enhanced by its sonic presentation.

GOOD SOUND IN A CLASSICAL-MUSIC RECORDING

As with pop music, classical music should sound clean, wide range, and tonally balanced. But because classical recordings are meant to sound realistic—like a live performance—they also require good acoustics, a natural balance, tonal accuracy, suitable perspective, and accurate stereo imaging (see Chapter 18).

Good Acoustics

The acoustics of the concert hall or recital hall should be appropriate for the style of music to be performed. Specifically, the reverberation time should be neither too short (dry) nor too long (cavernous). Too short a reverberation time results in a recording without spaciousness or grandeur. Too long a reverberation time blurs notes together, giving a muddy, washed-out effect. Ideal reverberation times are around 1.2 seconds for chamber music or soloists, 1.5 seconds for symphonic works, and 2 seconds for organ recitals. To get a rough idea of the reverb time of a room, clap your hands once, loudly, and count the seconds it takes for the reverb to fade to silence.

A Natural Balance

When a recording is well balanced, the relative loudness of instruments is similar to that heard in an ideal seat in the audience area. For example, violins are not too loud or soft compared to the rest of the orchestra; harmonizing or contrapuntal melody lines are in proportion.

Generally, the conductor, composer, and musicians balance the music acoustically, and you capture that balance with your stereo mic pair. But sometimes you need to mike certain instruments or sections to enhance definition or balance. Then you mix all the mics. In either case, consult the conductor for proper balances.

Tonal Accuracy

The reproduced timbre or tone quality should match that of live instruments. Fundamentals and harmonics should be reproduced in their original proportion.

Suitable Perspective

Perspective is the sense of distance of the performers from the listener—how far away the stage sounds. Do the performers sound like they're eight rows in front of you, in your lap, or in another room?

The style of music suggests a suitable perspective. Incisive, rhythmically motivated works (such as Stravinsky's "Rite of Spring") sound best with closer miking; lush, romantic pieces (a Bruckner symphony) are best served by more distant miking. The chosen perspective depends on the taste of the producer.

Closely related to perspective is the amount of recorded ambience or reverberation. A good miking distance yields a pleasing balance of direct sound from the orchestra and ambience from the concert hall.

Accurate Imaging

Reproduced instruments should appear in the same relative locations as they were in the live performance. Instruments in the center of the ensemble should be heard in the center between the speakers; instruments at the left or right side of the ensemble should be heard from the left or right speaker. Instruments halfway to one side should be heard halfway off-center, and so on. A large ensemble should spread from speaker to speaker, while a quartet or soloist can have a narrower spread.

It's important to sit equidistant from the speakers when judging stereo imaging, otherwise the images shift toward the side on which you're sitting. Sit as far from the speakers as they are spaced apart. Then the speakers appear to be 60 degrees apart, which is about the same angle an orchestra fills when viewed from the typical ideal seat in the audience (tenth row center, for example).

The reproduced size of an instrument or instrumental section should match its size in real life. A guitar should be a point source; a piano or string section should have some stereo spread. Each instrument's location should be as clearly defined as it was heard from the ideal seat in the concert hall.

Reproduced reverberation (concert-hall ambience) should surround the listener, or at least it should spread evenly between the speakers. Surround-sound technology is needed to make the recorded ambience surround the listener, although spaced-microphone recordings have some of this effect. Accurate imaging is illustrated in Figure 12.3.

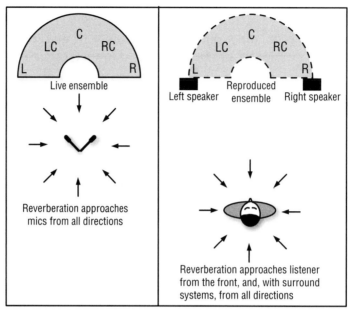

FIGURE 12.3
With accurate imaging, the sound-source location and size, and the reverberant field, are reproduced during playback.

There should be a sense of stage depth: front-row instruments sound closer than back-row instruments.

TRAINING YOUR HEARING

The critical process is easier if you focus on one aspect of sound reproduction at a time. You might concentrate first on the tonal balance—try to pinpoint what frequency ranges are being emphasized or slighted. Next listen to the mix, the clarity, and so on. Soon you have a lengthy description of the sound quality of your recording.

Developing an analytical ear is a continuing learning process. Train your hearing by listening carefully to recordings—both good and bad. Make a checklist of all the qualities mentioned in this chapter. Compare your own recordings to live instruments and to commercial recordings. Check out the Golden Ears ear-training CDs at www.moulton labs.com/full/product01. Audiofile Engineering makes a Mac & iOS app that includes feedback via quizzes: Quiztones (http://audiofile-engineering.com/quiztones).

A pop-music record that excels in all the attributes of good sound is The Sheffield Track Record (Sheffield Labs, Lab 20), engineered and produced by Bill Schnee. In effect, it's a course in state-of-the-art sound—required listening for any recording engineer or producer.

Another record with brilliant production is *The Nightfly* by Donald Fagen (Warner Brothers 23696-2), engineered by Roger Nichols, Daniel Lazerus, and Elliot Scheiner, produced by Gary Katz, and mastered by Bob Ludwig. This recording, and Steely Dan recordings by Roger Nichols, sound razor sharp, very tight and clear, elegant, and tasteful; and the music just pops out of the speakers.

The following listings are more examples of outstanding rock production, and set high standards:

Songs:

"I Need Somebody," by Bryan Adams; producer, Bob Clearmountain
"The Power of Love," by Huey Lewis & The News; producer, Huey Lewis & The News

Albums:

90125, by Yes; producer, Trevor Horn
Synchronicity, by The Police; producer, Hugh Padgham and The Police
Dark Side of the Moon, by Pink Floyd; producer, Alan Parsons
Thriller, by Michael Jackson; engineer, Bruce Swedien; producer, Quincy Jones
Avalon, by Roxy Music; engineer, Bob Clearmountain; producer, Roxy Music
Nevermind, by Nirvana; producer, Butch Vig
Come Away with Me, by Norah Jones; engineer, Jay Newland
Genius Loves Company, by Ray Charles; several engineers
Give, by The Bad Plus; engineer, Tchad Blake
Live in Paris and *The Look of Love*, by Diana Krall; engineer, Al Schmitt
Smile, by Brian Wilson; engineer, Mark Linett.

Some other recordings of note are those by Supertramp, Pat Metheny Group, Alison Krauss, The Wailin' Jennys, Heart, and Radiohead; and by producers George Martin and Andy Wallace.

Then there are the incredibly clean recordings of Tom Jung (with DMP records) and George Massenberg. Some classical-music recordings with outstanding sound are on the Telarc, Delos, and Chesky labels. You can learn a lot by emulating these superb recordings and many others.

Once you're making recordings that are competent technically—clean, natural, and well mixed—the next stage is to produce imaginative sounds. You're in command; you can tailor the mix to sound any way that pleases you or the band you're recording. The supreme achievement is to produce a recording that is a sonic knockout, beautiful or thrilling.

TROUBLESHOOTING BAD SOUND

Now you know how to recognize good sound, but can you recognize bad sound? Suppose you're monitoring a recording in progress, or listening to a recording you've already made. Something doesn't sound right. How can you pinpoint what's wrong, and how can you fix it?

The rest of this chapter includes step-by-step procedures to solve audio-related problems. Read down the list of "bad sound" descriptions until you find one matching what you hear. Then try the solutions until your problem disappears. Only the most common symptoms and cures are mentioned; console maintenance is not covered.

This troubleshooting guide is divided into four main sections:

   1. Bad sound on all recordings (including those from other studios)
   2. Bad sound on playback only (the mixer output sounds all right)
   3. Bad sound in a pop-music recording
   4. Bad sound in a classical-music recording

Before you start, check for faulty cables and connectors. Also check all control positions, rotate knobs and flip switches to clean the contacts, and clean connectors with DeoxIT from Caig Labs.

Bad Sound on All Recordings

If you have bad sound on all your recordings, including those from other studios, follow this checklist to find the problem:

   - Upgrade your monitor system.
   - Adjust tweeter and woofer controls on speakers.
   - Adjust the relative gains of tweeter and woofer amplifiers in a bi-amped system.
   - Relocate speakers.
   - Improve room acoustics.
   - Equalize the monitor system.
   - Try different speakers.
   - Upgrade the power amp and speaker cables.
   - Monitor at a moderate listening volume, such as 85 dBSPL. We hear less bass and treble in a program if it is monitored at a low volume, and vice versa. If we hear too little bass due to monitoring at a low level, we might mix in too much bass.

Bad Sound on Playback Only

If a hard-drive recording has glitches or drop-outs on playback, try this:

   - Increase the latency setting in your recording software.
   - Follow the tips in Appendix B, "Optimizing Your Computer for Multitrack Recording."

If a digital recording sounds distorted, these suggestions might help:

   - Keep the recording level as high as possible, but try not to exceed −6 dBFS (decibels full scale) in peak meter mode, not RMS mode.

- Avoid clipping in effects plug-ins.
- Record at a higher sampling rate or higher bit depth.
- Avoid sampling-rate conversion.
- Apply dither when going from a high bit depth to a lower bit depth.
- Try a higher-quality audio interface.

Bad Sound in a Pop-Music Recording Session

Sometimes you have bad sound in a pop-music recording session.

MUDDINESS (LEAKAGE)

If the sound is muddy from excessive leakage, try the following:

- Place the microphones closer to their sound sources.
- Spread the instruments farther apart to reduce the level of the leakage.
- Place the instruments closer together to reduce the delay of the leakage.
- Use directional microphones (such as cardioids).
- Overdub the instruments.
- Record the electric instruments direct.
- Use baffles (goboes) between instruments.
- Deaden the room acoustics (add absorptive material or flexible panels).
- Filter out frequencies above and below the spectral range of each instrument. Be careful or you'll change the sound of the instrument.
- Turn down the bass amp in the studio, or monitor the bass with headphones instead.

MUDDINESS (EXCESSIVE REVERBERATION)

If the sound is muddy due to excessive reverberation, try these steps:

- Reduce the effects-send levels or effects-return levels. Or don't use effects until you figure out what the real problem is.
- Place the microphones closer to their sound sources.
- Use directional microphones (such as cardioids).
- Deaden the room acoustics.
- Filter out frequencies below the fundamental frequency of each instrument.

MUDDINESS (LACKS HIGHS)

If your sound is muddy and lacks highs, or has a dull or muffled sound, try the following:

- Use microphones with better high-frequency response, or use condenser mics instead of dynamics.

- Change the mic placement. Put the mic in a spot where there are sufficient high frequencies. Keep the high-frequency sources (such as cymbals) on-axis to the microphones.
- Use small-diameter microphones, which generally have a flatter response off-axis.
- Boost the high-frequency equalization or cut slightly around 300 Hz.
- Change musical instruments; replace guitar strings; replace drum heads. (Ask the musicians first!)
- Use an enhancer signal processor, but watch out for noise.
- Use a direct box on the electric bass. Have the bassist play percussively or use a pick if the music requires it. When compressing the bass, use a long attack time to allow the note's attack to come through. (Some songs don't require sharp bass attacks—do whatever's right for the song.)
- Damp the kick drum with a pillow, folded towel, or blanket, and mike it next to the center of the head near the beater. Use a wooden or plastic beater if the song and the drummer allow it.
- Don't plug an electric guitar directly into a mic input. Use a direct box or a high-impedance input.
- Apply high-frequency boost after compression, not before.

MUDDINESS (LACKS CLARITY)

If your sound is muddy because it lacks clarity, try these steps:

- Consider using fewer instruments in the musical arrangement. Maybe turn down synth pads in the mix.
- Equalize instruments differently so that their spectra don't overlap.
- Try less reverberation.
- Using equalizers, boost the presence range of instruments that lack clarity. Or cut 1 to 2 dB around 300-500 Hz.
- In a reverb unit, add about 45 to 100 msec of predelay.
- Pan similar-sounding instruments to opposite sides.

DISTORTION

If you hear distortion when monitoring the mics in a pop-music recording, try the following:

- Increase input attenuation (reduce input gain), or plug in a pad between the microphone and mic input.
- Readjust gain-staging: set faders and pots to their design centers (shaded areas).
- If you still hear distortion, switch in the pad built into the microphone (if any).
- Check connectors for stray wires and bad solder joints.
- Unplug and plug-in connectors. Clean them with Caig Labs DeoxIT or Pro Gold.

TONAL IMBALANCE

If you have bad tonal balance—the sound is boomy, dull, or shrill, for example—follow these steps:

- Change musical instruments, change guitar strings, change reeds, etc.
- Change mic placement. If the sound is too bassy with a directional microphone, you may be getting proximity effect. Mike farther away or roll off the excess bass.
- Use the three-to-one rule of mic placement to avoid phase cancellations. When you mix two or more mics to the same channel, the distance between mics should be at least three times the mic-to-source distance.
- Try another microphone. If the proximity effect of a cardioid mic is causing a bass boost, try an omnidirectional mic instead.
- If you must place a microphone near a hard, reflective surface, try a boundary microphone on the surface to prevent phase cancellations.
- If you're recording a singer/guitarist, delay the vocal mic signal by about 1 msec.
- Change the equalization. Avoid excessive boost. Maybe cut slightly around 300 Hz if the sound is muddy, or cut around 3 kHz if the sound is harsh.
- Use equalizers with a broad bandwidth, rather than a narrow, peaked response.
- During mastering, try a spectrum analyzer/equalizer such as Harmonic Balancer (www.har-bal.com).

LIFELESSNESS

If your pop-music recording has a lifeless sound and is unexciting, these steps might help you solve it:

- Work on the live sound of the instruments in the studio to come up with unique effects.
- Add effects: reverberation, echo, exciter, doubling, equalization, etc.
- Use and combine recording equipment in unusual ways.
- Try overdubbing little vocal licks or synthesized sound effects.

If your sound seems lifeless due to dry or dead acoustics, try these:

- If leakage is not a problem, put microphones far enough from instruments to pick up wall reflections. If you don't like the sound this produces, try the next suggestion.
- Add reverb or echo to dry tracks. (Not all tracks require reverberation. Also, some songs may need very little reverberation so that they sound intimate.)
- Use omnidirectional microphones.

- Add hard, reflective surfaces in the studio, or record in a hard-walled room.
- Allow a little leakage between microphones. Put mics far enough from instruments to pick up off-mic sounds from other instruments. Don't overdo it, though, or the sound becomes muddy and track separation becomes poor.

NOISE (HISS)

Sometimes your pop-music recording has extra noise on it. If your sound has hiss, try these:

- Check for noisy guitar amps or keyboards.
- Switch out the pad built into the microphone (if any).
- Reduce mixer input attenuation (increase input gain).
- Use a more sensitive microphone.
- Use an impedance-matching adapter (a low- to high-Z step-up transformer) between microphones and phone-jack mic inputs.
- Use a quieter microphone (one with low self-noise).
- Increase the sound pressure level at the microphone by miking closer. If you're using PZMs, mount them on a large surface or in a corner.
- Apply any high-frequency boost during recording, rather than during mixdown.
- If possible, feed recorder tracks from mixer direct outs or insert sends instead of group or bus outputs.
- Use a lowpass filter (high-cut filter).
- As a last resort, use a noise gate.

NOISE (RUMBLE)

If the noise is a low-frequency rumble, follow these steps:

- Reduce air-conditioning noise or shut off the air conditioning temporarily.
- Use a highpass filter (low-cut filter) that is set around 40 to 100 Hz.
- Use microphones with limited low-frequency response.
- See the next section, "Noise (Thumps)."

NOISE (THUMPS)

- Change the microphone position.
- Change the musical instrument.
- Use a highpass filter set around 40 to 100 Hz.
- If the cause is mechanical vibration traveling up the mic stand, put the mic in a shock-mount stand adapter, or place the mic stand on some carpet padding. Try to use a microphone that is less susceptible to mechanical vibration, such as an omnidirectional mic or a unidirectional mic with a good internal shock mount.
- Use a microphone with a limited low-frequency response.
- If the cause is piano pedal thumps, also try working on the pedal mechanism.

HUM

Hum is a subject in itself. See the tips on hum prevention near the end of Chapter 4.

POP

Pops are explosive breath sounds in a vocalist's microphone. If your pop-music recording has pops, try these solutions:

- Place the microphone above or to the side of the mouth.
- Place a foam windscreen (pop filter) on the microphone.
- Stretch a nylon stocking over a crochet hoop, and mount it on a mic stand a few inches from the microphone (or use an equivalent commercial product).
- Place the microphone farther from the vocalist.
- Use a microphone with a built-in pop filter (ball grille).
- Use an omnidirectional microphone, because it is likely to pop less than a directional (cardioid) microphone.
- Switch in a highpass filter (low-cut filter) set around 80 to 100 Hz.

SIBILANCE

Sibilance is an overemphasis of "s" and "sh" sounds. If you are getting sibilance on your pop-music recording, try these steps:

- Use a de-esser signal processor or plug-in. Or use a multiband compressor, and compress only the range from 5 to 10 kHz.
- Place the microphone farther from the vocalist.
- Place the microphone toward one side of the vocalist, or at nose height, rather than directly in front.
- Cut equalization in the range from 5 to 10 kHz.
- Change to a duller sounding microphone such as a ribbon.

BAD MIX

Some instruments or voices are too loud or too quiet. To improve a bad mix, try the following:

- Change the mix. (Maybe change the mix engineer!)
- Compress vocals or instruments that occasionally get buried.
- Change the equalization on certain instruments to help them stand out.
- During mixdown, continuously change the mix to highlight certain instruments according to the demands of the music.
- Change the musical arrangement so that different musical parts don't play at the same time. That is, consider having a call-and-response arrangement (fill-in-the-holes) instead of everything playing at once, all the time.

UNNATURAL DYNAMICS

When your pop-music recording has unnatural dynamics, loud sounds don't get loud enough. If this happens, try these steps:

- Use less compression or limiting.
- Avoid overall compression.
- Use multiband compression on the stereo mix instead of wideband (full-range) compression.

ISOLATED SOUND

If some of the instruments on your recording sound too isolated, as if they are not in the same room as the others, follow these steps:

- In general, allow a little crosstalk between the left and right channels. If tracks are totally isolated, it's hard to achieve the illusion that all the instruments are playing in the same room at the same time. You need some crosstalk or correlation between channels. Some right-channel information should leak into the left channel, and vice versa.
- Place microphones farther from their sound sources to increase leakage.
- Use omnidirectional mics to increase leakage.
- Use stereo reverberation or echo.
- Pan effects returns to the channel opposite the channel of the dry sound source.
- Pan extreme left-and-right tracks slightly toward center.
- Make the effects-send levels more similar for various tracks.
- To give a lead-guitar solo a fat, spacious sound, use a stereo chorus. Or send its signal through a delay unit, pan the direct sound hard left, and pan the delayed sound hard right.

LACK OF DEPTH

If the mix lacks depth, try these steps:

- Achieve depth by miking instruments at different distances.
- Use varied amounts of reverberation on each instrument. The higher the ratio of reverberant sound to direct sound, the more distant the track sounds.
- To make instruments sound closer, use a longer predelay setting in your reverb (maybe 40-100 msec). Use a shorter predelay (under 30 msec) to make instruments sound farther away.

Bad Sound in a Classical-Music Recording

Check the following procedures if you have problems recording classical music:

TOO DEAD

If the sound in your classical recording is too dead—there is not enough ambience or reverberation—try these measures to solve the problem:

- Place the microphones farther from the performers.
- Use omnidirectional microphones.
- Record in a concert hall with better acoustics (longer reverberation time).
- Turn up the hall mics (if used).
- Add artificial reverberation.

TOO CLOSE

If the sound is too detailed, too close, or too edgy, follow these steps:

- Place the microphones farther from the performers.
- Place the microphones lower or on the floor (as with a boundary microphone).
- Roll off the high frequencies.
- Use mellow-sounding microphones (many ribbon mics have this quality).
- Turn up the hall mics (if used).
- Increase the reverb-send level.

TOO DISTANT

If the sound is distant and there is too much reverberation, these steps might help:

- Place the microphones closer to the performers.
- Use directional microphones (such as cardioids).
- Record with a spaced pair of directional mics aiming straight ahead.
- Record in a concert hall that is less live (reverberant).
- Turn down the hall mics (if used).
- Decrease the reverb-send level.

STEREO SPREAD TOO NARROW OR TOO WIDE

If your classical-music recording has a narrow stereo spread, try these steps:

- Angle or space the main microphone pair farther apart.
- If you're doing mid-side stereo recording, turn up the side output of the stereo microphone.
- Place the main microphone pair closer to the ensemble.

If the sound has excessive stereo spread (or "hole-in-the-middle"), try the following:

- Angle or space the main microphone pair closer together.
- If you're doing mid-side stereo recording, turn down the side output of the stereo microphone.

- In spaced-pair recording, add a microphone midway between the outer pair, and pan its signal to the center.
- Place the microphones farther from the performers.

LACK OF DEPTH

Try the following to bring more depth into your classical-music recording:

- Use only a single pair of microphones out front. Avoid multimiking.
- If you must use spot mics, keep their level low in the mix.
- Add more artificial reverberation to the distant instruments than to the close instruments.

BAD BALANCE

If your classical-music recording has bad balance, try the following:

- Place the microphones higher or farther from the performer.
- Ask the conductor or performers to change the instruments' written dynamics. Be tactful!
- Add spot microphones close to instruments or sections needing reinforcement. Mix them in subtly with the main microphones' signals.

MUDDY BASS

If your recording has a muddy bass sound, follow these steps:

- Aim the bass-drum head at the microphones.
- Put the microphone stands and bass-drum stand on resilient isolation mounts (such as a carpet pad), or place the microphones in shock-mount stand adapters.
- Roll off the low frequencies or use a highpass filter set around 40 to 80 Hz.
- Use artificial reverb with a shorter decay time at low frequencies.
- Record in a concert hall with less low-frequency reverberation.

RUMBLE

Sometimes your classical-music recording picks up rumble from air conditioning, trucks, and other sources. Try the following to clear this up:

- Check the hall for background rumble problems.
- Temporarily shut off the air conditioning.
- Record in a quieter location.
- Use a highpass filter set around 40 to 80 Hz.
- Use microphones with limited low-frequency response.
- Mike closer and add artificial reverb.

DISTORTION

If your classical-music recording has distortion, try the following:

- Switch in the pads built into the microphones (if any).
- Increase the mixer input attenuation (turn down the input trim).
- Check connectors for stray wires or bad solder joints.
- Avoid sample-rate conversion.
- Apply dithering when going from 24 to 16 bits.

BAD TONAL BALANCE

Bad tonal balance expresses itself in a sound that is too dull, too bright, or colored. If your recording has this problem, follow these steps:

- Change the microphones. Generally, use flat-response microphones with minimal off-axis coloration.
- Follow the three-to-one rule mentioned in Chapter 7.
- If a microphone must be placed near a hard, reflective surface, use a boundary microphone on the surface to prevent phase cancellations between direct and reflected sounds.
- Adjust equalization. Try a spectrum analyzer/equalizer such as Harmonic Balancer (www.har-bal.com).
- Place the mics at a reasonable distance from the ensemble (too-close miking sounds shrill).
- Avoid mic positions that pick up standing waves or room modes. Experiment with small changes in mic position.

This chapter described a set of standards for good sound quality in both popular- and classical-music recordings. These standards are somewhat arbitrary, but the engineer and producer need guidelines to judge the effectiveness of the recording. The next time you hear something you don't like in a recording, the lists in this chapter will help you define the problem and find a solution.

CHAPTER 13
Digital Recording

Before 1976, all music was recorded on analog tape recorders. They record the audio signal as magnetic patterns that rise and fall in strength as the signal waves do. Digital recorders, on the other hand, store the audio signal as a numerical code of ones and zeros.

Let's venture into the world of digital audio. We'll overview how digital recording works, explore 2-track digital recorders, and explain multitrack digital recorders. Chapter 14 covers computer recording in detail.

DIFFERENCES BETWEEN ANALOG AND DIGITAL

Both analog and digital recorders can sound accurate—the playback signal sounds like the input signal. But there are subtle differences. A digital recording usually is called "clean" because it adds almost no noise or distortion to the input signal. Some analog tape recorders add a little "warmth" to the sound. It's likely due to slight third harmonic distortion, head bumps (bass boost), and tape compression at high recording levels. Compared to analog tape recordings, digital recordings have measurably less hiss, frequency-response errors, modulation noise, distortion, unsteady pitch, and print-through. However, those characteristics are inaudible in the best analog tape recorders when they are properly maintained and aligned. Although older digital recorders sounded harsh compared to analog, they have improved with each generation.

Compared to analog tape recorders, digital recorders and their media tend to be smaller and lower cost. They can record timing information. Also, they have random access so you can locate a particular section of the recording very quickly.

DIGITAL RECORDING

Digital recording does not chop up your music. Many people have the misconception that digital recording breaks up an audio signal into little slices, so that some of the signal is missing. Nope—all of the analog signal is captured and reproduced, and I'll explain how.

FIGURE 13.1
Analog to digital conversion.

Like an analog tape deck, a digital recorder puts audio on a magnetic medium, but in a different way. Here's what happens in the most common digital recording method— pulse code modulation or PCM:

1. The signal from your mixer, preamp, or audio interface (Figure 13.1A) is run through a lowpass filter (anti-aliasing filter), which removes all frequencies above 20 kHz.
2. Next, the filtered signal passes through an analog-to-digital (A/D) converter. This converter measures (samples) the voltage of the audio waveform several thousand times a second (Figure 13.1B).
3. Each time the waveform is measured, a binary number (made of 1s and 0s) is generated that represents the voltage of the waveform at the instant it is measured (Figure 13.1C). This process is called quantization. Each 1 and 0 is called a bit, which stands for binary digit. The more bits that are used to represent each measurement (the higher the bit depth), the more accurate the measurement is.
4. These binary numbers are stored on the recording medium as a modulated square wave recorded at maximum level (Figure 13.1D). For example, the numbers can be stored magnetically on a hard disk.

The playback process is the reverse:

1. The binary numbers are read from the recording medium, such as a hard disk (Figure 13.2A)

First value

1000 1011 1010 1100 1111 1100

0000 −1011 −1110 −1100 −1010 0000

Last value

(A) The numbers are read from the recording medium.

(B) The binary numbers are converted to voltage steps.

(C) The voltage steps are smoothed out (filtered out).

FIGURE 13.2
Digital to analog conversion.

2. A digital-to-analog (D/A) converter translates the numbers back into an analog signal made of voltage steps (Figure 13.2B).

3. An anti-imaging filter (lowpass filter, smoothing filter) smoothes the steps in the analog signal, resulting in the original analog signal (Figure 13.2C).

The curve or shape of the analog waveform between samples is re-created by the anti-imaging filter. Nothing is lost…digital recording does not slice up your music! With recordings made on a CD, the process does filter out signals above 22 kHz, but we can't hear that high anyway.

As an analogy, suppose you want to reproduce the circular shape of a wooden disk.

1. You lay a ruler across the disk every ¼ inch and measure how wide the disk is at each point (Figure 13.3A). That's like sampling.

2. You cut a bunch of wood strips with the same lengths as those measurements, and put the strips side-by-side. They look like stair steps going up and down (Figure 13.3B).

3. You file the raggedy edges of the wood strips so they join together smoothly. That's like the anti-imaging filter. You've just reproduced the shape of the original wooden disk (Figure 13.3C).

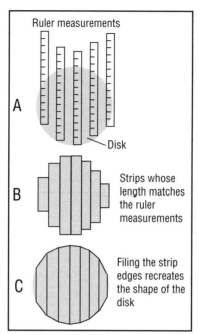

Ruler measurements

A

Disk

B

Strips whose length matches the ruler measurements

C

Filing the strip edges recreates the shape of the disk

FIGURE 13.3
An analogy of digital recording and playback: re-creating a disk shape using wooden strips.

As we said, digital recording reduces noise, distortion, speed variations, and data errors. Because the digital playback head reads only 1s and 0s, it is insensitive to the magnetic medium's noise and distortion. During recording and playback, numbers are read into a buffer memory and read out at a constant rate, eliminating speed variations in the rotating media. Reed-Solomon coding during recording, and decoding during playback, corrects for missing bits by using redundant data.

If a digital recording is on a defective medium such as a scratched compact disc, errors (missing samples) can occur. Usually these errors can be corrected by interpolation. This algorithm looks at data before and after the blank sample and "guesses" what its value should be. If the errors are too extended to correct, the resulting audio has a silent spot or a burst of noise.

Almost all digital recording devices employ the same A/D, D/A conversion process, but use different storage media: a hard-disk drive records on magnetic hard disk, a compact disc and DVD recorder record on an optical disc, a memory recorder records onto a flash memory card, and a sampler records into computer memory. The sound quality of any of these devices depends mainly on its A/D and D/A converters.

Digital audio is recorded on a computer hard drive as a WAV file or AIFF file. Both are standard formats for audio files. WAV is preferred for both PC and Apple; AIFF (Audio

Interchange File Format) is an Apple format. Both formats use linear PCM encoding, with no data compression (explained in Chapter 19). Two WAV formats are Riff and Broadcast Wave, which facilitates interchange of program material between audio workstations.

Bit Depth

As we said, the audio signal is measured many thousand times a second to generate a string of binary numbers (called words). The longer each word is (the more bits it has), the greater is the accuracy of each measurement. Short words give poor resolution of the signal voltage (high distortion); long words give good resolution (low distortion). Bit depth and resolution are other terms for word length.

In our wooden disk analogy, bit depth (word length) corresponds to the accuracy of the ruler measurement: whether we measured to the closest 1/8″, 1/16″, 1/32″, etc.

A word length of 16 bits is adequate (but not optimum) for hi-fi reproduction. It is the current standard for CDs. Most digital recorders offer 24-bit word lengths. More bits sound smoother and more transparent, but need more disk storage space.

Even though CDs are a 16-bit format, they sound better when made from 24-bit recordings. During mastering, you add dither (low-level noise) to the 24-bit recording, then export or save it as a 16-bit recording. You copy the 16-bit recording to CD. The dither helps the 16-bit recording sound more like the 24-bit recording. In other words, dither lets you retain most of the quality and resolution that you recorded at 24 bits, even though the recording ends up on a 16-bit CD. We explain dither later in this chapter.

Suppose you record with a 24-bit word length. All signal levels in that recording have 24-bit resolution. Even low-level signals use all the bits—but most of the bits are zeroes.

Sampling Rate

Sampling rate or sampling frequency is the rate at which the A/D converter samples or measures the analog signal while recording. For example, a rate of 48 kHz is 48,000 samples per second; that is, 48,000 measurements are generated for each second of sound. The higher the sampling rate, the wider the frequency response of the recording.

In our wooden disk analogy, sampling rate corresponds to the number of wood strips you cut to form the disk. If you measure the disk every 1/2″ across, that is like using a low sampling rate. If you measure the disk every 1/8″ across, that is like using a high sampling rate.

The sampling rate for high-quality audio can be 44.1 kHz, 48 kHz, 88.2 kHz, 96 kHz, or 192 kHz. CD quality is 44.1 kHz/16 bits. A 96-kHz sampling rate should be used on DVD. State-of-the-art formats include DXD (Digital eXtreme Definition) at

352.8 kHz/24 bits, and Super Audio CD and linear PCM at 192 kHz/24 bits (but you're more likely to see 96 kHz/24 bits).

According to some sources, higher sampling rates sound smoother and more transparent but need more disk storage space and a faster hard drive. One study in the September 2007 issue of the *Journal of the Audio Engineering Society* has shown that rates higher than 44.1 kHz yield no audible improvement over 44.1 kHz.

Digital recording is based on the Nyquist-Shannon theorem. It defines the "Nyquist frequency" as half the sampling rate. If the sampling rate is 44.1 kHz, the Nyquist frequency is 22.05 kHz.

The theorem proves that the original analog signal can be reconstructed exactly from its samples—if the highest frequency in the signal is less than the Nyquist frequency. That happens when the original signal is sent through a lowpass (high-cut) filter to remove frequencies above the Nyquist frequency.

If the analog signal is not filtered, aliasing occurs: high frequencies above the Nyquist frequency appear as lower frequencies. In other words, inaudible ultrasonic frequencies are converted to audible tones. That's why an anti-aliasing filter is needed.

If you lowpass filter the analog signal to the Nyquist frequency before it is sampled, there is only one waveform that can pass through the sample points, and that is the original analog waveform. Figure 13.1B shows the sample points and the waveform that passes through them.

The harshness of some early digital recorders was not caused by slicing and dicing the signal. It was due to excessive phase shift of the anti-aliasing and anti-imaging filters. Those filters have been much improved, so current digital audio is generally much smoother and more like analog.

Current A/D and D/A converters use a process called oversampling to improve the sound. Oversampling is sampling an audio signal at a higher rate than is needed to reproduce the highest frequency in the signal. For example, sampling a 20 kHz audio signal at 8 times 44.1 kHz is called "8x oversampling." This process is followed by a digital lowpass filter and a gentle-slope analog anti-alias filter. The result is less phase shift and less harshness compared to a steep, "brick-wall" analog filter used alone.

In summary, a digital audio system samples the analog signal several thousand times a second, and quantizes (assigns a value to) each sample. Sampling rate affects the high-frequency response. Bit depth affects the dynamic range, noise, and distortion.

An alternative to PCM is 1-bit (bitstream) encoding with a 2.8224 MHz or higher sampling rate. This is the Direct Stream Digital (DSD) process used in the Super Audio CD and in some Korg portable digital recorders. DSD offers a frequency response from DC to 100 kHz with 120-dB dynamic range and a very smooth, analog-like sound.

Most recent digital devices use delta-sigma A/D converters, which produce 1-bit streams at a very high sample rate (typically using 64x oversampling). Then a digital lowpass filter removes the aliasing and quantizing noise. Finally, a decimation circuit downsamples (converts) the bitstream to 16-bit or 24-bit PCM samples at the desired sample rate (44.1 kHz, 96 kHz, and so on).

With delta-sigma A/D conversion, recorded sample rates above 44.1 kHz do not improve the sound quality because the analog input signal is 64x oversampled no matter what the recorded sample rate is.

In a digital transmission, the two channels of a stereo program are multiplexed. That is, one word from channel 1 is followed by one word from channel 2, which is followed by one word from channel 1, and so on.

Data Rate and Storage Requirements

The data rate of digital audio (in bytes per second) is

$$\text{Bit depth}/8 \times \text{Sampling rate} \times \text{Number of tracks}$$

Divide by 1,048,576 to get megabytes (MB) per second. For example, the data rate of a 24-bit/44.1-kHz recording of 16 mono tracks is

$$(24/8 \times 44{,}100 \times 16)/1{,}048{,}576 = 2\text{MB/sec}$$

Recording digital audio on a hard drive consumes a lot of space. The storage required is

$$\text{Bit depth}/8 \times \text{Sampling rate} \times \text{Number of tracks} \times 60 \times \text{Number of minutes}$$

Divide by 1,048,576 to get MB. Divide MB by 1024 to get gigabytes (GB). For example, suppose you record a concert at 24 bits, 44.1 kHz, 16 tracks, for 2 hours. The hard-drive space needed is

$$(24/8 \times 44{,}100 \times 16 \times 60 \times 120)/1{,}048{,}576 = 14{,}534.9\text{MB or } 14.2\text{GB}$$

Note that the operating system might report the capacity of the disk drive up to 7% lower than it really is.

Table 15.1 in Chapter 15 shows the amount of storage needed for a 1-hour recording. Not shown in Table 15.1, a 1-bit stereo recording at 2.8 MHz consumes about 40 MB per minute or 2.4 GB per hour.

Digital Recording Level

In a digital recorder, the record-level meter is a peak-reading LED or LCD bar graph meter (Figure 13.4) that reads up to 0 dBFS (FS means full scale). In a 16-bit digital recorder, 0 dBFS means all 16 bits are on (1) at the waveform peak. In a 24-bit digital recorder, 0 dBFS means that all 24 bits are on or off.

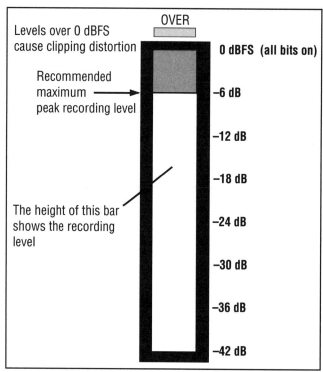

FIGURE 13.4
Digital record-level meter.

The OVER indication means that the input level exceeded the voltage needed to produce 0 dBFS, and there is some short-duration clipping of the output analog waveform. This clipping can sound really nasty. Some manufacturers calibrate their meters so that 0 dBFS is less than 16 or 24 bits on; this allows a little headroom. When you set the recording level, it's a good idea to aim for −6 dB maximum so that unexpected peaks don't exceed 0 dBFS. If you're making a 24-bit recording, the recording level is not very critical because a 16-bit signal is at −48 dBFS!

THE CLOCK

Each digital audio device has a clock or internal oscillator that sets the timing of its samples. The clock signal is a series of pulses running at the sampling rate. When you transfer digital audio from one device to another, their clocks must be synchronized. One device must provide the master clock and the other must be the slave. This clock comes embedded in the digital signal, or comes on a separate wire or connector as a word clock signal.

In your digital devices or DAW, look for a switch or software menu that selects the clock source. The three choices are internal clock, external clock, or word clock.

Internal clock is the A/D converter's clock. External clock (also called digital input) is the clock signal embedded in another device's digital audio signal. Word clock is a separate cable with timing signals. Set the master device to internal clock and set the slave devices to external clock (digital input). If you're using a master word-clock generator feeding other devices' word-clock connectors, set the slave devices to word clock.

Some receiving devices synchronize automatically to the embedded clock from the source device. In that case you don't need to set the clock source.

Your system's A/D converter or interface is usually the best master clock. Connect its word clock output (if any) to the input of a word-clock splitter or distribution unit. Connect a cable from each output of the distribution unit to the word-clock input of each of the other devices. That way all the devices will be synchronized.

Word-clock cables under 25 feet should be unbalanced 75-ohm cables with BNC connectors. Word-clock cables over 25 feet should be balanced 110-ohm AES digital cables with XLR connectors, although regular mic cables can work.

In large recording systems where you mix or assemble digital audio signals from a variety of sources, it's convenient to use a separate master clock such as Apogee's Big Ben, Mutec's iClock, or Drawmer's M-Clock. Most master clocks have several word-clock outputs. Connect each one to each slave device in your studio to keep them in sync. The receiving device's word-clock connector must be terminated properly—refer to the user's manual. If you daisy-chain the word-clock connectors between audio devices (not recommended), only the last device in the chain should be terminated.

DIGITAL AUDIO SIGNAL FORMATS

There are eight major formats for carrying digital-audio signals between various devices: AES3, MADI, S/PDIF, ADAT Lightpipe, TDIF, FireWire, USB, and Thunderbolt. With all these formats you connect a single cable between digital out and digital in. Let's look at each format.

- AES3 (also called AES/EBU or IEC-60958 Type I): This professional format transfers 2 channels of digital audio over a 110-ohm shielded twisted-pair cable with XLR connectors. That is not the same as a mic cable. An AES3 cable can be up to 100 meters long. AES3id uses 75-ohm coaxial cable and BNC connectors for runs up to 1,000 meters. You need one AES3 interface for each pair of channels.
- MADI (Multichannel Audio Digital Interface or AES10-2003): This professional format transports 28, 56, or 64 channels up to 24 bit/96 kHz on 75-ohm coaxial cable or fiber-optic cable. It's often used for linking large mixing consoles to digital multitrack recorders. The cable may be up to 3000 meters long.
- S/PDIF Sony/Philips Digital Interface (also called IRC-0958 Type II): This is a 2-channel consumer or semipro format. It uses a single 75-ohm coaxial cable

up to 10 meters long with RCA or BNC connectors, or a fiber-optic cable with TOSLINK connectors. Optical interfaces prevent ground loops and cable losses.

- ADAT Lightpipe: The Alesis ADAT digital multitracks use a Lightpipe, which sends 8 channels of digital audio on a single optical cable with TOSLINK connectors. Every 8 channels of transfer require a separate Lightpipe cable. Data transfer is up to 24 bit/48 kHz or 24 bit/96 kHz with half the number of channels.
- TASCAM TDIF (TASCAM digital interface): The TASCAM DA-88 and similar modular digital multitracks use a multiconductor cable with standard DB-25 connectors. TDIF sends 8 channels of digital audio in and out on a single cable, which can be up to 5 m long.
- USB (Universal Serial Bus), FireWire (IEEE 1394) and Thunderbolt are standard Mac/PC protocols for high-speed serial data transfer between a computer and an external device, such as a hard drive, USB drive, MIDI interface or audio interface. More details are in Chapter 14 under the heading "FireWire, USB, or Thunderbolt Audio Interface."

AES3, S/PDIF, and ADAT Lightpipe are self-clocking systems: the clock is embedded in the signal and marks the start time of each sample. MADI, TDIF, and ADAT Sync carry a separate word-clock signal on a separate connector or wire. Note that FireWire and USB signals do not have clock information; they are reclocked at the receiving device.

Converting Signal Formats

AES and S/PDIF signals are similar but not necessarily compatible. You can convert one to the other using a format converter. Some sound cards and digital mixers do this conversion. Lightpipe and TDIF signals can be converted as well.

Some digital audio devices do not implement AES or S/PDIF correctly, so they do not interface with some other devices.

DITHER

Your HD recorder, software, and digital mixer may run at 24 bits, but the result ends up on a 16-bit CD. When you save a 24-bit audio file as a 16-bit file to transfer to CD, those last eight bits are truncated or cut off. The result may be a grainy static sound at very low levels. This distortion can be prevented by adding low-level random noise (dither) to the signal. Let's explain.

A 24-bit resolution can accurately capture the quietest parts of a musical program: very low-level signals such as the end of long fades and reverb tails. But truncation of that signal to 16 bits makes those low-level signals sound grainy or fuzzy, because 16 bits is a less accurate measurement of the analog waveform than 24 bits. This fuzzy sound, called quantization distortion, doesn't exist at normal high levels.

What causes this distortion? Each digital word is made of a certain number of bits. During quantization, the A/D converter assigns the closest possible digital number to represent the measured voltage of each sample. The last or rightmost bit (least significant bit or LSB) switches on or off depending on whether the converter rounds the word value up or down. If this switching occurs in the 16th bit, it may be faintly audible as a fuzzy noise during quiet passages.

Also, a 24-bit recording has 256 possible levels in the lower 8 bits. But after the signal is truncated to 16 bits, that resolution is lost.

To solve this problem, dithering adds random noise (random 1s and 0s) to the lowest 8 bits of the 24-bit signal (at about −100 dB) before they are truncated to 16 bits. That noise modulates the 16th bit with some 24-bit information (bits 17 through 24) in the form of pulse-density modulation. The average value of that modulated square wave is recovered by a lowpass filter. Then most of the 24-bit sound quality is restored, and the quantization distortion changes to a smooth hiss.

To make that added hiss less obvious, noise shaping is used. Noise shaping applies an oversampling filter to the noise, which reduces its level in the midrange where our ears are most sensitive, and increases its level in the high frequencies where it's less audible.

Compared to a truncated signal, a truncated-and-dithered signal sounds slightly more transparent. Fades and reverb tails sound smoother, and there's more sonic detail and depth. Signals below the noise floor become audible.

For best sound quality, apply dither only once when you convert a high bit-depth source to its 16-bit CD format. For example, record at 24 bits and stay there through the entire project. Do not redither material that has already been dithered—switch off any dithering. Add dither as the very last step when you export your stereo mix to 16 bits.

To hear the effects of dithering, start with a clean 24-bit recording, reduce its level 50 dB in your editing software, and export it as a 16-bit file. Export it in three ways: without dither, with dither, and with noise shaping added. Next, normalize the exported recordings so that the highest peak hits 0 dBFS. Then listen to the resulting 16-bit files at high level over headphones. Compare the processing and use whatever sounds best. Generally, a signal that is truncated with no dither is accompanied by a rough, grainy noise. A signal that is truncated and dithered is accompanied by smooth, quiet hiss or silence.

With good dithering algorithms, it's possible to preserve most of the 24-bit quality (ultra-low distortion, fine detail, and ambience) when converting to 16 bits. One such system is the POW-r™ Psychoacoustically Optimized Word-Length Reduction algorithm from the POW-r Consortium LLC. It reduces the high-resolution, higher word lengths (20 to 32 bit) to a CD-standard, 16-bit format while retaining the transparency of the high-resolution recording. In other words, the 16-bit CD sounds like the original 24-bit recording.

JITTER

Accurate A/D and D/A conversions rely on the clock precisely sampling the analog signal at equal time intervals. Jitter is unstable timing of samples that occurs in A/D and D/A conversion. Any change between the sample times, even nanoseconds, creates amplitude errors—small changes in the audio waveform's shape—resulting in a slight veiling of the sound (low-level distortion or noise). Some researchers claim that a jitter spec under 50 to 500 picoseconds is inaudible.

During A/D and D/A conversion, one cause of jitter is analog noise and crosstalk in the recording system. They affect the switching times and switching threshold of the clock, causing frequency modulation of the clock. They also affect the analog filters and oscillators used in the clock's phase locked loops (PLLs).

Jitter is also caused by inadequate cables. They pick up hum and noise, and introduce phase shift and high-frequency attenuation, which degrade the timing of the digital signal. However, the AES3 and ADAT interfaces can handle a lot of timing variation, so cable-induced jitter is almost never a problem.

In theory, jitter does not occur during real-time data transmission over FireWire or USB because they are clockless systems. In other words, FireWire and USB connections do not introduce jitter on their own. FireWire data transfers are "isochronous," which means the data must be delivered at a certain minimum data rate. Digital audio and video signal streams require isochronous data flow to ensure that audio data is delivered as fast as it is played and recorded.

However, there's a potential problem. Part of the FireWire and USB signals are sync-message packets. FireWire and USB interfaces can slightly vary the timing of those messages. Then, when the receiving device extracts the clock with a phase-locked loop, jitter can be created.

To reduce jitter:

- Use high-quality clock sources with low jitter specs (under 50 picoseconds). Some audio interfaces and A/D converters are specially designed to reduce jitter, and usually include a PLL.
- Use high-quality, well-shielded cables designed for digital signals, and as short as possible. Don't use mic cables for AES3; don't use hi-fi cables for S/PDIF.
- Keep analog and digital cabling separate.
- Use your audio interface or A/D converter as the master clock. Do not drive it from an external clock source.

DIGITAL TRANSFERS AND COPIES

When you send a digital audio signal in real time from one device to another, they must be set to the same sampling rate. For example, if the digital signal from a multitrack recorder is 48 kHz, the digital mixer it feeds must also be set to 48 kHz. In a digital transfer, the sending device usually is the master, and the receiving device is the slave. Normally you can set the slave to automatically lock onto the sampling rate of the master. Set the master device to Internal Clock, and set the slave device to External Clock. Or set the slave's clock source to whatever device the master is.

The word length (number of bits) of digital transfers is less critical. As long as the two machines handle the same digital format (AES, S/PDIF, etc.), transfers of any word length can be done.

It's okay to send a lower bit signal to a higher bit device. For instance, feed a 16-bit signal from a DAT into a 24-bit digital mixer. The mixer will add more zeroes to fill out the digital word, with no effect on sound quality. It's also no problem to feed a device more bits than it can hold. The receiving device ignores the extra bits. For example, if you send a 24-bit signal to a 16-bit recorder, the last 8 bits will be truncated or cut off.

Truncation adds slight distortion, but this can be reduced by adding dither—low-level noise—to the signal before truncating. So it's best to keep the word length as long as possible as you're working on a project. Always apply dither before you reduce the word length, and only then.

What if you have a 48 kHz recording that you want to release on CD? The CD mastering engineer can (1) convert the 48 kHz digital signal to 44.1 kHz, which might degrade the sound, or (2) use the analog output of his or her playback machine, and convert the analog signal to digital at 44.1 kHz. This may or may not sound better.

Digital audio transfers to or from DAT or MDM tape, and to or from CD-Rs, might or might not create perfect clones because of inadequate error correction or interpolation errors.

When you copy a WAV or AIFF file rather than a digital audio signal, the copy is a perfect clone of the original file. The file transfers much faster than the real-time playback of the signal. Flawless file copies can be made in these ways:

- Inside a computer from one hard drive location to another
- From one computer's hard drive to another computer's hard drive
- Via Ethernet, USB, FireWire, or the Internet
- Between computers via data CD-R, Jaz, or Zip drive; floppy disks; or flash memory cards

The following data compression algorithms don't lose any data, so they don't degrade audio quality:

- PKWare's PKZIP
- WinZip Computing's WinZip
- Aladdin Systems' StuffIt

2-TRACK DIGITAL RECORDERS

Now that we have overviewed digital recording, let's look at digital recorders. They come in 2-track and multitrack formats. The 2-track formats are

- Digital audio workstation; this is a computer (usually a laptop) with a sound card or other audio interface, plus recording software
- CD-R recorder
- Flash memory recorder (handheld SD recorder)
- Apple iOS device running a recording app (covered in Chapter 4)

We'll look at each one. No matter which format you use, the sound quality depends on the A/D converter and mic preamp (if any).

The Digital Audio Workstation

The digital audio workstation (DAW) is a computer running recording software with a connected audio interface such as a sound card (Figure 13.5). A DAW allows you to record, edit, and mix audio programs entirely in digital form. It can store up to several hours of digital audio or MIDI data. You can edit this data with great precision on a computer monitor screen. What's more, you can add digital effects and perform automated mixdowns. We'll cover DAWs in detail in Chapter 14.

A laptop computer can make a great portable 2-track digital recorder. Just add some recording software and an audio interface connected by USB or FireWire.

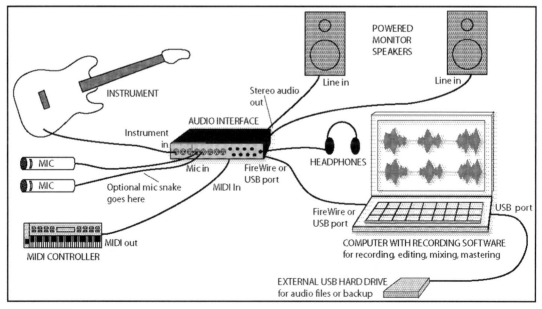

FIGURE 13.5
A computer digital audio workstation (DAW).

CD Recordable (CD-R)

Another form of digital recording is the compact disc. On your own desktop, you can burn a CD by using a CD-R recorder (CD burner). It's exciting to hear one of these CDs playing your music with the purity of digital sound. The sound quality meets or exceeds CD standards.

CD-R stands for Compact Disc-Recordable. This optical medium is a write-once (non-erasable) format. CD-RW stands for Compact Disc ReWritable; you can erase it and record a new program. Although any modern CD player or CD-ROM drive will play a CD-R, many CD players—especially older ones—can't play CD-RW discs. Most new CD-ROM drives support CD-RW, but not all will read CD-RW discs at full speed. CD-RW blank discs cost more than CD-R blank discs.

How can you use the CD-R format? You could make demos for your own band, or make a one-off copy of your stereo mixes for clients. Use the CD-R as a premaster to send to a CD replicator. Another function is to compile sound libraries of production music, samples, and sound effects. In addition to recording audio, CD-Rs can store data files such as WAV, AIFF, or MP3 files to send to other studios. If handled and stored with care, the CD format is a dependable storage medium, so it's a great way to archive your recordings.

COMPUTER CD-R BURNER

Most computers come with a CD burner built in, or you can buy one and install it. Cost is about $35. The unit plugs into an EIDE or SATA connector in your computer. It can write audio or computer data. Its blank discs, called "CD-R" or "Data CD-R," cost less than 35 cents in quantity. (Music CD-Rs will work in a computer CD burner as well.) Disc length is 74 minutes (650 MB) or 80 minutes (700 MB). Some discs permit recording at up to 52 times real time if your computer and hard drive are fast enough.

The CD burner also requires a CD recording program, usually packaged with the recorder or computer. You don't have to use that program; other ones are available that you might prefer. Be sure the program is compatible with your recorder. You'll also need a sound card or audio interface and some software to record audio onto hard disk.

Multidrive CD burners let you record or copy several CDs at once.

CD-R TECHNOLOGY

While conventional CD players follow the Sony-Philips Red Book standard, CD-Rs conform to the Orange Book Part II standard. Once recorded, a CD-R disc meets the Red Book standard.

A recordable CD is the same size as a standard CD, but it is more colorful. On top is a metal reflective layer; on the bottom is a recording layer made of blue cyanine dye or yellow (gold) phthalocyanine dye. The blue layer appears green because of the gold

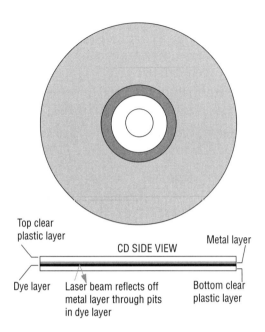

Top clear
plastic layer

CD SIDE VIEW

Metal layer

Dye layer   Laser beam reflects off
metal layer through pits
in dye layer

Bottom clear
plastic layer

FIGURE 13.6
CD-R construction.

layer behind it. Yellow dye lasts a little longer in accelerated aging tests, and it may work better with high-speed CD-R drives. A few other colors are available as well.

A blank CD-R is made of four layers (Figure 13.6):

1. Clear plastic (protects the metal layer)
2. Metal layer (gold, silver, or silver alloy, which reflects the laser light)
3. Dye (for the recording)
4. Clear plastic (protects the dye layer)

The dye fills a spiral groove which is etched in the bottom clear-plastic layer. This groove guides the laser.

To record data on a disc, the laser melts holes in the dye layer. The plastic layer flows into the holes to form pits. During playback, the same laser reads the disc at lower power. At each pit, laser light reflects off the metal layer. The reflected light enters the laser reader, which detects the varying reflectance as the pits go by.

In contrast with a standard CD, a CD-R disc has two more data areas:

- The program calibration area (PCA). The CD recorder uses this area to make a test recording, which determines the right amount of laser power to burn the disc (4 to 8 milliwatts).
- The program memory area (PMA). This area stores a temporary table of contents (TOC) as the CD-R tracks are assembled. The TOC is a list of the tracks, their start times, and the total program time. The recorder uses the PMA for this information until it writes the final TOC.

According to CD-R manufacturers, the expected lifetime of a CD-R is about 70 to 100 years with careful handling. CD-RW life expectancy is claimed to be about 25 years. Avoid sunlight and high temperatures. Be sure not to damage the label side of the disc by writing on it with a ballpoint pen or pencil—use a soft felt-tip pen that is water-based, such as a Sharpie CD/DVD marker. Store CDs vertically and keep them in their cases. Avoid labels because the adhesive can attack the protective plastic layer, and the paper might warp over time.

CD-R SESSIONS, DISC-AT-ONCE, AND TRACK-AT-ONCE

A session on disc is made up of a lead-in, program area, and lead-out. Each session has its own TOC. Each lead-in and lead-out consumes 13 MB of disc space.

With the multisession feature, you can write several sessions on a disc at different times. This feature comes in handy when you need to add information to a disc a little at a time. Only the first session on disc will play on an audio CD player, so the discs are just for your own use and not for distribution. On a multisession disc, you lose about 23 MB of space when you close the first session and about 14 MB for each following session.

CD-R recorders also permit Disc-At-Once recording, in which the entire disc must be recorded nonstop. You can't add new material once you write to the disc. With some software, Disc-At-Once lets you set the length of silence between tracks (pause length), and lets you control how the tracks are laid out on disc. Use Disc-At-Once for all pro audio work.

Most CD-R recorders allow Track-At-Once recording. They can record one track (or a few tracks) at a time—up to 99 tracks. You can play a partly recorded disc on a CD-R recorder, but the disc will not play on a regular CD player until the final TOC is written. Track-At-Once is not recommended for audio because it puts 2-second spaces and (sometimes) clicks between audio tracks.

If you want no pauses between tracks (as on a live album), get a CD-R writer with Disc-At-Once. Also get some software that can adjust the pause length down to zero (such as Express Burn Disc from NCH Software). Another option is software with PQ code editing, which lets you set the start time of each song. Note that a standalone CD-R recorder will copy your edited program as it is, with or without pauses.

A good CD-R burner has a buffer of at least 2 MB. Some burners include SCMS copy code. Be sure that you can return the CD-R burner if it proves to be unreliable.

Using a computer CD-R burner is covered in Chapter 15. For more information on CD-R technology, see www.cdrfaq.org.

Handheld Flash Memory Recorder

This is a portable 2-track digital recorder that records onto a flash-memory card such as a Compact Flash or Secure Digital (SD) card (see Figure 13.7). It records MP3 or

FIGURE 13.7
Zoom H4n flash memory recorder.

uncompressed PCM WAV files (which are CD quality or better). You can copy the recorded files to a computer hard drive through a USB cable.

Available mic connectors are XLR, 1/4″ phone or 1/8″ phone, with or without 48 V phantom power or plug-in power for the mics.

Options are sample rate (44.1 kHz to 96 kHz), battery life (4 to 19 hours), battery type (alkaline, lithium, or NiMH), rechargeable or standard battery, memory size (up to 32 GB), USB version (1, 2, or 3), mics (built-in or not), and extra features such as a guitar tuner, vocal cancellation, pitch change, overdubbing, and real-time audio interfacing to a computer.

Examples of models are in Chapter 4 under the heading "Portable 2-Track Recorder."

MULTITRACK DIGITAL RECORDERS

So far we've covered 2-track digital recorders. Now we'll consider the multitrack digital recorders listed below:

- Computer DAW (covered earlier)
- Hard-disk recorder
- Recorder-mixer

Hard-Disk (HD) Recorder

This digital multitrack recorder was described in Chapter 4 under the heading "Separate Multitrack Recorder and Mixer."

Recorder-Mixer (Digital Personal Studio)

This device was explained in Chapter 4 under the heading "Digital Multitracker (Recorder-Mixer)." An example is shown in Figure 13.8. Listed below are some features to look for.

- Type of analog inputs and outputs. Balanced XLR or TRS phone-jack mic connectors are preferred over unbalanced (TS) phone-jack connectors. Balanced connections let you run longer mic cables without picking up hum.
- Type of digital inputs and outputs. S/PDIF is a common format.
- Number of mic inputs: 2 and up. If you're recording an entire band or drum set at once, you'll need at least 8 mic inputs.
- Number of mixer channels: 8 and up.

FIGURE 13.8
TASCAM DP-03, an example of a digital personal studio.

- Number of tracks: 4 to 36. The more tracks you have, the more instruments and vocals you can record on individual tracks.
- Number of tracks that can be recorded simultaneously: 8 to 24. If you want to record an entire band at once, you need to record many tracks at the same time.
- Analog-to-digital converter's bit depth: 16 to 24 bit. The more bits, the higher the sound quality.
- Analog-to-digital converter's sampling rate: 44.1 to 96 kHz. The higher the sampling rate, the higher the sound quality, but 44.1 kHz is often good enough for projects that will end up on a CD. The latest units record with 24-bit resolution and up to 96-kHz sampling rate. Some record 24 tracks of 24-bit programs at sample rates of 44.1 and 48 kHz. They record 12 tracks of 24-bit recording at 88.2 and 96 kHz.
- Number of built-in digital effects: Up to 8 stereo or 16 mono. Examples of effects are reverberation, echo, flanging, chorus, and compression. They add sonic excitement and a professional touch to your productions. Some recorder-mixers have an expansion card that accepts downloadable plug-in effects.
- Types of EQ: 2- or 3-band, fixed or parametric. EQ means tone control. A 2-band EQ adjusts bass and treble, and a 3-band adjusts bass, midrange, and treble. A fixed-frequency EQ is less flexible than a parametric or sweepable EQ, which lets you adjust the frequency range you want to work on.
- Phantom power for condenser mics.
- Backlit LCD display (bigger is better) with a waveform display.
- Built-in CD burner.
- Built-in microphone.
- Backup or transfer recordings to computer via USB.
- Data compression: recordings can be made in MP3 format as well as WAV format. The WAV format is preferred for higher sound quality.
- Footswitch so you can record over mistakes (punch-in) with your foot while playing your instrument.
- Pitch control. Adjust the recording speed up or down so you can match the pitch of recorded tracks to the pitch of new instruments to be recorded.
- Backup options: removable hard drive or CD-R.
- Levels of undo: up to 1000. The Undo function lets you undo an editing change you just did. If you don't like how an edit sounded, you can press Undo and go back to the way things were before you made the change.
- Number of locate points: up to 1000. A locate point is a point in time in the recording that you can have the recorder memorize and locate later. For example, you might want to mark the time where the beginning of a song is, so you can tap a button and go there instantly.
- Expansion ports for SMPTE and MTC sync, MDM interface, and extra ins and outs. Beginners probably don't need these features, but they let you connect your multitracker to other devices.

- MIDI implementation: MIDI Machine Code (MMC), Control Change (CC). These features let you control the mixer via MIDI commands.
- Types of synchronization: SMPTE, word clock, MTC, MIDI Clock with Song Position Pointer, ADAT, DA-88, and RS-422. These are various formats that you might want your multitracker to sync with.
- Jog/shuttle wheel to "scrub" audio—play it slowly forward and backward to locate an edit point.
- Remote control.
- Automated mixing. A hard-drive or memory recording in the multitracker remembers your song-mix settings and mix changes. The next time you play the song, the mixer resets to those settings automatically. Some multitrackers have motorized faders, so that the faders move up and down just as they did when you adjusted them while mixing.
- Editing (cut and paste, etc.). Editing lets you do all sorts of things with the song arrangement. You might remove parts of songs that you don't want to keep, loop a drum part so that it repeats continuously, or copy a chorus to several points in a song.
- Video output that plugs into a computer monitor screen. During editing, it's a lot easier to see details in the sound waveform if you can plug a large monitor into your multitracker.
- Mastering tools and CD-burning ability. In mastering you compile a playlist of song mixes from which you can create an album. Some multitrackers include programs that let you record your songs on a CD-R recorder, either external or built-in.
- Number of virtual tracks (recordings of separate takes of the same instrument): 8 and up.

A virtual track is a 1-channel recording of a single take or performance on a random-access medium. Most random-access recorders let you record several virtual tracks or takes of a single instrument, then select which take you want to hear during mixdown. For example, suppose you have a 16-track hard-disk recorder. You might record 8 or more virtual tracks (takes) of a vocal on hard disk. Then during mixdown, choose which virtual track will play as one of those 16 tracks. Some recorders can be set up to select parts of different virtual tracks during playback—a process called comping. For example, play vocal take 15 during the verse, play vocal take 3 during the chorus, and so on.

In other words, you can associate several virtual tracks, or takes, with a single "real" track. You might play 8 tracks at once, and you can choose which virtual track (take) plays on each track. For example, you can record several takes of a guitar solo—keeping each one—and choose the best take during mixdown. You also can create a composite track, which is made of the best parts of several takes. All hard-drive recorder-mixers let you record virtual tracks.

Pros and Cons of Three Multitrack Recording Systems

Let's compare a hard-drive recorder, recorder-mixer, and computer DAW.

Multitrack Hard-Drive Recorder and Separate Mixer

Pros:

- The recorder is portable—great for on-location recording.
- Real faders and knobs.
- Can export audio data to a computer for editing, or can use its own editing firmware.
- Stable: Less likely to crash than a computer recording system. This is a big advantage both in on-location recording and studio recording.
- The mixer allows easy setup of monitoring and cue mixes.
- The recorder might have higher-quality converters than a computer recording system.

Cons:

- If you want to do automated mixes, this system requires a mixing console with automation, or a computer recording program with automation.
- If you mix with a hardware mixer rather than a computer, you'll also need some hardware effects.
- Lacks the advantages of a computer recording system.

Recorder-Mixer

Pros:

- Easy to use and portable for on-location recording.
- Simple. No need for a computer, external multitrack recorder, or outboard effects.
- Portable—good for on-location recording, but bigger than an HD recorder.
- Less likely to crash than a computer recording system.
- Can export audio data to a computer for editing, or use its own editing firmware.
- All-in-one system: no external cables needed between recorder and mixer or between mixer and effects.
- Low-cost CD backup.
- Automated mixing in some models.

Cons:

- Lacks the advantages of a computer recording system.
- Less flexible than recording software. Not as upgradable. However, some recorder-mixers have expansion cards that accept downloadable plug-in effects.
- May require copying data to a computer for sophisticated editing and plug-ins.
- Sophisticated units can be very difficult for a beginner to figure out.

Computer Recording System (DAW)

Pros:

- Can be low cost if you already have a fast computer (except for the top-level Pro Tools systems). Software effects (plug-ins) cost much less than hardware effects.
- Flexible (due to software upgrades and downloadable plug-ins).
- Everything can be "in the box." The mixer, recorder, effects, and CD burner are all in the computer. Fewer cables are needed.
- Fast backup of files to CD or external hard drive.
- Automated mixing.
- When you quit a project, the computer remembers all your mixer settings, plug-ins, and edits. It's easy to recall a project.
- Sophisticated editing.
- Can be used with your existing analog or digital mixer.
- You record tracks directly to the computer. You don't need to copy external recordings to the computer before editing and mastering them.
- It's easy to find parts of a song by looking at the tracks on screen.

Cons:

- Computer may crash.
- Not as "hands on" as a recorder-mixer, but a controller surface can help with this.
- Not easily portable except in laptop systems.
- Requires a fast computer with a large hard drive optimized for digital audio.
- Requires a lot of computer tweaks and strong computer skills (but this is a plus for some users).

As we've seen, there is a wide variety of digital recording formats. Learn all you can about digital technology, then choose the formats that meet your needs.

BACKUP

The hard drive on which you store your projects will eventually crash. So it's vital to back up your recordings' WAV files and session files after each session. That is, copy them to another storage medium.

Some backup media are CD-R, CD-RW, hard drive (internal or external FireWire/USB), DVD-R, and DVD-RW. Some engineers prefer to archive audio programs on analog tape as well as digital media. For additional backup recommendations, see the NARAS Technical Guidelines for Producers and Engineers at http://www.grammy.org/recording-academy/producers-and-engineers/guidelines.

Also, see the backup article at www.emusician.com/news/0766/backup-your-backup/144301. It covers RAID drives as well as other backup systems. RAID stands for

"Redundant Array of Independent Disks." It's a group of two or more hard drives that can be configured in a variety of ways. For example, you can record on two hard drives simultaneously so you always have a backup if one drive crashes.

Clone your system drive using a program such as Casper. If your hard drive crashes during a session, you can replace it with the clone drive, reboot, and be back in business.

CHAPTER 14

Computer Recording

With recording software and an audio interface, you can turn your computer into a powerful digital recording studio. This digital audio workstation (DAW) lets you record dozens of audio tracks, edit them, add effects, do a mixdown with automation, and burn a professional-quality CD—all in your computer. The cost is only a few hundred dollars.

In addition to recording audio, most DAW software can act as a sequencer by recording MIDI data. You can record, edit, and play both audio and MIDI tracks in the same program.

Another function of a DAW is to edit tracks that were originally recorded on a hardware multitrack recorder. You can transfer eight or more tracks at once to your computer by using a sound card with Alesis Lightpipe or Tascam TDIF connectors, or with an Alesis FirePort that works with an Alesis HD24 hard-disk recorder.

A DAW has three parts (Figure 14.1):

1. A fast computer with lots of memory and a large hard drive
2. An audio interface to get audio and MIDI into and out of your computer
3. Recording software

You also need some powered monitor speakers and at least one mic. Some optional extras are a control surface and DSP cards. These are explained later under those headings.

BASIC OPERATION

When you launch the recording software, you see simulated tracks and recorder transport buttons such as record, stop, and play. You also see a mixer with virtual controls: simulated faders, knobs, buttons, and meters.

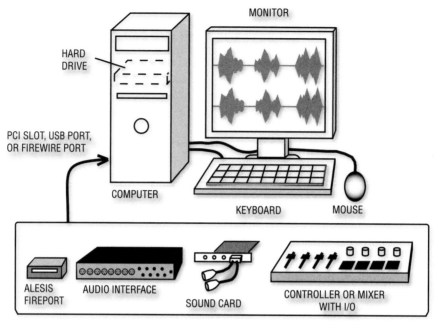

FIGURE 14.1
A computer DAW.

Most DAW software has a few windows or views (Figure 14.2). You view and manipulate the tracks in the Track window, do edits in the Track or Edit window, and adjust mixer controls in the Mixer window. Each window can be opened to fill the entire screen.

Some engineers prefer to work only in the Track window. That's possible if you can view each track's fader along with the tracks. You can control effects and automation in the track view as well.

Recording and Playback

Here's how the system operates. Audio from your mic, instrument, mixer, or mic preamp goes to the inputs on the audio interface, which converts the audio to computer data and sends it to the computer. The software lets you record this data on the computer's hard drive. During playback, the recorded data streams from the hard drive into the interface, which converts the data back into audio at the interface outputs. Monitor speakers connected to those outputs play the audio.

Audio data on the hard drive is read by an electromagnetic head. Because the head can be controlled to jump to any location on disk, it has random access: it can instantly

FIGURE 14.2
An example of DAW software windows.

locate any part of the audio program. As the head is jumping around, the data pieces it picks up are read into a buffer memory, then read out at a constant rate.

The waveform of the recorded audio appears on your monitor screen (Figure 14.3). You can zoom out to see the entire program, or zoom in to see individual samples.

Like a multitrack tape recorder or hard-disk recorder, a DAW has several tracks to record on. You might record a MIDI drum pattern on track 1, a bass audio signal on track 2, guitar on 3, keys on 4 and 5, vocals on 6, and so on. During playback those tracks mix or combine into a stereo signal that plays through your audio interface.

Editing

Editing is a wonderful feature of all computer recording programs. When you edit audio, you select a part of the audio waveform on a track (or tracks). Then you cut, copy, or paste that audio, or slide it in time. It's like a word processor for audio.

FIGURE 14.3
A waveform editing screen (SONAR Producer).

Editing lets you clean up noises, replace bad notes with good ones, fix bad timing, or create a stuttering effect. With editing you can rearrange song sections or make choruses repeat without having to record each one. You might fade out the end of a song or crossfade between two songs. We cover editing in detail later in this chapter under the heading "Editing Tips."

Mixdown

Once all your tracks are recorded and edited, it's time for mixdown. Here is the general procedure:

1. Adjust each on-screen fader with your mouse to set the level of each track until you create a good balance among tracks.
2. Make selections with the mouse to add panning, EQ, compression, and effects to various tracks.
3. Set up automation (envelopes) so that the computer remembers your mix moves and reproduces them each time you play the mix.
4. Once your mix is perfected, export it to a stereo WAV file or AIFF file on your hard drive. Within your project folder, you might create two folders called "mixes-24bit" and "mixes-16bit," and export all your mixes there for easy locating. Name them according to the song title. Use the 16-bit mixes to burn reference CDs; use the 24-bit mixes for mastering.
5. Repeat steps 1 to 4 for all the songs in a demo or album.
6. Burn a CD of the finished mixes or convert them to MP3 files.
7. Alternatively, import the mixes one after the other into a stereo track. Put the mixes in order with some silence between them. Burn a CD of the ordered mixes.

Table 14.1	Hard-Drive Storage Required for a 1-Hour Recording		
No. of tracks	Bit depth	Sampling rate	Storage needed
2	16	44.1 kHz	606 MB
2	24	44.1 kHz	909 MB
2	24	96 kHz	1.9 GB
8	16	44.1 kHz	2.4 GB
8	24	44.1 kHz	3.6 GB
8	24	96 kHz	7.7 GB
16	16	44.1 kHz	4.8 GB
16	24	44.1 kHz	7.1 GB
16	24	96 kHz	15.4 GB
24	16	44.1 kHz	7.1 GB
24	24	44.1 kHz	10.7 GB
24	24	96 kHz	23.3 GB

With this basic understanding, let's take a closer look at DAW components: the computer, audio interface, and recording software.

THE COMPUTER

Either Mac or PC will give great results with audio software. Just be sure that the software you want to use is compatible with your computer platform.

In order to play a lot of tracks and effects in real time, you need a computer with a fast central processing unit (CPU); lots of RAM; and a large, fast hard drive. A minimum system would be a 2 GHz CPU, 2 GB RAM, and an 80 GB hard drive. Two hard drives are faster: one for system files and programs, and another for audio data. The drive should be capable of high sustained transfer rate (throughput). Current SATA revision 2 drives can sustain up to 300 MB/s. SATA revision 3 drives can sustain up to 600 MB/s which are both more than fast enough for multitrack recording.

Multitrack audio consumes a lot of disk storage space. Table 14.1 shows the amount of hard-drive space needed for a 1-hour recording with various recording formats.

If you plan to record a 1-hour album with about four takes per song, multiply the "Storage Needed" by four. In addition to the disk storage for each song's tracks, you will need about 30 to 200 MB per song mix, and up to 700 MB for an 80-minute CD album of the song mixes.

AUDIO INTERFACE

Once you have a fast computer with a large hard drive, you need a way to get audio signals into and out of the computer. An audio interface does the job. Four types of interface are listed below, and we'll look at each one (see Figure 14.1).

- Sound card (also called a PCIe audio interface)
- FireWire, USB, or Thunderbolt audio interface.
- Control surface with I/O (input/output connectors).
- Alesis FirePort. This converts WAV files from an Alesis HD24 FST-formatted hard drive to FireWire, and sends the data to your computer. FireWire (IEEE-1394) is a high-speed data link between digital devices.

Sound Card (PCIe audio interface)

The simplest form of interface is a 2-channel sound card, which plugs into a PCIe user slot in your computer's motherboard. Low-cost sound cards have unbalanced 1/8-inch (mini) phone jack connectors, which include a mic input, stereo line input, and stereo line output. Generally, the sound quality and connectors of low-cost cards are not up to professional standards.

The next step up is a sound card with 1/4-inch TRS (tip-ring-sleeve) connectors (Figure 14.4) or XLR connectors on cables. It offers 2 to 8 balanced inputs. Current high-quality sound cards can record with 24-bit resolution. Many sound cards have MIDI connectors and an onboard synthesizer. Some have a FireWire port. Examples of high-quality sound cards are RME, Frontier Design, M-Audio, Lynx Studio Technology, E-MU, and Echo Digital Audio.

FIGURE 14.4
A sound card.

Choose either a 2-channel or multichannel card. A 2-channel sound card is adequate if you are recording one instrument at a time (such as a vocal, sax, keyboard, guitar, or a mix of several drum mics). You can overdub more parts using the same stereo input on different tracks. Also, a 2-channel sound card is sufficient if you are using an Alesis FirePort, which copies several Alesis HD24 recorder tracks to your hard drive, bypassing the sound card. In that case, you'd use the sound card just for monitoring.

You'll need a multichannel sound card (or a multitrack hard-disk recorder) if you want to record several instruments at once, such as a band or individual mics on a drum set. This type of card has several connectors on short cables, one per channel. Each instrument's signal goes to a separate channel in the sound card, and each channel's signal is recorded on a separate track. You can install two or more sound cards side-by-side to get more channels.

Check the sound card's website to make sure that it is compatible with your computer and recording software.

Here are some desirable features to look for when shopping for a sound card:

- 24 bit, 44.1 kHz minimum, full duplex recording. A full duplex card can record and play back simultaneously. The card works on two DMA channels.
- 85 dB or greater signal-to-noise ratio, 20 Hz to 20 kHz frequency response (+/–0.5 dB or less).
- XLR connectors, 1/4-inch TRS phone jacks, RCA phono jacks, MIDI connectors.
- Operates at +4 dBu as well as -10 dBV.
- Comes with drivers that work with your operating system and your recording software.

If the sound card includes an onboard synthesizer chip, look for these features:

- General MIDI (GM) compatible
- MPU-401 MIDI interface compatible
- Wavetable synthesis (better sound than FM synthesis)
- Programmable synth patches
- 24-note polyphony minimum
- 2 MB of wavetable ROM or RAM minimum

Some PCIe cards have an I/O box, in which all the connectors are in a common chassis. See Figure 14.5. Because the interface's analog circuits are outside of the computer, they tend to pick up less computer electrical noise than analog sound cards do. This interface accepts analog audio or MIDI signals, converts them to computer data, and sends the data to the computer via a card in a PCIe slot.

FireWire, USB, or Thunderbolt Audio Interface

This is a chassis with 2 to 16 mic preamps, an A/D converter, and a USB, FireWire, or Thunderbolt port that sends digital audio on a single cable to your computer

FIGURE 14.5
E-MU 1616 M, an example of a PCIe audio interface I/O box.

FIGURE 14.6
PreSonus FireStudio Project, an example of a FireWire audio interface.

(Figure 14.6). On the same cable, the interface also receives digital audio from the computer. Some units have MIDI ins and out, and a high-impedance instrument input for an electric guitar, electric bass, acoustic guitar pickup or synth.

These audio interfaces connect to a mating connector on the outside of your computer. To install a PCIe interface you must open the computer, although PCIe devices can be connected via Thunderbolt through the use of an external chassis such as the Magma ExpressBox 3T.

Some makers of USB/FireWire audio interfaces are PreSonus, MOTU, Apogee, RME, M-Audio, Focusrite, Roland, Akai, Tascam, Mackie, and E-MU.

A few mixers could be called audio interfaces because they include a FireWire or USB port. It sends the output signal of each mic preamp to a separate track in your recording software, and returns the stereo mix back to the mixer for monitoring. This can be a good system for recording a band because it offers easy setup of monitor mixes with effects. Chapter 4 lists some models of audio interfaces and mixers under the heading "Computer Digital Audio Workstation."

A low-cost alternative to an audio interface is a USB microphone, or else a standard microphone plugged into an XLR-to-USB mic adapter. It lets you record the signal of one microphone to your computer via USB.

FIGURE 14.7
USB port and cable connector.

FIGURE 14.8
FireWire port and cable connector.

Figure 14.7 shows a USB port and icon.

FireWire comes in two speeds: FireWire 400 (IEEE 1394), which runs at 400 Mbps (megabits per second), and FireWire 800 (IEEE 1394b) which runs at 800 Mbps (100 MB per second). In contrast, USB 2.0 high-speed is 480 Mbps. USB 3.0 (SuperSpeed USB) is 10x faster than USB 2.0, running at 5 Gbps. It might replace FireWire, although Thunderbolt is a stronger contender.

You can add a FireWire port to your computer with a $35 FireWire card, and some computers have FireWire built in. To prevent data errors, get a FireWire card with a Texas Instruments (TI) chipset. Figure 14.8 shows a FireWire port and icon.

Thunderbolt is Intel's new ultra high-speed data transfer cable running at 10 Gbps (Figure 14.9). That's 20 times faster than USB 2.0 and 12 times faster than FireWire 800. Thunderbolt adapts to PCI Express controllers and provides bidirectional data flow (as do USB and FireWire).

USB, FireWire, and Thunderbolt devices are hot-swappable: you can insert or remove the connector while the computer is on. All protocols are compatible with Mac or PC.

Those ports are also available in PCMCIA, CardBus, and ExpressCards, which fit into laptops. A PCMCIA card is a credit-card-size memory card or I/O device that connects to a slot in a computer. CardBus is an advanced PCMCIA card with faster speed due to its direct memory access (DMA) and 32-bit data transfer. An ExpressCard is the latest

FIGURE 14.9
Thunderbolt port and cable connector.

iteration of the PCMCIA standard and transmits at either 480 Mbps in USB mode or 2.5 Gbps in PCIe mode.

Audio Interface Features

When deciding which interface to purchase, consider the features below.

DIGITAL I/O

Some interfaces have digital inputs and outputs as well as analog. Digital I/O comes in various formats (AES3, S/PDIF, ADAT Lightpipe), so make sure that the interface's format matches that of your digital recorder, digital mixer, or outboard A/D converter. See Chapter 13 under the heading "Digital Audio Signal Formats."

ANALOG I/O

Inputs: Decide how many XLR connectors you need for microphones, line inputs you need for line-level signals, and instrument inputs you need for electric instruments. Outputs: You may need more than one output for multiple studio monitors or headphones. Check whether the analog I/O on the interface is balanced or unbalanced. Balanced connections reduce electrical interference picked up by cables. However, unbalanced connections cost less, and usually are adequate for cable runs under 10 feet. Balanced connectors are XLR or 1/4-inch TRS phone jacks. Unbalanced connectors are RCA phono jacks or 1/4-inch TS (tip-sleeve) phone jacks.

SAMPLING RATE AND BIT DEPTH

Available sampling rates in audio interfaces are 44.1 to 192 kHz. Recordings made with higher sampling rates give better sound quality but consume more hard-drive space. If you are making CDs, a sampling rate of 44.1 kHz does the job.

Most interfaces can handle 24-bit as well as 16-bit signals. Recordings made with 24-bit resolution sound a little smoother and cleaner (less distorted) than 16-bit recordings. Even if your final product is a 16-bit CD, it will sound better with a 24-bit recording that is dithered down to 16 bits during mastering. See Chapter 13 under the heading "Dither."

MIDI PORTS

Many interfaces have MIDI ports. A MIDI IN port accepts MIDI signals from a MIDI controller and sends them to your computer's sequencing software. A MIDI OUT port sends MIDI signals from your computer's sequencing software to a synthesizer or sound generator. If your audio interface has MIDI ports, you don't need a separate MIDI card in your computer.

WORD CLOCK

Some interfaces offer word clock connectors, which send and receive timing signals for digital audio. A word clock is used to synchronize the digital devices in a large studio for real-time transfer of digital audio. See Chapter 13 under the headings "Word Clock" and "Jitter".

DRIVER SUPPORT

An audio driver is a program that allows recording software to transfer audio to and from an audio interface. Many interfaces are sold with several types of drivers, or are supported by the core audio driver of your operating system. Be sure that your interface comes with drivers that work with (or are supported by) your operating system and your recording software.

A good driver has a minimum latency spec under about 4 msec. Latency is the signal delay through the driver and interface to the monitor output. This can be a problem in recording a soft synth, because you hear the playback of the synth a few msec after striking the keys on the MIDI controller. See the section "Minimizing Latency" later in this chapter.

The most popular audio driver formats are ASIO, WDM, DAE, Direct I/O, GSIF, MAS, SoundManager, Wave, and Core Audio. Here's a brief description of each format:

- ASIO (Audio Streaming Input and Output) (Mac OS 9+, PC): A very popular driver developed by Steinberg. Allows multiple channels of simultaneous input and output, and low latency with software synthesizers.
- WDM (PC): Windows Driver Model with Kernel Streaming, a multichannel driver. Allows low latency with WDM-compatible audio hardware and DXi software instruments. DXi stands for DirectX Instruments, Cakewalk's virtual instrument integration standard. DirectX audio effects can be used live on input signals, not just during playback. This lets you monitor and record effects in real time as you're recording.

- DAE (Mac, PC): Used only with Digidesign audio interfaces. It's a multichannel driver that runs with a compatible host such as Pro Tools, Logic, and Digital Performer. DAE lets you use RTAS and/or TDM plug-ins (explained later under the heading "Avid Pro Tools").
- Direct I/O (Mac, PC): Works with Digidesign interfaces as a multichannel driver only. Does not let you run RTAS or TDM effects.
- GSIF (PC): Permits very low latency when playing samples from hard disk with TASCAM's GigaStudio software sampler.
- MAS (Mac): Developed by Mark of the Unicorn (MOTU). Offers resolutions up to 24 bit/96 kHz and multiple simultaneous input and output channels. It's also a format for plug-ins (software audio effects).
- SoundManager (Mac): Macintosh's standard audio driver. It lets you record and play mono and stereo files up to 16 bit and 44.1 kHz. It has a moderate amount of latency.
- Wave (PC): The PC standard audio driver. Wave can be used with a variety of audio interfaces (like Sound Blaster-type sound cards) to record and play mono or stereo audio. It has a moderate amount of latency.
- Core Audio: Apple's low-latency driver used in Mac OS X.
- MME is an older driver that offers lower performance than newer ones.

ASIO or WDM are the main choices for PC. Cubase seems to work best with ASIO; Cakewalk Sonar works with WDM or ASIO. Choose drivers recommended by (or supplied by) the audio interface manufacturer. If you hear glitches, drop-outs, or buzzing with a particular driver in use, try switching to another one—that often stops the problem.

CAUTION: If you have multiple drivers installed, they may conflict. Then the computer might crash or the recording software might not access the audio interface. Disable or delete unused drivers.

OTHER OPTIONS

Listed below are several features that some interfaces offer.

- Zero-latency monitoring: The input signal to the interface is fed to its output, so you hear the input signal without any signal-processing delay (latency).
- Pro Tools compatible: The interface works with Pro Tools recording software and hardware.
- Surround sound: Provides 5.1 or 7.1 surround sound monitoring.
- Bus powering: The FireWire or USB connection powers the interface; you don't need another power supply.
- Battery powering: This makes the interface portable for on-location recording with a laptop.
- Supplied recording software: The interface is packaged with recording software, so you might not need to buy other software.

- Built-in effects: reverb, compression, EQ. Musicians can hear these effects without any latency (delay) in their headphones while playing or singing. That's because the effects are generated inside the interface, rather than through computer software and drivers, which create latency. An example of such an interface is the PreSonus AudioBox VSL.
- DSP: Onboard chips that take some of the processing load off the CPU.
- SMPTE sync: The interface will synchronize with a SMPTE time code signal (explained later in this chapter).

Control Surface

So far we've covered the computer and audio interface. Let's look at other hardware for a computer studio.

Using a mouse to adjust the controls on-screen is slow and can lead to repetitive stress syndrome. An alternative is a DAW control surface or controller (Figure 14.10). It's a chassis with physical controls for software functions. Resembling a mixer with real faders, knobs, and transport buttons, it lets you adjust the software's virtual controls that you see on the monitor screen. The controller attaches to the computer through a MIDI connector, Ethernet connector, USB, or FireWire port.

FIGURE 14.10
M-Audio Project Mix, an example of a control surface.

Many control surfaces also act as audio interfaces: they include mic/line inputs and outputs (I/O) and MIDI connectors.

Some control surfaces are dedicated to one DAW recording program, such as Pro Tools 003 Series, Avid C|24 and Command|8, Cakewalk V-Studio, and Steinberg Cubase CC121 and WK Audio ID controller for Nuendo. Others are universal: they work with several different DAW programs. Examples are the Euphonix Artist series controllers, M-Audio Project Mix, Alesis Master Control, and Tascam FW-1028. Be sure to check whether the controller you want to buy will work with your recording software.

What if your controller has 8 faders, but your project has 16 virtual faders on screen? You can make the controller access groups (banks) of virtual faders, 8 at a time, by pressing a bank switch. Normally, your controller faders will affect virtual faders 1 through 8. Press the bank switch to make the controller faders affect virtual faders 9 through 16. Press it again to access virtual faders 17 through 24, and so on. Controllers that have 4 faders can access banks of virtual faders 4 at a time.

Advanced controllers offer these features:

- Motorized faders: The faders in the control surface are motorized, so they move like the virtual faders on screen.
- Standalone mixer mode: You can use the control surface as a regular mixer.
- Footswitch jack: Accepts a footswitch for punch-in/out. The footswitch works only if your recording software supports this function.
- Meters.
- Monitoring section.
- Aux send/return: Allows you to connect an external analog signal processor.
- Insert jacks: An insert jack in series with each channel's signal allows you to plug in an analog compressor.
- Expandability: You can add more controllers, side-by-side, to control more virtual faders at once.

Note that a MIDI controller or a MIDI fader box can control a sequencer's MIDI tracks, but not audio virtual controls. A control surface can control both.

The Frontier Design Group AlphaTrack (Figure 14.11) and the PreSonus Faderport are controllers with several buttons and one long-throw fader. They are easier to use than a mouse and have a small footprint.

Alesis FirePort

This is a special type of audio interface that transfers audio tracks from an Alesis HD24 multitrack recorder to a computer. The HD24 records tracks as WAV files on an Alesis FST-formatted hard drive that is plugged into the recorder. After recording tracks, you can remove the hard drive and plug it into an Alesis FirePort (Figure 14.1), a small

FIGURE 14.11
Frontier Design AlphaTrack, an example of a single-fader controller.

device that converts those tracks' WAV files to FireWire format and sends them to your computer hard drive for editing and mixing.

As we've seen, there are a wide range of features and connectors among different interface models. A good interface is well worth the price because of its top-quality sound, convenient connections, and easy control of audio signals.

DIGITAL SIGNAL PROCESSING (DSP) CARD

Another DAW option is a digital signal processing (DSP) card that you plug into a user slot in your computer's motherboard. Why is it helpful? The more software effects (plug-ins) you use, the fewer tracks you can mix, because effects put a big load on the CPU. Normally this is not a problem except in mixes with more than 24 tracks and loads of plug-ins in use. You can resolve this issue by installing a DSP card, which handles the processing for the plug-ins. The plug-ins access the card rather than the CPU. You can install multiple cards to expand the system. Examples are TC Works Powercore (www.tcelectronic.com), Avid HDX-based Pro Tools cards (www.avid.com), and Universal Audio UAD-2 PCIe cards (www.uaudio.com).

ANALOG SUMMING AMPLIFIER

This device is a mixer without faders, used to mix tracks or submixes from a DAW through analog circuitry. By using a summing amplifier, you bypass the DAW's internal mix bus, which might create slight distortion due to rounding errors, numeric overload, and bit truncation. (Some DAW software claims not to produce those errors, such as Pro Tools TDM systems.) You connect each DAW track's analog output to a summing amplifier input, and use the DAW's faders and automation to set mix levels. Examples of summing box manufacturers include Dangerous, SPL, InnerTube Audio, AMS-Neve, Boutique Audio/Inward Connections, Radial Workhorse, and Tube-Tech.

Some engineers claim they avoid the distortion of mixing entirely inside the computer by recording and mixing at lower levels.

RECORDING SOFTWARE

Now let's turn from hardware to software. Some popular programs for recording, editing, and mixing are listed in Chapter 4 under the heading "Computer Digital Audio Workstation." They share a lot of the same features but implement them differently. Also, each product has unique features and its own GUI (Figure 14.12).

FIGURE 14.12
Cakewalk Sonar Producer X1 track view.

Check out the products online to see how you might like working with them. Some websites have interactive demos; others have free trial versions that you can download. Many sites have tutorials.

When you first install recording software or hardware, expect some frustration. Your software may not be compatible with your hardware, or may not perform as expected. You might need to modify certain settings in your computer system or recording software. Read the manual and any readme files that came with your product. Also, check out the products' websites for knowledge bases, FAQs, etc. Each DAW program has online discussion groups or forums which can be very helpful. If you're still having problems, call or e-mail the tech support for your product.

For each recording program, books are available that provide operating details and power-user tips. To find them, search www.amazon.com or the references in Appendix E. Type the name of the program and manufacturer in the search field.

This chapter is not meant to duplicate the information in those specialized books or in manufacturers' websites. Instead, it provides an overview of recording software.

Features

Here are some features to look for in recording programs:

- Multitrack digital recording from 2 tracks on up: Some programs can record and play back an unlimited number of tracks depending on the speed of your computer, RAM, and hard drive.
- MIDI sequencer: This lets you import or record several tracks of MIDI performance data, edit it, and play it. MIDI sequencer tracks can be mixed with audio tracks. Some programs include MIDI effects, MIDI plug-ins, and input quantizing.
- A freeze function mixes down effects and edits on a track to another track with embedded effects. This frees up CPU resources. Freeze also converts MIDI soft-synth tracks to audio tracks, which also reduces CPU loading. Unfreeze returns to the original tracks for editing.
- Peak and RMS metering.
- Compatibility with control surfaces.
- Surround mixing (in high-end programs).
- Customizable GUI: Change the graphical user interface to suit the way you work. Create configurations to use as templates for similar projects.
- Keyboard shortcuts: Instead of constantly using a mouse, tap certain keys on your computer keyboard. This speeds operation and gives your mouse hand a rest.
- Virtual tracks: Extra recorded takes of a musical performance. During mixdown you can choose the best takes to use, or the best parts of each take.
- Automated mixing: All settings for a song project can be saved and automated. The computer remembers your mix moves, and sets the mixer faders and pan

pots accordingly during subsequent mixdowns. Some programs let you automate effects settings as well. For more information, see the section on automated mixing in Chapter 11.

- Loop-based composition, construction, and editing tools, and ACID-loop and MIDI Groove Clip support. All these are covered later in this chapter. Loops and samples are included in high-end programs.
- Locate and marker points: Mark several points in a song (intro, verse, chorus, solos) so that you can go to them instantly.
- Routing or virtual patch bay: Assign any input to any track.
- Automatic pitch correction of audio (by Autotune, Melodyne, or V-Vocal).
- Automatic timing correction of notes by Beat Detective (Pro Tools) or Audiosnap (Cakewalk Sonar Producer) lets you lock the tempo of audio tracks, such as drums, to a tempo map. You can also make one track follow another track's tempo, or extract timing from audio tracks and apply them to MIDI tracks.
- Audio-to-MIDI conversion: For example, have a singer's notes play the same pitches on a synthesized cello.
- Video display: An on-screen window that shows video clips. Found in high-end software, this feature lets you synchronize music and sound effects to a video program.
- CD recording: Record your mixes on a CD burner.
- PQ editing: Set up a list of CD song-start times before burning a CD of your mixes.
- Import and export various file types (such as .wav, .mp3, .wma, .omf), sampling rates, and bit depths.
- 32-bit or 64-bit floating-point resolution in the DSP.
- Support for 64-bit operating systems and multicore CPUs.
- ASIO and WDM compatibility.
- Ability to learn the function of each controller encoder or button, mapping the controller to the on-screen controls.
- Spectral analysis: A display of level versus frequency of the audio program as it progresses in time.
- Notation application: Converts a MIDI file to musical notes on a treble and bass clef.
- Sync: Synchronization to SMPTE time code or MIDI time code (described later in this chapter).
- Support for DirectX and VST audio effects (DirectX is a standard audio driver).
- Support for DXi and VSTi soft synths (VSTi stands for Virtual Studio Technology for Instruments, Steinberg's virtual instrument integration standard).
- Included soft synths and samplers.
- Convolution reverb, vintage compressors, audio restoration, mastering tools.
- Plug-ins with delay compensation (described below).

Plug-ins

As we described in Chapter 9, a plug-in is a software module that adds digital signal processing, effects, or soft synths (virtual instruments) to a DAW recording program (Figures 14.13 and 14.14). Recording software comes with several plug-ins already installed. You also can purchase and download plug-ins from the Web. Unlike hardware effects, plug-ins are instantly upgradable—just download the latest version.

When you click on the name of a plug-in, a screen pops up that looks like a hardware processor with knobs, faders, lights, and meters. Parameters (such as reverb time, chorus depth, or compression ratio) can be adjusted, and most parameters can be saved and automated.

Every effect mentioned in Chapter 9 is available as a software plug-in: reverb, compression, EQ, chorus, flanging, distortion, tape saturation, lo-fi, restoration, pitch and time stretching, echo, mastering, and much more. Some plug-ins are bundles (suites) or collections of effects, such as Waves bundles. Virtual instruments include acoustic piano, voices, ethnic instruments, choir, guitar, bass, drums, strings, classic keyboards, and so on.

FIGURE.14.13
Lexicon Pantheon Reverb plug-in.

FIGURE 14.14
McDSP Channel G Compressor plug-in.

A number of plug-ins can simulate or model the sound of specific devices. For example, Amp Farm by Line 6 is a guitar-amp modeling plug-in, while Antares Microphone Modeler can simulate the sound of popular studio mics. Vintage compressors, tube mic preamps, concert halls, analog tape recorders, and console strips (Figure 14.15) have been modeled as well.

Some plug-ins run in real time as the audio is playing. Others process a sound file and create a new file as with normalization or time compression/expansion.

Plug-in effects can be applied in three ways:

1. Master effect: The effect is on the master output bus, and processes the entire mix. Multiband compression is a typical master effect.
2. Aux-send or loop effect: The effect is on an aux bus. Set the plug-in's dry/wet mix control all the way to "wet" or "100% mix." Adjust the amount of effect on each track by turning up the track's aux send control. This option reduces the number of plug-ins needed for a mix, and so reduces the load on your CPU. Delay effects—reverb, echo, chorus and flanging—are aux effects.
3. Insert effect: This effect is on a specific track and affects only that track. EQ and compression are insert effects. If you insert a delay effect into a track, you adjust the dry/wet mix control in the effect to control the amount of effect that you hear.

FIGURE 14.15
Waves SSL G-Channel plug-in.

Plug-ins come in several formats; the most popular are Core Audio's AU (Audio Units) for Mac, RTAS (Real Time Audio Suite) for Pro Tools, and Steinberg's Virtual Studio Technology (VST) and DirectX for Windows. DirectX is a group of application-program interfaces that enhance multimedia (video and audio) on Windows systems.

There are way too many plug-ins to mention here, but here are a few prominent plug-in companies: Waves (Plug-In Bundles, Signature Series Collections, Musicians 2, Native Power Pack, One Knob, AudioTrack Native, API Collection, Aphex, Vocal Rider, L3 Maximizer, and many more), McDSP, SoundToys, Sonnox, Softube, Slate Digital, Submersible Music, Avid Strike, Propellerhead Reason, Mixosaurus, FXpansion, Toontrack, URS, Focusrite, Lexicon, Antares, Mu Technologies, Serato, Celemony, Altiverb, Speakerphone 2, Crane Song, Avid, Eventide, David Hill, Synchro Arts, iZotope, T-Racks, Steinberg, Bias, Sony, and Native Instruments.

You can find plug-ins on their manufacturers' websites and on the websites of audio retailers. For example, Avid's website lists dozens of plug-in partners. Another great resource is http://www.kvraudio.com with its searchable plug-in database.

Avid Pro Tools

Pro Tools is the most popular DAW software and hardware system—an industry standard used by the pros. Pro Tools sessions recorded anywhere will load in any other Pro Tools-equipped studio. It runs on Mac or PC, and includes both songwriting and recording tools for audio, MIDI, loops, samples, and soft synths. Details are at www.avid.com/us/products/Pro-Tools-Software/.

Other excellent DAW programs are available. People often use the term "Pro Tools" generically to mean "digital audio workstation." Engineers prefer various DAWs based on their work style.

Because Pro Tools is so popular, and has a number of configurations that can be confusing, I'll describe them here.

- **Pro Tools 9** is software that uses your computer's CPU for all processing. Your CPU speed, hard-drive speed, and amount of RAM determine how many plug-ins and soft synths you can use simultaneously. Pro Tools 9 works with any audio interface. **Pro Tools 10**, like version 9, runs entirely off your computer's CPU (Figure 14.16). Price: $599.
- **Pro Tools|HD Native**: Like Pro Tools 9 or 10, the Native version uses your computer's CPU for processing. That's the definition of "native" or "host-based." It consists of a Native PCIe core card and Pro Tools Native HDX software, and is claimed to provide the lowest latency, most stability, and highest performance of any native digital audio workstation. It provides up to 192 tracks with automatic delay compensation for plug-ins. Price: $3400.

FIGURE 14.16
Avid Pro Tools 10 screen.

- **Pro Tools HDX** is the high-end professional system with DSP cards, up to 192 kHz sample rate, expandable I/O, and many options. The dedicated DSP cards free up your CPU for other tasks. The result is you can run more tracks, more plug-ins, and more soft synths at the same time. Price: $10,000 to $125,000 or more.
- **Pro Tools LE** has been replaced by Pro Tools 9 and 10, but you should be familiar with the many LE systems already installed. LE components included Pro Tools LE software and a choice of audio interface: Mbox Mini, Mbox, Mbox Pro, 003 Rack Factory, 003 Factory control surface, or Command|8 USB control surface. Price: $500 to $3000.
- **Pro Tools M-Powered** also has been replaced by Pro Tools 9 and 10. Its software is designed to work with certain M-Audio® audio interfaces and control surfaces. The software costs $250 while the interfaces range from $100 to $549.

Each system is compatible with the high-end Pro Tools HD system, so pros can use any system at home and import their projects into Pro Tools HD.

Let's look at each system in more detail.

Pro Tools 9 or 10 is intended for singer/songwriters, project and personal studios, remote recording engineers, and bands. It comes with seven virtual instruments, 20 effects, and dozens of processing plug-ins. Other features include a score editor (compatible with Sibelius music notation software) and easy MIDI editing. The program offers up to 96 or 192 simultaneous audio tracks, 128 instrument tracks, 512 MIDI tracks, 256 internal busses, and 160 aux tracks.

Pro Tools 9 and 10 work with "EUCON-aware" control surfaces. EUCON is a high-speed Ethernet protocol that allows EUCON-compatible controllers to operate software such as Pro Tools.

OMF/AAF/MXF file interchange support lets you can exchange projects created in other audio and video software with Pro Tools and vice-versa. That lets you collaborate easily with other musicians and recording engineers.

Pro Tools 9 and 10 have automatic delay compensation, so when you add a plug-in or an external effects unit, you don't need to compensate manually for the delay that they introduce.

Here are a few of the new features introduced in Pro Tools 10:

- Ability to use Avid's new AAX (Avid Audio Extension) plug-in format, which is 64-bit ready and lets you use native or DSP processing.
- Clip gain: Each clip now has its own volume automation which acts independently from the track volume.
- New plug-ins: channel strip, down mixer (convert 7.1 surround to 5.1, 2.1, stereo, etc.), and Mod Delay III time-based effects.
- 500 new EUCON-controllable functions
- Real-time fades.
- Higher sound quality due to 32-bit floating-point processing, which prevents dithering down to 24 bits between plug-ins.
- Functions to export selected tracks as a new session, or export multiple file types.

Pro Tools|HD Native offers the lowest latency of any host-based DAW. It supports up to 64 channels of I/O and 192 audio tracks, is very stable, and is said to provide the best performance possible using a host computer.

Unlike USB- or FireWire-based DAWs which have latency, Pro Tools|HD Native employs a high-speed PCIe core card to virtually eliminate latency. You can combine it with one or more Pro Tools HD Series interfaces (such as HD OMNI or HD I/O, sold separately) to achieve the highest sound quality.

If your projects don't require the extreme DSP acceleration of Pro Tools|HD, you can use HD Native's PCIe core card to save money. It works with Pro Tools HD software.

Features include advanced automation, surround mixing, input monitoring, AFL/PFL solo, destructive punch, RTAS plug-ins (explained later), and other professional features. Automatic delay compensation is supported on hardware and software insert.

HD Native will synchronize to time code and video reference with frame-edge accuracy when using SYNC HD (sold separately). It works with ICON, C|24, Artist Series, or Pro Series consoles and control surfaces, Mac or Windows.

Pro Tools HD, the high-end pro version, comes in four Core systems (each with more DSP power and more I/O) and one or more audio interfaces. Each system requires the Pro Tools 192 I/O or the 96 I/O audio interface, and supports legacy Pro Tools audio interfaces for additional inputs and outputs. With the release of ProTools 10, Avid has introduced their new HDX system. Pro Tools HD systems are compatible through Pro Tools 10; however, HDX systems will be required by Pro Tools 11 and beyond.

HD offers a wide range of sample rates, hi-res audio, and up to 192 tracks. An HD system can have 8 to 96 ins and outs. The four basic Core systems have one, two, or three DSP cards, and a system can use up to seven DSP cards.

Listed below are the four Core systems. Each can be expanded by adding more HD Accel cards or HDX cards. (Accel cards are becoming out of date):

- Pro Tools HD 1 includes the HD Core DSP card, which handles up to 32 channels of I/O and up to 96 tracks.
- Pro Tools HD 2 Accel includes the HD Core card and an HD Accel card, which provide more than four times the mixing and processing power of HD 1 systems, 64 channels of I/O, and 192 tracks.
- Pro Tools HD 3 Accel, a more powerful system, includes the HD Core card and two HD Accel cards, which provide up to 96 channels of I/O and 192 tracks.
- Pro Tools HDX: Each HDX DSP card has 3–7 times the processing power of a single HD Accel card.

Pro Tools HD systems also include these interface options.

- The 96 I/O Audio Interface converts audio at up to 24 bit/96 kHz. It includes 8 channels of analog I/O, 8 channels of ADAT optical I/O, 2 channels of AES/EBU and S/PDIF I/O, and a word clock. It can link to other 96 I/O Audio Interfaces and to Avid 888/24, 882/20, 1622, or 24-bit ADAT Bridge.
- The 192 I/O Audio Interface converts audio at up to 24 bit/192 kHz. It includes 8 channels of analog I/O, 8 channels of AES/EBU, 8 channels of TDIF, 16 channels of ADAT, and 2 additional channels of ADAT or S/PDIF digital I/O. A 192 Digital Expansion Card provides 8 more channels of AES/EBU, TDIF, and ADAT I/O connections and a legacy port to link to older Digidesign interfaces.
- The ICON D-Control work surface is a large mixing console/controller surface designed to work with Pro Tools software. C|24 is another powerful control surface.
- The Avid System 5-MC is the premier large-format Pro Tools workstation controller. It's EUCON compatible.
- Avid Artist Series and Pro Series consoles and controllers also work with Pro Tools software. They offer EUCON integration.

New features of the latest Pro Tools software are on the www.avid.com website.

Plug-ins for Pro Tools come in four formats: TDM, HTDM, RTAS (Real Time Audio Suite), and AudioSuite. TDM plug-ins work only with Pro Tools' professional TDM systems, and use dedicated cards that plug into your computer. HTDM is the same for the Pro Tools HD (high definition) system. Many RTAS plug-ins are the same as TDM plug-ins, except that they use your computer's processor instead of specialized TDM DSP cards. All the above plug-ins work in real time.

AudioSuite plug-ins don't work in real time. Instead, they process a track's signal and create a new track in which the effect is part of the recorded sound. That is, the effects are embedded in the track. Playing a track with embedded effects is less work for the CPU than playing a track with real-time effects. Less load on the CPU means you can play more tracks without drop-outs.

Loads of plug-ins are available for Pro Tools. Many are included, and some can be downloaded from third-party vendors. Use the plug-in finder on the Avid website (www.avid.com).

Included with every Pro Tools system is a Music Production Enhancement Suite, which includes soft synths, looping, guitar-amp modeling, a sample player, and tube EQ.

OPTIMIZING YOUR COMPUTER FOR MULTITRACK RECORDING

Once you've chosen some recording software and installed it, you'll want to make your computer run as fast and glitch-free as possible. You need to optimize its settings for best results. Appendix B suggests some ways to speed up your hard drive, reduce software interruptions, and reduce CPU usage. If you follow those tips, you will have a faster system that handles more plug-ins and more tracks at once. Also, when you play tracks or burn CDs, clicks and drop-outs in the audio will be less likely.

USING A DAW

Let's say you purchased some DAW hardware and software, and tweaked your computer to make it fast and reliable. It's ready to rock. Here are some tips on setting up and using your DAW.

Connections

First, connect your audio equipment to the DAW as shown in Figure 14.17. You may need some adapter cables between devices. The following adapter cable connects a sound card's unbalanced I/O to a power amp, powered speakers, or mixer: stereo 1/8-inch phone plug to two unbalanced phone plugs, or two RCA phono plugs (whatever matches your equipment).

Software Settings

Starting from a PC's desktop view, select Start > Settings > Control Panel. Double-click Sound and Audio Devices > Audio tab. Set the Playback and the Recording

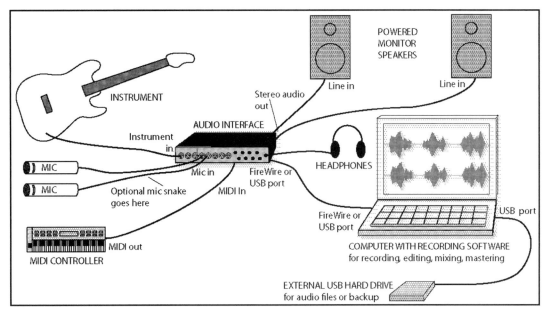

FIGURE 14.17
Connecting audio equipment to a DAW.

default devices to the sound card or interface you are using for audio editing. In Device Manager, disable the computer's built-in sound card to avoid conflicts.

Before you record a new song, create a separate session project for that song. Each song should have its own session project and its own folder of audio files. Set a bit depth and sampling rate for the project, such as 24 bit/44.1 kHz. Name each track before recording.

If you want to record from a mic or instrument, you will record their amplified signals on an audio track. If you want to record from a MIDI controller, you will record its MIDI signal on a MIDI track. Details on recording MIDI sequences are in Chapter 16.

Set up the audio drivers in your recording software. These settings might be called "Record Timing Master" and "Playback Timing Master." Specify the input and output devices for each audio track, such as "sound-card left channel input" and "sound-card stereo outputs." Also, specify the input and output devices for each MIDI track, such as "sound-card MIDI IN" and "sound-card stereo outputs."

If you are recording many tracks simultaneously, or at a high sample rate, this may generate more data than your CPU and buffers can handle—causing drop-outs. A drop-out is a short silence or a noise burst (glitch) in the recorded audio. To prevent drop-outs while recording, turn off the track playback metering, effects, and waveform preview. Increase the buffer size. Turn off or reduce video acceleration. Do not zoom or scroll while recording. Also, follow the tips in Appendix B on optimizing your computer for multitrack recording.

The buffer setting is critical. A large buffer lets you use many tracks and plug-ins without drop-outs, but it creates a lot of latency (monitoring delay). If latency is high, and you press a key on a MIDI keyboard, a soft synth will respond after a noticeable delay. A small buffer minimizes latency so soft synths respond faster, but can cause drop-outs. Generally it's best to set the buffer/latency small (under 4 msec) while tracking soft synths and set it large otherwise (maybe 25 msec). See Appendix B under the heading "Minimizing Latency."

To monitor a live mic signal without latency while overdubbing, listen to a mix of the live mic signal and the previously recorded tracks. Mute the mic input signal in the sound card's software and disable Echo Input Monitoring in the mic's track.

Editing Tips

Now that your DAW is set up, you can record and edit audio. When you select a portion of the audio program, you create a **clip** or **region**. Examples of clips are an entire song, the chorus of a song, a guitar track, a drum riff, or a single note. Using a mouse, you can create a clip by marking its beginning and end points in the waveform. In Figure 14.18 a clip is selected and highlighted.

A clip is actually a pair of pointers to part of an audio file on your hard drive. One pointer is the data address for the beginning of the clip, and the other pointer is the address for the end of the clip. Rather than containing audio data, clips tell the software which section(s) of the audio file to play.

When you set up the sequence of clips in a mix, you're telling the hard-drive head which pointers to play in what order. When you delete a clip, you're telling the hard-drive head to skip the clip's pointers during playback. If you copy an audio clip, the

FIGURE 14.18
Selecting and highlighting a clip.

software does not make copies of the audio file it points to—instead, it plays the same audio file each time it sees a clip pointing to that file. When you split a clip into parts and put them in a different order, the audio file itself is not split. Instead, the software plays the sections of the audio file that the clips point to, in the desired order.

These types of edits are called nondestructive edits. Only the pointers change; the data on disk is not changed or destroyed. Nondestructive edits are not permanent. If you don't like an edit, you can undo it and try it again.

Some types of edits or processing are destructive: they write over the data on disk. However, some recording programs save the data before you edit it. Then you can undo the change by reverting to the saved data.

I'll give examples of several things you can do with editing.

DELETE NOISES

Suppose you recorded an acoustic guitar track, but it has a chair squeak between the verse and chorus. Zoom in on that noise, play it, highlight it, and delete it (type Ctrl-X with most DAWs). You might have the option to close up the space you created—don't do that in this case. Figure 14.19 shows a guitar track with a noise deleted between notes.

CLEAN UP TRACKS

Before the beginning of a recorded song, you'll typically hear noises and a count-off. To remove them, define the group of noises as a clip and delete it (Ctrl-X). Some software lets you **slip-edit** or drag the beginning of each track to the right to remove noises there.

FIGURE 14.19
A guitar track with a noise deleted.

FIGURE 14.20
Edited tom-tom tracks with only the hits remaining.

You can cut out silent areas in each track to reduce leakage and background noise. Say you've recorded a real drum set. Look at the tom-tom tracks. The tom mics pick up a lot of unwanted leakage from the snare drum and cymbals. In the tom-tom tracks, you can cut out the space between tom hits. This prevents the leakage and tightens up the sound. Figure 14.20 shows some edited tom-tom tracks with only the hits remaining. You could use a noise gate instead of doing this editing.

DUPLICATE MUSICAL PARTS

Suppose you have a perfect take of the background vocals in a chorus. Rather than recording more parts for the other choruses, you could copy the vocals from one chorus and paste them at the measures where the chorus is repeated. Take care to align the vocal clips in time with the other tracks.

One way to duplicate several tracks in another part of a song is to bounce them to a stereo WAV file, then import that WAV file where you want it. Here's how: Solo and balance the tracks you want to bounce. Set the master fader to make the mix level near 0 dBFS. Export the tracks as a 32-bit stereo WAV file, then import that file into a new stereo track and align it with other tracks.

REPLACE WRONG NOTES

Does the bass track have a few fluffed notes? Simply delete a wrong note, copy an in-tune note and paste it where the wrong note was. Zoom in to adjust the timing with precision.

MAKE SPECIAL EFFECTS

Editing can create unusual effects. For instance, copy a syllable in a vocal track and paste it several times in a row to create stuttering. Suppose you want to double a guitar that is in the left channel to make it stereo. Copy the guitar track, paste it to another track, slide the pasted guitar track to the right about 20-30 msec (which delays the guitar signal), and pan the delayed guitar track to the right. Adjust the track faders so that you hear the guitar spread evenly between your monitor speakers.

FADE-OUT, FADE-IN

A fade-out at the end of a song is a slow drop in volume to silence. DAW programs let you insert a fade-out at the end of a finished mix (or on individual tracks). If you want to fade in a synth-pad intro, you can do that too. Fade-out and fade-in curves come in various shapes, so use whatever sounds best.

CROSSFADE

This means, "fade out of one song while fading into another." Also, you can crossfade across an edit point to make it sound smoother. Figure 14.21 shows two songs overlapping. The first song fades out as the other song fades in, and the two songs blend where they overlap.

FIX TIMING ERRORS

With most DAW programs you can slip each track forward or backward in time (left or right on the screen) independent of other tracks. You also can move single notes in

FIGURE 14.21
A crossfade between two songs.

FIGURE 14.22
(A) Two musical parts out of sync. (B) Two parts in sync.

time. If a note in a bass track comes in too late, highlight that note as a separate clip, and slide it or nudge it to the left until it is on the beat. By visually aligning the start times (attacks) of notes in various tracks, you can make a recorded band sound very tight.

Figure 14.22 shows two tracks that I wanted to synchronize. I set the DAW to display only those two tracks. In Figure 14.22A, the notes in the bottom track play later than the notes in the top track. In Figure 14.22B, the notes in the bottom track have been aligned with the notes in the top track—they play in sync.

REMOVE OR REARRANGE SONG SECTIONS

You do this type of edit on a finished stereo mix, not necessarily on individual tracks.

Suppose you want to shorten a song by removing a section. Highlight the section you want to delete, cut it (Ctrl-X), and close up the space. Zoom into the edit point and slide the beginning or end of the clips there to make a smooth transition. Try to create a zero-crossing at the edit point, so that the waveform crosses the 0 dB line there (as in Figure 16.8). This method also works for editing mistakes out of spoken-word recordings. Just be careful not to leave double breaths during pauses.

Now suppose you want to remove a section of a song and put it somewhere else in the song. If you think the bridge should come earlier, you can remove it from its current location and put it where you want it.

1. Highlight the bridge waveform.
2. Cut it (Ctrl-X) and close up the space.
3. Split the song where you want the bridge to go.
4. Slide the right-side part of the song to the right to create a space.
5. Paste the bridge waveform in that space.
6. Slide the waveform segments to create a seamless edit.

CHANGE THE SONG ORDER WHILE MASTERING

When mastering a program for CD, you can select entire songs and move them around to change the order they will play in.

Those are just a few ways that editing can improve your mixes and song arrangements.

Maintaining Audio Quality

Do your computer mixes sound harsh, grainy, or "digital"? Here are some tips to make your in-the-box mixes sound as smooth as analog.

Suppose you're recording and mixing with a DAW set to 24-bit resolution. You think you're doing everything right, but the final product still sounds vaguely distorted. You record tracks near 0 dBFS but no higher, and your meters don't show any clipping. Still, your mixes lack that smooth, effortless sound you used to get when recording analog. The sound is harsh or fatiguing.

What's going on? Chances are the signal is clipping but the meters don't indicate it. We'll explain this phenomenon, and will suggest ways to solve that problem. We'll also give other tips on helping your audio to sound less processed and more musical.

BACKGROUND

First we need to go over some digital audio basics. The A-to-D converter in your audio interface measures or samples the incoming analog signal many thousand times a second and converts those samples to binary numbers, which you record on your hard drive.

Figure 14.23 shows an analog sine wave that is sampled periodically. The digital meters in your DAW show those sample levels, but not necessarily the peak signal levels that

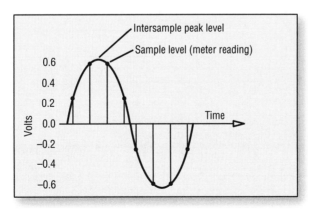

FIGURE 14.23
Analog signal levels can exceed digital sample levels.

appear between samples. In your D-to-A converter, a lowpass filter (reconstruction filter) creates occasional peaks between the samples that can be up to 3 dB higher than the measured level at certain high frequencies. If your recorded levels are near 0 dBFS, this can cause clipping (overs or overloads) that the meters don't display. An "over" is a segment of audio in which three or more consecutive samples are at 0 dBFS (all bits on).

Similarly, some plug-ins can create larger peaks than those that went in—without any increase in volume or audible change in the program. Even an EQ cut or rolloff (subtractive EQ) can increase the signal level by producing intersample peaks due to ringing at the cut frequency.

With these facts in mind, let's explore some ways to prevent harsh digital sound.

USE METERS CORRECTLY

First, let's review the concept of average levels and peak levels. As shown in Figure 14.24 a musical signal changes in level (voltage) continuously as it plays. Imagine a musical passage with a low-level synth pad, but with high-level drum hits. The average level or volume of the passage is low, but the transient peak levels are high.

Peak levels may be up to 24 dB above average levels depending on the type of signal. Percussive sounds have much higher peaks than continuous sounds do (synth pads, organ, flute)—even if the two signals have similar average levels.

The meters in your DAW can show signal levels in two modes: RMS and peak. RMS (or root-mean-square) readings correspond to the average levels, while peak readings show the level of peaks or short transients. The average or RMS level indicates approximately how loud the sound is, while the peak level shows how close the signal is to clipping.

Since you don't want to clip or distort the signal while recording or mixing, use peak metering, not RMS.

FIGURE 14.24
Average levels and peak levels in a musical signal.

REDUCE RECORDING LEVELS

To prevent intersample clipping, reduce your recording levels by turning down the gain in your mic preamps. Record at about −6 dBFS maximum in peak meter mode. Make sure that the recording meters are set to pre-fader.

Some DAWs, such as Cakewalk Sonar Producer, color the top 6 dB of the meters red as an aid to setting levels correctly.

One benefit of lower recording levels is that you won't overdrive your mic preamp. The distortion of most analog gear increases as you approach maximum gain. The mic pre-amp in your audio interface is less likely to distort its analog signal at 6 dB below maximum than at 3 to 0 dB below maximum.

Another benefit of reduced levels is that it creates some headroom for your plug-ins. Six dB of headroom should eliminate any invisible overs. Then you can set up your mix balances without having to adjust levels every time you insert a plug-in. With lower track levels, you won't hear distortion from plug-ins and your D/A converter while you are mixing.

Signals can clip during any gain boost, such as by a fader, compressor, or EQ. So keep recorded levels low (−6 dBFS max) so you don't have to change plug-in levels during a mix. Each track's waveform may appear small, but most DAWs let you zoom in vertically.

Remember, the recorded level on each track drives the plug-ins—not the track fader, which comes after the plugs. If a track was recorded at 0 dBFS maximum, use your DAW's trim control (or insert a −6 dB trim plug-in) to reduce the signal level going into your plugs. Do this before setting up a compressor or gate, because trim affects their gain reduction.

Doesn't a low recording level increase noise? Yes, but in practice it's not a problem. A 24-bit recording has a theoretical signal-to-noise ratio of 144 dB. So even if you record at −6 dB max, you're still far above the noise floor and you won't hear any increase in hiss.

Low noise is important because the more tracks that are in use, the higher the noise. For instance adding 16 tracks together at equal fader settings will increase noise by about 12 dB. Still, the S/N is more than you need to have a clean recording. So you can use that extra S/N to create headroom by recording and mixing at lower levels. If your peak recording level is −6 dBFS, you create 6 dB of headroom. If your peaks are −10 dBFS, you create 10 dB of headroom.

An alternative is to get an oversampling meter plug-in, which shows the actual inter-sample signal levels. By modeling the D/A conversion process, this type of meter reconstructs intersample peaks that regular meters miss. Examples are the TL MasterMeter for

Pro Tools, Sony Oxford Limiter, TCE 6000 MD3 algorithm, TCE MD3 PowerCore plug-in, and Voxengo Elephant.

If you just keep your levels down, you don't have to worry about intersample peaks.

REDUCE LEVELS AT OTHER POINTS

Listed below are other spots in the signal path to set your level to −6 dB maximum:

- On any stem (group or submix).
- At the output of each plug-in. Check for distortion (clipping) in each plug-in, especially in an equalizer set to boost. Find the plug's output gain control and turn it down until clipping stops. You might have several plugs in series, and you don't want the output of one plug to overload the input of another.
- At the output of any upsampling plug-ins. These DSP processes upsample the signal (increase its sample rate), do the effect, then downsample the signal at the output. The downsampler acts like a partial reconstruction filter. So even with the effect turned off, the up- and downsampling can create overs.

Level setting is important during mixdown, too. Set the maximum peak level of the master output bus meters to −3 to −6 dBFS to allow for signal peaks that do not display on sample-reading meters. You can make up the gain later during mastering, where you can boost the overall level or peak-limit/normalize to make a hot CD.

Try to keep the master faders at or near 0 dB. If the master faders are set low, you will turn up the channel faders to get a good mix level. That results in high-level signals that can overload the mix bus. Still, if a master-fader meter clips, it's okay to lower the master fader a few dB until the clipping stops.

Note that mix-bus overload is not a serious problem if the DAW uses 48-bit or 64-bit floating-point math in the mix bus, because float processing can pass signals above 0 dBFS without clipping. Still, it's good practice to control the levels hitting the mix bus. Summing calculations become less accurate when signals exceed full scale.

Consider not using makeup gain in compressors. Makeup gain can cause the initial uncompressed transient to clip by boosting its level. Raise the track fader instead.

If your mix is going to a mastering engineer, omit any stereo bus processing. Record the file at −6 to −3 dBFS max so that the mix doesn't overload the mastering engineer's D/A converter. If you are mastering your own mixes (say, for customer approval), use an oversampling meter that displays the actual reconstructed signal, or set the limiter ceiling at −3 dB to −0.3 dB, not 0 dB or higher. This avoids making files that sound distorted on your customer's playback systems. Many CD players distort with samples above the clip level.

So far we looked at level reduction as a path to smoother sound. Now let's consider other methods to make digital audio easier to listen to.

WORK ON THE SOUND SOURCE

Sometimes a musical instrument or vocal sounds edgy, even when recorded at low levels. The cause might be the source itself, the microphone frequency response, or mic placement. Here are some ways to make the source sound better:

- With EQ, consider using less high-frequency boost. Consider cutting just a few dB around 3 to 7 kHz, which tends to reduce harshness. Try a 4 kHz lowpass (high-cut) filter on distorted guitar amps to take the edge off.
- Try multiband compression on voices that get raspy at high volume. Set the compressor so it kicks in above 3 kHz or so during loud passages.
- Many condenser mics have high-frequency peaks in their response, which can sound brittle or sibilant. Prevent that effect by applying a high-frequency rolloff or shelf in your EQ. Better yet, use ribbon mics or flat-response condenser mics. Many ribbon microphones have reduced output at high frequencies, which can give a rounder tone.
- Move the mic to a spot that sounds mellower. For example, mike a vocal partly off to the side if it sounds too edgy when miked in front. Mike a trumpet off-axis. Mike a guitar amp near the edge of the speaker cone.
- Try using a tube mic preamp to warm up the sound. Not all tube preamplifiers do that, but some have euphonic even-order distortion that some people call added "warmth."

OTHER TIPS

- Dirty contacts can cause signal rectification and distortion. Clean connectors with agents such as Caig Labs DeoxIT (www.caig.com). Consider using high-quality cables.
- Use better converters.
- Use audio interfaces designed for low jitter.
- During mastering, add dither when truncating from 24 bit to 16 bit. This eliminates quantization distortion at low signal levels.
- Minimize the number of calculations that your DAW has to do. Each calculation creates some error or distortion by increasing the word length beyond 24 bits. Rather than boosting the gain of a musical section and dropping it later, undo the changes and apply only the correct amount of gain change.
- Consider using volume envelopes (fader moves) rather than compression. Compression adds distortion because it changes the waveform. Too-fast release times cause distortion in low-frequency notes and also cause pumping. A fader-setting change on a series of notes is less audible than a compressor working on each note.
- Limiting adds distortion. When you peak-limit and normalize a stereo mix to make a hot CD, try not to exceed 6 to 7 dB of peak reduction. Use even less

limiting on mixes that don't have loud transients. Rather than normalizing the limited signal to 0 dBFS, normalize to −3 to −1 dB to allow for intersample peaks.

- Experiment with an analog tape plug-in to warm up a track (e.g., Crane Song Phoenix). Or record to analog multitrack tape and dump the tracks to your DAW for editing/mixing.
- To reduce graininess in a reverb plug-in, use a high-density setting or use a convolution reverb.
- For best quality, record at 24 bits and stay there throughout the project. Then turn on dither and export to a 16-bit file that you will master for a CD. If you are sending 24-bit data files to a mastering engineer, do not turn on dither—let the engineer do it.
- Normalization is a DAW process that raises the gain of the entire program so that the highest peak reaches 0 dBFS (in peak meter mode, not RMS). Do not normalize the program until final mastering.
- Use cables designed for digital audio, and use short cables. Use internal sync on your A/D converter to reduce jitter.
- Avoid A/D and D/A conversions. Once your signal is digital, try to keep it that way. After converting the analog mic signal to digital in your audio interface, do all your processing in the computer if possible, then burn a CD of the final product. The only D/A conversion will be when the CD plays back.

Give all these suggestions a try—your music will sound better.

As we've seen, there is a whole world of computer music applications. Download the free trial versions to see what you like. Set up your DAW system, read the software manual, and have fun creating music.

AUDIO FOR VIDEO

You can use your DAW to create a sound track for a video or film production.

First we need to define SMPTE time code. SMPTE stands for the Society of Motion Picture and Television Engineers, who standardized the time-code signal for use in video productions. SMPTE time code is something like a digital tape counter, where the counter time is recorded as a signal on tape or hard disk.

Pictures on a video screen are updated approximately 30 frames per second, where a frame is a still picture. SMPTE time code assigns a unique number (address) to each video frame—8 digits that specify hours:minutes:seconds:frames.

Suppose you want to create a sound track for a video program. First you copy the video onto your computer's hard drive. When you open the video file in your audio editing program, it shows up in a window on your computer screen (Figure 14.25, top). While watching the video, you create the sound track by editing audio and music clips.

FIGURE 14.25
A video window and sound clips in a DAW screen.

Here is the process:

1. First, you're handed a DVD or sent a file that is a video recording of the program you're working on, plus location audio. Copy the file(s) to your hard drive.

2. Launch your audio software, import the video file, and watch the picture. Using soft synths, compose and record musical parts related to the video scenes. Slide the audio clips so they sync with the video.

3. Missing or poorly recorded dialog is replaced during a process called looping or automatic dialog replacement (ADR). Have actors watch the video and lip-sync their lines using the DAW.

4. Audition several sound effects from CD libraries, pick the ones you like, and import them into tracks in the DAW.

5. Using slow motion or freeze frame, go through the video and note where each sound effect should occur. Nudge audio clips in time to align with video events.

Below is a more detailed procedure. I'm using these terms in the steps below:

AUDIO EDITOR: The person who performs the audio edits and maybe composes the music.

AUDIO MIXER: The person who mixes the end product for a feature film. There might be several mixers: one for dialog, one for sound effects, and one for music.

1. During a video shoot, the audio from the location audio mixer is recorded directly to camera.
2. The audio editor receives a video file with the location sound track embedded (but extractable). In addition the audio editor gets either a WAV, OMF or, AAF file of the audio to mix. The audio track that is embedded in the video file is usually a scratch mix to make sure the audio from the WAV, OMF, or AAF file is in sync with the video. You line them up by ear (and eye).
3. In Hollywood film productions, the audio editor delivers to the audio mixer either an OMF or AAF file that contains all the audio edits in sync with the video. The audio in this file is in multitrack format so the audio mixer has maximum flexibility in mixing.
4. The audio mixer lines up the video and audio file at SMPTE time 01:00:00:00, the traditional project start time, and starts mixing. The video and audio in the computer are synchronized by the audio interface at the resolution of the sample rate of the mixer's system (typically every 48,000th of a second).
5. After the audio mixer finishes the mix using the audio elements from the audio editor, he or she sends the final mixed file (WAV or AIFF) to the video editor. The video editor marries the mixed track to the video on a video editing system.

In smaller productions, the audio editor/mixer follows this procedure:

1. Receive a WAV, OMF, or AAF file from the video editor.
2. Import location audio, narration, and sound effects. Compose and record music. Sync these elements to picture and do a mix.
3. Export the mix (usually as a Broadcast WAV or AIFF file). Put a time code stamp (01:00:00:00) at the beginning of the file.
4. Send the mix to the video editor.

Thanks to Matthew Gellert of bonesteelfilms.com and David Schmidt of acapellaaudio.com for their expert help with this section.

CHAPTER 15

Session Procedures, Mastering, and CD Burning

"We're rolling. Take One." These words begin the recording session. It can be an exhilarating or an exasperating experience, depending on how smoothly you run it.

The musicians need an engineer who works quickly yet carefully. Otherwise, they may lose their creative inspiration while waiting for the engineer to get it together. And the client, paying by the hour, wastes money unless the engineer has prepared for the session in advance.

This chapter describes how to conduct a multitrack recording session. These procedures should help you keep track of things and run the session efficiently.

There are some spontaneous sessions, especially in home studios, that just "grow organically" without advance planning. The instrumentation is not known until the song is done! You just try out different musical ideas and instruments until you find a pleasing combination.

In this way, a band that has its own recording gear can afford to take the time to find out what works musically before going into a professional studio. In addition, if the band is recording itself where it practices, the microphone setup and some of the console settings can be more or less permanent. This chapter, however, describes procedures usually followed at professional studios, where time is money.

PREPRODUCTION

Long before the session starts, you're involved in preproduction—planning what you're going to do at the session, in terms of overdubbing, track assignments, instrument layout, and mic selection.

Instrumentation

The first step is to find out from the producer or the band what the instrumentation will be and how many tracks will be needed. Make a list of the instruments and vocals that will be used in each song. Include such details as the number of tom toms, whether acoustic or electric guitars will be used, and so on.

Recording Order

Next, decide which of these instruments will be recorded at the same time and which will be overdubbed one at a time. It's common to record the instruments in the following order, but there are always exceptions:

1. Loud rhythm instruments—bass, drums, electric guitar, electric keyboards
2. Quiet rhythm instruments—acoustic guitar, piano
3. Lead vocal and doubled lead vocal (if desired)
4. Backup vocals (in stereo)
5. Overdubs—solos, percussion, synthesizer, sound effects
6. Sweetening—horns, strings

The lead vocalist usually sings a guide vocal or scratch vocal along with the rhythm section so that the musicians can get a feel for the tune and keep track of where they are in the song. You record the vocalist's performance but you will probably rerecord it later. That eliminates leakage and lets you focus on the lead vocal.

In a MIDI studio, a typical order might be:

1. Drum machine or soft synth drum set (playing programmed patterns)
2. Synthesizer bass sound
3. Synthesizer chords
4. Synth melody
5. Synth solos, extra parts
6. Vocals and miked solos

Track Assignments

Now you can plan your track assignments. Decide what instruments will go on which tracks of the multitrack recorder. The producer may have a fixed plan already.

What if you have more instruments than tracks? Decide what groups of instruments to put on each track. In a 4-track recording, for example, you might put guitars on track 1, bass and drums on track 2, vocals on track 3, and keyboards on track 4.

Remember that when several instruments are assigned to the same track, you can't separate their images in the stereo stage. That is, you can't pan them to different positions—all the instruments on one track sound as if they're occupying the same point in space. For this reason, you may want to do a stereo mix of the rhythm section on tracks 1 and 2, for instance, and then overdub vocals and solos on tracks 3 and 4.

It's possible to overdub more than four parts on a 4-track recorder. To do this, bounce (mix) several tracks onto one, then record new parts over the original tracks. Your recorder manual describes this procedure.

If you have many tracks available, leave several tracks open for experimentation. For example, you can record several takes of a vocal part using a separate track for each take, so that no take is lost. Then combine the best parts of each take into a single final performance on one track. Most recorder-mixers let you do these extra takes on virtual tracks.

It's also a good idea to record the monitor mix on one or two unused tracks. The recorded monitor mix can be used as a cue mix for overdubs, or to make a recording for the client to take home and evaluate.

Session Sheet

Once you know what you're going to record and when, you can fill out a session sheet (Figure 15.1). This simple document is adequate for home studios. "OD" indicates an overdub. Note the recorder-counter time for each take, and circle the best take.

Production Schedule

In a professional recording studio, the planned sequence of recording basic tracks and overdubs is listed on a production schedule (Figure 15.2).

```
SONG:  Escape to Air Island

TRACK      INSTRUMENT        MICROPHONE
1          BASS              DIRECT
2          KICK              AKG D-112
3          DRUMS             CROWN GLM-100
4          LEAD VOC OD       STUDIO PROJECTS B1
5          HARM. VOC OD      STUDIO PROJECTS B1
6          LEAD GUIT OD      SHURE SM57
7          KEYS L            DIRECT
8          KEYS R            DIRECT

TAKE 1   03:21 - 06:18   FS
     2   06:25 - 09:24   INC
     3   10:01 - 13:02
```

FIGURE 15.1
A session sheet for a home studio.

WEST WIND STUDIOS
PRODUCTION SCHEDULE

Artist: Steve Mills
Album: Long Distance Music
Producer: B. Brauning
Engineer: D. Scriven
Date: 1-17-09

Project files: c:\cakewalk projects\mills\
mr_potato_head.cwp   and   sambatina.cwp

Audio data folders: d:\mills\mr_potato_head\   and   \sambatina\

1. Record Mr. Potato Head
Instrumentation: Electric bass, drums, electric rhythm guitar, electric
lead guitar, acoustic piano, sax, lead vocal.
Comments: Record rhythm section with scratch vocal. OD sax, piano,
lead vocal. Double lead guitar in stereo.

2. Record Sambatina
Instrumentation: Electric bass, drums, acoustic guitar, percussion, synth
Comments: Record rhythm section with scratch acoustic guitar.
Record synth MIDI. OD acoustic guitar, percussion and synth.

3. Mix Mr. Potato Head
Comments: Add 40msec echo to toms. Increase reverb on sax solo.

4. Mix Sambatina
Comments: Add flanger to bass on intro only. Do automated flange
on percussion.

FIGURE 15.2
A production schedule.

Track Sheet

Another document used in a pro studio is the track sheet or multitrack log (Figure 15.3). Write down which instrument or vocal goes on which track. The track sheet also has blanks for other information such as take numbers. If you are using a DAW, you can enter this information by typing it in.

Microphone Input List

Make up a microphone input list similar to that seen in Table 15.1. Later you will place this list by the mic snake box and by the mixing console.

Be flexible in your microphone choices—you may need to experiment with various mics during the session to find one giving the best sound with the least console equalization. During lead-guitar overdubs, for example, you can set up a direct box, three

```
                        WEST WIND STUDIOS
                          TRACK SHEET

      Artist: S. Mills              Album: Long Distance Music
      Producer: B. Brauning         Engineer: D. Scriven
      Date: 1-17-09                 Song title: Dig Up Nebraska


      Take # 1              Start 00:31        Stop 03:31
      Take # 2  NC          Start 03:54        Stop 06:49
      Take # 3  LFS         Start 07:21        Stop 08:38
      Take # 4  FS          Start 09:01        Stop 09:12
      Take #(5)             Start 09:25        Stop 12:27


      Track 1:  elec bass  (Fender Precision  DI)
      Track 2:  kick
      Track 3:  snare
      Track 4:  rack tom
      Track 5:  floor tom
      Track 6:  overhead L
      Track 7:  overhead R
      Track 8:  rhythm gtr L  (Martin HD-28)
      Track 9:  rhythm gtr R  (Taylor 814ce)
      Track 10:  lead gtr  (Parker PM-20PRO, Roland Cube 60)
      Track 11:  keys L (Prophet '08  DI)
      Track 12:  keys R (Prophet '08  DI)
      Track 13:  scratch lead vocal
      Track 14:  lead vocal
      Track 15:  background vocals  1
      Track 16:  background vocals  2
```

FIGURE 15.3
A track sheet (multitrack log).

close-up microphones, and one distant microphone—then find a combination that sounds best.

Find out what sound the producer wants—a "tight" sound; a "loose, live" sound; an accurate, realistic sound, etc. Ask to hear recordings having the kind of sound the producer desires. Try to figure out what techniques were used to create those sounds, and plan your mic techniques and effects accordingly. Tips on choosing a microphone are given in Chapter 6.

Instrument Layout Chart

Work out an instrument layout chart, indicating where each instrument will be located in the studio, and where baffles and isolation booths will be used (if any). In planning the layout, make sure that all the musicians can see each other and are close enough together to play as an ensemble. Often a circular arrangement works well.

Table 15.1	A Microphone Input List	
Input	**Instrument**	**Microphone**
1	Bass	Direct
2	Kick	EV N/D868
3	Snare	AKG C451
4	Overhead L	Shure SM81
5	Overhead R	Shure SM81
6	High toms	Sennheiser MD421-II
7	Floor tom	Sennheiser MD421
8	Electric lead guitar	Shure SM57
9	Electric lead guitar	Shure SM57
10	Piano treble	Crown PZM-6D
11	Piano bass	Crown PZM-6D
12	Scratch vocal	Beyer M88

Keep all these documents in a single folder or notebook that is labeled with the band's name and recording date. Also include contact information, time sheets, invoices, and all the correspondence about the project.

SETTING UP THE STUDIO

About an hour before the session starts, clean up the studio to promote a professional atmosphere. Lay down rugs and place AC power boxes according to your layout chart.

Now position the baffles (if any). Put out chairs, stools, and music stands according to the layout. Run a headphone extension cable from each artist's location to the headphone junction box in the studio.

Place mic stands approximately where they will be used. Wrap one end of a microphone cable around each microphone-stand boom, leaving a few extra coils of cable near the mic-stand base to allow slack for moving. Run the rest of the cable back to the mic input panel or snake box. Plug each cable into the appropriate wall-panel or snake-box input according to your mic input list.

Some engineers prefer to run cables in reverse order, connecting to the input panel first and running the cable out to the microphone stand. That procedure leaves less of a confusing tangle at the input panel where connections might be changed.

Now set up the microphones. Check each mic to make sure its switches are in the desired positions. Put the mics in their stand adapters, connect the cables, and balance the weight of each boom against the microphone.

Finally, connect the musicians' headphones for cueing. Set up a spare cue line and microphone for last-minute changes.

SETTING UP THE CONTROL ROOM

Having prepared the studio, run through the "Session Preparation" section of Chapter 11 to make sure the control room is ready for the session. Then turn up the monitor system. Carefully bring up each fader one at a time and listen to each microphone. You should hear normal studio noise. If you hear any problems such as dead or noisy microphones, hum, bad cables, or faulty power supplies, correct them before the session.

Verify the mic input list. Have an assistant scratch each mic grille with a fingernail and identify the instrument the microphone is intended to pick up. If you have no assistant, listen on headphones as you scratch the grilles.

Check all the cue headphones by playing a tone or music through them and listening while wiggling each cable. Make sure they play at a reasonable level, not too loud.

SESSION OVERVIEW

This is the typical sequence of events:

1. For efficiency, record the basic rhythm tracks for several songs on the first session.
2. Do the overdubs for all the songs in a dubbing session.
3. Mix all the tunes in a mixdown session.
4. Edit the tunes and master the album.

Some musicians prefer instead to record all the parts for one song at a time.

After the musicians arrive, allow them 1/2 hour to 1 hour free setup time for seating, tuning, and mic placement. Show them where to sit, and work out new seating arrangements if necessary to make them more comfortable.

Once the instruments are set up, listen to their live sound in the studio and do what you can to improve it. A dull-sounding guitar may need new strings, a noisy guitar amp may need new tubes, the drums may need tuning, and so on. Adjust the studio lighting for the desired mood.

RECORDING

Before you start recording, you might want to make connections to record the monitor mix. This recording is for the producer to take home to evaluate the performance.

Many DAW programs have session templates: a group of tracks with EQ, compression, and effects plug-ins already in place. You simply load in the template and you're ready to go. You can create your own session templates, such as "16 tracks with aux 1 and 2" or "drum set."

Some DAWs let you create a track template, which is a format for a single track. You might have a vocal track template with the input channel, reverb, EQ, and compression already set up. Each time you're about to record another vocal, import the vocal track template.

Follow the mixer recording procedures described in Chapter 11. Set recording levels, then set cue mixes for the musicians' headphones. The monitor mix affects only what is heard, not what is recorded. To clarify the beat of the cue mix, you might create a mix of only a few instruments such as drums, bass, chords, and vocal.

When you're ready to record the tune, briefly play a metronome to the group at the desired tempo, or play a click track (an electronic metronome) through the cue system. Or just let the drummer set the tempo with stick clicks. A click track is helpful if you are starting with an acoustic guitar and are adding drums and bass later.

Start recording. Note the recorder counter time. Hit the slate button (if any) and announce the name of the tune and the take number. Then the group leader or the drummer counts off the beat, and the group starts playing.

The producer listens to the musical performance while the engineer watches levels and listens for audio problems. As the song progresses, you may need to make small level adjustments.

The assistant engineer (if any) runs the multitrack recorder and keeps track of the takes on the track sheet, noting the name of the tune, the take number, and whether the take was complete (Figure 15.3). Use a code to indicate whether the take was a false start, long false start, nearly completed, a "keeper," and so on.

While the song is in progress, don't use the solo function, because the abrupt monitoring change may disturb the producer. The producer should stop the performance if a major flub (mistake) occurs but should let the minor ones pass.

At the end of the song, the musicians should be silent for several seconds after the last note. Or, if the song ends in a fade-out, the musicians should continue playing for about 30 seconds so there is enough material for a fade-out during mixdown.

After the tune is done, you can either play it back or go on to a second take. If you connected your multitrack recorder to the insert jacks, use the faders to set a rough mix with EQ and effects. The musicians will catch their flubbed notes during playback; you just listen for audio quality.

Now record other takes or tunes. Pick the best takes. Punch in to correct errors on each best take, or fix errors with editing.

To protect your hearing and to prevent fatigue, try to limit tracking sessions to four hours or less. Take breaks to give your ears and body a rest.

Relating to the Musicians

During a session, the engineer needs not only technical skills, but also people skills. It's important to respect artistic personalities and to keep the creative energy flowing.

First, learn the names of the band members and refer to them by name during the session.

Musicians are often nervous at the beginning of a session. Having a sense of humor and exuding confidence helps to put the artists at ease. Talk about the band's music, origin, or instruments. Be sure that the artists are comfortable—adjust the layout and lighting, offer beverages and snacks. Tell new clients that mistakes are normal and are easy to fix. Ask band members, "Is the mix okay in your headphones? Is the volume all right? Does your recorded instrument sound okay?"

Don't badmouth other musicians, and respect the privacy of your clients. Each musician you record should feel confident that you're not telling others about their mistakes.

When you want to try something that could enhance a song, don't just do it. Ask first. Say, "What would you think about doubling the guitar in stereo?" or "I want your input on this."

During a playback, try not to point out the errors. The musicians will hear them and will try to do better on the next take. Don't say, "That sucked." Say "It's almost there—do you want to punch in a few spots?" Or say "That was good, but how about another take?" The goal is to make the best record you can, not to make clients feel bad about themselves. Give a musician time to practice during overdubs, too.

If the band members are getting tired or are having trouble getting through a particular song, you might say "Let's take a break" or "Let's try another song and come back to this one later."

Finally, if a piece of gear breaks or your software develops a glitch, try to work around it quietly without making it obvious. Musicians need an engineer who handles technical problems in a professional way.

OVERDUBBING

After recording the basic or rhythm tracks for all the tunes, add overdubs. A musician listens to recorded tracks over headphones and records a new part on an open track. See the Chapter 11 sections "Overdubbing" and "Composite Tracks."

You might have the musician play in the control room while overdubbing. You can patch a synth or electric guitar into the console through a direct box, and feed the direct signal to a guitar amp in the studio via a cue line. Pick up the amp with a microphone, and record and monitor the mic signal.

Do any drum overdubs right after the rhythm session because the microphones are already set up, and the overdubbed sound will match the sound of the original drum track.

BREAKING DOWN

When the session is over, tear down the microphones, mic stands, and cables. Put the microphones back in their protective boxes and bags. Wind the mic cables onto a power-cable spool, connecting one cable to the next. Wipe off the cables with a damp rag if necessary. Some engineers hang each cable in big loops on each mic stand. Others wrap the cable "lasso style" with every other loop reversed. You learn this on the job.

Put the labeled recording and session documents in a labeled box or file folder. Normally the studio keeps the multitrack recording for possible future work unless the group wants to buy or rent it. Backup the multitrack recordings (audio files) and session files (track setups for each song) to another medium, such as an external hard drive, CD-R, or DVD-R. Glyph makes pro-quality external hard drives for audio and video storage.

If you're using a hardware mixing console, log the console and effects settings by writing them in the track sheet or reading them slowly into a portable recorder. At a future session you can play back the recording and reset the console the way it was for the original session. Automated consoles can store and recall the console's control adjustments. DAWs remember both the mixer settings and the plug-in settings.

MIXDOWN

After all the parts are recorded, you're ready for mixdown. Follow the mixdown procedures in Chapter 11 and repeat for all the song mixes. Again, be sure to backup your session files (the mix and edit settings) and audio files (multitrack audio recordings and stereo mixes).

MASTERING

After recording the mixes, you can burn a CD of the mixes as they are, or you can master an album or demo of those mixes. Mastering is the last creative step before burning the final CD used for duplication or replication. In mastering you edit out noises or false starts before the beginning of each song, put the songs in the desired order, insert a few seconds of silence between songs, and make each song the same loudness and the same tonal balance. The goal is a consistent sound from track to track, so that everything flows better and the album sounds unified. You also might try to make the CD as loud or "hot" as possible (without destroying its sound quality).

At this point, you have three choices:

1. If the CD will be just a reference, not a demo or album, you can burn a CD of the unedited mixes and stop there.
2. If the CD will be a demo or album, burn a CD of the unedited mixes, then send it to a mastering engineer for mastering.
3. Or you can edit and master the mixes yourself, then burn a CD. It will be the CD premaster for duplication or replication. "Duplication" is making copies with a CD-R burner, while "replication" is stamping CDs from a glass master (the professional standard).

Let's look at each option above.

Burning a Reference CD

You can use the CD-burning software that came with your computer CD burner, or use other software. Some examples of CD-burning programs are Express Burn Disc from NCH Software, Roxio's Creator 2012 and Toast 11 Titanium, Sony CD Architect, and Nero Burning ROM. Some DAW software (such as WaveLab) includes a CD-burning application.

Before you start, note that most CD-burning software puts 2 seconds of silence between songs. If you want to have extra seconds of silence at the end of a song, record the silence as part of the song's WAV file. You might want to normalize each song's WAV file. This raises the level of the song so that the highest peak in the song reaches maximum level: 0 dBFS (decibels full scale) or whatever level you specify. Normalizing does not make the tracks the same loudness, because loudness depends on the average signal level, not the peak level.

Here's a CD-burning procedure:

1. Launch the CD-R burning software.
2. Select which WAV files you want to put on the CD-R, drag them to the playlist, and arrange them in the desired order. The total playing time must be less than the CD-R length (74 or 80 minutes).
3. Choose "On the Fly" or "Disk Image." When recording on the fly, the computer grabs the sound files from random locations on hard disk and puts them in order as the CD-R does a burn. When recording a disk image, the computer rewrites the sound files to a single, contiguous space on your hard disk and puts them in order. Then you copy the disk image from hard disk to CD-R. Disk Image is less likely to produce CD errors than On the Fly.

4. Set the burn speed. Some CD recorders and blank CDs can record up to 52 times normal speed. Normal speed is 172 kBps (Bytes is capital B), double speed is 344 kBps, etc. A 52× recorder can burn a 74-minute CD in less than 2 minutes. Some CD burners automatically select the optimum speed based on what the blank CD-R can handle. High speeds do not degrade sound quality seriously, but they tend to increase errors—which the CD player may or may not correct accurately. The recommended speed to prevent errors is 2× to 4× when making a final master CD to be sent out for replication. But some CD recorders and blank CD-Rs have fewer errors at higher speeds.

5. Set the software to Disc-at-Once Mode. Start recording the CD-R. The WAV files will transfer in order to the CD-R disc. To prevent glitches, do not multitask while the CD is recording.

CD-R discs cannot be erased and used over again, so try to make everything right before you burn a disc. You could do a practice burn on a CD-RW.

6. As soon as the recording is done, the display will indicate that the table of contents is being written. Eventually, the system will beep and eject the disc. To prevent error-causing fingerprints, be sure to handle the disc only by the edges. Pop the disc in an audio CD player, press Play, and check that all the tracks play correctly.

Sending Out Your CD for Mastering

You might prefer to send your CD of mixes to a good mastering engineer. This person can listen to your program with fresh ears, then suggest processing for your album that will make it sound more commercial. He or she is likely to have a better monitor system and better equipment than yours. They have heard hundreds of recording projects done by others, and know how to make your CD sonically competitive.

If you plan to have your program mastered outside, do not apply any signal processing to your finished mixes such as editing, level changes, compression, normalization, fades, or EQ. Let the mastering house do it with their better equipment and software. Also, leave some headroom by recording the finished mixes at about −3 dBFS maximum in peak meter mode.

Deliver your mixes in the highest possible resolution that the mastering engineer can accept, such as 96 to 192 kHz and 24 to 32 bits on a data CD (ISO 9660 format). However, a 24 bit/44.1 kHz format is usually adequate. Copy the mixes' WAV files to CD, rather than creating CD audio tracks (.cda files). The resulting data CD can be read by a CD-ROM but cannot be played on a CD player. If your CD-burning program does not support data CDs, either get one that does, or create an audio or music CD at 16 bits/44.1 kHz.

Also send the mastering engineer the sequence of songs and any suggestions on how you would like certain songs to be improved or edited. The mastering engineer might suggest that you need to redo some mixes.

Mastering Your Own Album

You can master an album with your multitrack recording software, or with a mastering program such as Steinberg Nuendo, Sony Sound Forge, BIAS Peak Pro, IK Multimedia T-RackS, or Magix Samplitude. A suite of plug-ins is Sonalksis Mastering Suite.

First, discuss the order of the songs on the recording. The first track on the album should be a strong, accessible, up-tempo piece. Change keys from song to song. The last track should be as good as or better than the first to leave a good final impression.

Mastering engineer Bob Katz offers this advice for song sequencing: Create the album in sets of one to three songs of the same tempo. Make sure the songs flow from one to the next. Here's a suggested song order:

1. An up-tempo, exciting song that hooks the listener.
2. After a short space, an up-tempo or mid-tempo song.
3. After three or four songs, slow down the tempo.
4. Reach a climax near the end of the album.
5. The last song should sound relaxed and intimate, perhaps using fewer members of the band.

You can drag the WAV files of your song mixes into iTunes and experiment with various sequences.

To get your mastered songs onto a CD, you need CD-burning software, either consumer or pro. With a consumer CD-burning program—such as Roxio's Creator 2012, Roxio's Record Now Music Lab, or Nero—you drag the WAV files of each of your song mixes into a playlist, then burn the CD.

With a professional program, you create one long WAV file of all the song mixes. You can adjust the timing between each pair of songs. You specify when each song starts, and the software creates a table of contents on the CD made of song-start IDs. This feature is called PQ subcode editing. Some programs that do this are Sony CD Architect and Goldenhawk Technology CDRWIN (www.cdrwin.de/cdrwin-en.html). Some mastering software includes a CD-burning application that lets you set start IDs. Examples are Steinberg WaveLab, Sony Sound Forge, Magix Samplitude and Sequoia, Sadie, and Amadeus Pro (for Mac).

Here's a suggested procedure to master an album:

1. If you already recorded your mixes directly onto hard disk as stereo WAV files, skip to step 4.
2. If your song mixes are WAV files on a CD-R, put the CD-R in your CD-ROM drive and copy each song to your hard drive. Then go to step 4. If your song mixes are audio tracks (.cda files) on a CD-R, convert each track to a WAV file on your hard drive using ripper software. Skip to step 4.

3. If your mixes are on a standalone 2-track recorder without a USB or FireWire port, plug the recorder's digital output into the digital input of your audio interface. If your interface lacks a digital input, use analog out to analog in. Launch the recording software. Set levels if necessary, start recording on your hard drive, and play the 2-track recording containing your mixes. All the mixes will copy in real time to your hard drive to create a single long WAV file.

If your 2-track recorder has a USB or FireWire port, plug it into your computer's matching port. The recorder will appear as a hard drive to your computer. Copy the WAV file(s) of the mixes to your computer hard drive.

4. Now that all the song mixes are on your hard drive, launch your DAW software. Import the first song's WAV file to a track at time zero. The waveform of the song appears as a clip or segment of audio. Slip-edit or trim the beginning and end of the song clip to remove extra space and noises. Repeat for the other songs.

Click on the trimmed Song 1 and slide it 10 frames to the right in your mastering software. That way the CD player won't miss the beginning of Song 1.

It's convenient to put each song clip on a different track, one after another (Figure 15.4). That way you can easily adjust the spacing between songs, and apply different fader settings and processing to each song as needed. Assign all the tracks to the same stereo bus so you can apply bus processing to them later.

FIGURE 15.4
Placing song clips on successive tracks makes mastering easier.

5. Now that all the songs are in place and trimmed, add a fade-out at the end of certain songs if desired. If your CD-burning program can write PQ sub-codes (start IDs), you can crossfade between two songs: overlap their clips near the transition point, and use the crossfade function in your DAW. Or fade out the end of one song and fade in the beginning of the next (as shown in Figure 14.17).

6. If your CD-burning program can write PQ subcodes (start IDs), you can adjust the spacing or gap between songs. Two to three seconds of silence between songs is typical, but go by ear. Use a longer space if you want to change moods between songs. Use a shorter space to make similar songs flow together. A short space also works well after a long fade-out because the fade-out acts like a long pause between songs.

7. Click on and play part of each song's waveform to check for loudness. Make all the songs equally loud by adjusting the track's fader for each song. If you put all the songs on the same stereo track, set up an automation volume envelope.

Here's one method to match song levels. If most songs are equally loud, set the fader to zero before each of those songs in order to avoid processing. Then turn down any louder songs and turn up quieter songs to match the rest of the songs. Do this by ear. You might use post-fader RMS metering on the stereo mix bus as a rough indicator of average levels.

CAUTION: If you increase a song's level, make sure the peaks in the song don't clip.

Another way to match song levels is to find the song with the highest peaks, and leave its level alone Adjust the levels of other songs so they match the loudness of the song with the highest peaks. Again, do this by ear. You may need to turn up the intro of a song so it works in context with the previous song. Consider using Harmonic Balancer (www.har-bal.com), a tool to match song levels and tonal balances.

8. Apply EQ to songs that need it. Put songs with different EQs on different tracks. If necessary, touch up levels after adding EQ.

9. If desired, apply multiband compression, limiting, and normalization to the mix bus in order to get a "hot" or loud CD. Don't overdo it, or your project will be distorted and fatiguing to listen to. Squashed dynamics can suck the life out of music. If you apply only peak limiting and normalization, you'll get a hot CD without altering the musical dynamics. *Play audio clip 40 at www.elsevierdirect. com/companions/9780240811444.*

The idea is to knock down the peaks in the waveform because they do not contribute to perceived loudness—the average level does. Once you limit the peaks about 6 dB, you can normalize (raise the overall level) and thus create a louder program.

Normalizing to 100% of full scale can create errors with some D/A converters, so you might normalize to 96% or −0.3 dB.

THE LOUDNESS WARS

Musicians and record labels, more than recording engineers or mastering engineers, are requesting hotter levels on CDs. One reason is that a low-level CD played with others in a CD changer or MP3 player tends to sound less exciting. Musicians also want their songs to stand out on the radio.

Loudness depends mostly on average (RMS) level rather than peak level. So in an attempt to raise average levels, engineers limit or clip the peaks. Overall compression of the mix is common too. Unfortunately, the song loses its dynamic range and its musical excitement.

When I master a pop-music CD, I often compromise by limiting peaks by no more than 7 dB, then normalizing to −0.3 dB without overall compression or clipping. That retains the dynamic range of the music, has no clipping, and sounds pretty competitive in loudness. But it sacrifices some of the clarity and impact of percussive transients.

If playback equipment had switchable automatic volume control (compression), then we wouldn't need to maximize levels in the recording. The iTunes Sound Check feature is a step in the right direction: it adds gain and limits peaks of quiet recordings to match the loudness of overly compressed ones. But that reduces the sound quality of the quieter recordings. It would be better to turn down the overly compressed recordings to match the average level of more dynamic recordings. That way, all songs would sound equally loud, and mixes with a wide dynamic range would keep their clarity and impact.

According to mastering engineer Bob Katz, hot CDs and quiet CDs end up at the same level when processed with radio-station compressor/limiters. Overcompressed CDs do not sound louder on the air; they just sound more distorted. So ignore the "loudness wars" and avoid excessive compression. Bob recommends that pop music be mastered at an average or RMS level of −14 dBFS while listening at 83 dBSPL C-weighted, slow meter reading. He suggests an average level of −20 dBFS for classical music.

10. After your musical program is edited, you have two options:
    1. If you have professional CD burning software that can write PQ subcodes, export (save) the mastered song mixes to a single 16 bit/44.1 kHz stereo wave file. If your recording was 24 bit, first turn on dithering. That retains much of the 24-bit sound quality when converting to 16-bit format.
    2. If you have consumer software that burns a CD from a list of song WAV files, click on one song's clip in your DAW to highlight it, then export the mix to a single 16 bit/44.1 kHz stereo wave file. If your recording was 24 bit, first turn on dithering. Repeat for all the song clips.

TRANSFERRING THE MASTERED PROGRAM TO CD-R

At this point your hard drive contains either one long WAV file of all the mastered songs, or several WAV files, each one a mastered song. If the latter, launch your CD-burning application and drag the WAV files in order to the playlist. To prevent errors on the CD, burn it at 4x speed and don't multitask. Check the finished CD in a CD player.

Suppose you have pro CD-burning software. It creates CD tracks from a single WAV file by following a cue sheet (a text file that lists the start time of each song). When the CD is recorded, the software creates a start ID for each song based on the cue sheet. With this method you can create CD song-start IDs that occur even during a continuous program, such as a live concert recording.

CDRWIN at www.goldenhawk.com is such a program. I'll describe how to use it as an example.

1. Note the start time of each song in the DAW editing software's edit decision list or playlist. Or check the start time of each song clip by right-clicking on it or by looking at its beginning on the project time line. The start time of each song is specified in your DAW in minutes:seconds:frames, and there are 30 frames per second. Write down the start time of each song.
2. In the cue sheet, make each song's start time about one-third of a second (10 frames) before the actual start time of each song. That way the CD player's laser will have time to find the track before playing audio. For example, if a song actually starts at 12:47:28, make its start time in your cue sheet 12:47:18. List Song 1's start time as 00:00:00. The cue sheet should also include the name and file path of the mastered program's WAV file.
3. Place a blank CD-R in your CD burner. Do not put a paper label on the CD before recording because the label can cause jitter.
4. Open your CD-burning program and load in the cue sheet text file. Set the recording speed to 2× to 4× speed to reduce errors. Some CD recorders and blank CD-Rs have fewer errors at higher speeds.
5. Start recording the CD. The CD-burning program writes the cue sheet's start times in the CD-R's table of contents, and copies the mastered program's WAV file to the CD-R. Each CD track starts at the time you specified in your cue sheet. To prevent glitches, do not multitask while the CD is burning.
6. When the CD is finished, handle it by its edges to prevent error-causing fingerprints. Play the CD from start to finish to check for glitches. Press the track-advance button to make sure each song starts at the right time. If you look at the CD's contents in your computer, you'll see a separate.cda (CD audio) file for each song.

You can identify the CD either by writing on the label side with a felt-tip marker or by sticking on a paper label. To avoid damaging the CD plastic layer, use water-based

Sharpie or Memorex CD marking pens, available in the CD-R section of many stores. Use a label applicator device and label design software such as Stomp's CD Stomper, Neato's MediaFACE, or Roxio's Record Now Music Lab. Some software can even print liner notes, including the track numbers, titles, and timing. If you are duplicating CDs yourself, you might want to use a CD printer rather than labels. Caution: Adhesive CD labels can damage the data on disc over several years.

Another way to create master CDs is to use the Tascam DV-RA1000HD.

CD-Text and ISRC Codes

You can add nonaudio data to your CD when you create the cue sheet. If you want the listener to see on-screen each song's title and performer as the CD plays, create a CD-Text file using the CD-Text editor that comes with most CD-burning software. Include the file path of the CD-Text file in the cue sheet. Not all CD players support CD-Text.

Some players get CD-Text information from an online database rather than from the CD. You upload the CD-Text information to the database so that the player can display it. For example, iTunes needs to access an online CD database called Gracenote (formerly CDDB) to get the artist and song titles. Gracenote is at www.gracenote.com.

Every database has a different procedure for uploading CD-Text. When using the iTunes database, follow this procedure:

1. Insert your CD and open iTunes.
2. Select all tracks in the iTunes window.
3. Select Get Info and enter the album name, artist, year, genre, copyright info, and so on.
4. Select each track individually, select Get Info, and enter the track's correct name.
5. Go to Advanced > Submit CD Track Names. Your information will uploaded to the Gracenote database. When someone puts your CD into iTunes, your info will show up.

When using the Windows Media Player database, send a copy of your CD to AMG (All Media Guide) at www.rovio.com.

Many players access the freedb database at www.freedb.org. Go there to see a list of supported players you can use to submit CD information. Popular applications are Exact Audio Copy and Nero. Another database is Muse at www.rovio.com.

iTunes prior to version 7 did not display CD-Text. The current version of iTunes can display CD-Text information, but only if it has been uploaded to the Gracenote database.

The International Standard Recording Code (ISRC) is a 12-character code that identifies each track. It's a tool for royalty collection and anti-piracy. Information on applying for ISRC codes is at www.recordlabelresource.com/isrc-registration.html and www.riaa.com. If the CD tracks have ISRC codes, include them in the cue sheet as well. Some online music providers, such as CDBaby, can create the ISRC codes for you.

MASTER LOG

Type or print a log describing the CD-R master (such as shown in Table 15.2) and include it with your master.

Also include the CD liner notes such as song lyrics, instrumentation, composers, credits, arrangers, publishers, copyright dates, and artwork. Include contact information for the artist, engineer, and mastering engineer. Stay in contact with the duplication or replication house, especially about artwork.

Before you send out the master, be sure to make another copy that you keep in your studio. This copy can be used if the master is lost or damaged.

Note that the CD-R premaster doesn't leave the studio until all studio time is paid for! When this is done, send the CD-R to the duplication house.

The document "Recommendations for Delivery of Recorded Music Projects" provides examples of session sheets, tracks sheets, and so on. It also recommends media for delivering and backing up master recordings. It can be found at www.aes.org/technical/documents. Search for AESTD1002.1.03-10_1.pdf. Also see the NARAS guidelines at http://www.grammy.org/recording-academy/producers-and-engineers/guidelines.

It's amazing how the long hours of work with lots of complex equipment have been concentrated into that little CD-R—but it's been fun. You crafted a product you can be proud of. When played, it recreates a musical experience in the ears and mind of the listener—no small achievement.

Table 15.2	A CD Master Log		
Track	**Start MM:SS:FF**	**Duration**	**Title**
1	00:00:00	03:23	60 Gigabytes to Go
2	03:24:20	03:15	Unplugged Plug-In
3	06:41:00	02:49	Digital Dropout Blues
4	09:32:05	02:55	Late and See
5	12:29:11	03:07	Buffer the Vampire

Total running time: 15:47
CD Master Log
Album title: Don't Press That Button (demo)
Artist: Puff Daddy Bartlett
Mastering date: 2-26-06

COPYRIGHTS AND ROYALTIES

It's important to copyright original songs so that others cannot claim them as their own. Go to www.copyright.gov/forms to download form SR (for Sound Recordings) and form CON (continuation of form SR to add more songs). The filing fee is $45 for submission by regular mail or $35 by electronic submission. That fee lets you copyright several songs by the same composer(s) as a single work on one CD. Just be sure to list all the song titles. Send the forms, CD, and fee in a box to the U.S. Copyright Office, address at that website.

If you record and sell cover tunes (songs that have been recorded by other artists), or if you sample other artists' music in your own recording, you must obtain a "mechanical license" and pay a royalty fee to BMI or ASCAP. The Harry Fox Agency (HFA) is the premier musical industry resource for licensing. They offer a service called Songfile, a simple, fast way to obtain licenses for 25 to 2,500 copies of a recording as a CD, cassette, LP, or digital download. If the cover song is under 5 minutes long, the royalty fee is about $45 per song for every 500 CDs that you distribute. For details check out www.harryfox.com, www.bmi.com, and www.ascap.com.

CHAPTER 16

MIDI and Looping

Welcome to the computerized world of MIDI equipment—samplers, synthesizers, sampling keyboards, drum machines, and sequencers. Because other texts explain this equipment in detail, we'll offer brief descriptions here.

Musical Instrument Digital Interface (MIDI) is a standard connection between electronic musical instruments and computers that allows them to communicate with each other.

Here are some things that MIDI lets you do:

- Play a piano-style keyboard, drum pads, MIDI guitar, or breath controller to produce a sound like any instrument.
- Record and play back a musical performance that is independent of the sound of that performance. You can hear your performance played by any musical instrument or synthesized sound, or change the instrument sound after your performance is recorded.
- Record a difficult part slowly, or one note at a time, and play it back at any tempo you wish.
- Enter the notes on a musical scale or on a piano-roll grid, one at a time, and play them back at any tempo.
- Edit any note of the performance—change its timing, length, or pitch.
- Create the effect of a band playing. Record a performance that is played by drum samples, overdub another performance that is played by a synthesized bass, overdub another performance played by piano samples, and so on.
- Combine or layer the sounds of two electronic musical instruments by playing them both with the same performance.
- Quantize the notes to make them rhythmically precise.

The MIDI signal is a stream of digital data—not an audio signal—running at 31,250 bits per second. It sends information about the notes you play on a MIDI controller, such as a piano-style keyboard or drum pads. Up to 16 channels of information can be sent on a single MIDI cable.

There are three types of MIDI ports on MIDI devices:

1. MIDI IN receives data going into the device.
2. MIDI OUT sends out data generated by the device.
3. MIDI THRU is like MIDI OUT, but duplicates the data that is at the MIDI IN port.

Connect MIDI OUT from the sending device to MIDI IN of the receiving device. For example, connect a keyboard controller's MIDI OUT to the MIDI IN connector of a MIDI/audio interface, a sequencer, or a sound module. Use MIDI THRU to connect two or more receiving devices in a row. For example, connect a keyboard controller's MIDI OUT to sequencer MIDI IN, and connect sequencer MIDI THRU to sound module MIDI IN.

You can compose and record music using MIDI without knowing the codes or data that MIDI equipment uses. Those codes are described in the MIDI 1.0 Specification. If you want to explore those codes, see these resources:

www.ibiblio.org/emusic-l/info-docs-FAQs/MIDI-doc/index.html
www.midi.org/techspecs/midispec.php
www.midi.org/techspecs/gm.php
http://en.wikipedia.org/wiki/MIDI_1.0

Also see MIDI Clock and MIDI Time Code in the Glossary. See Appendix E for books on MIDI.

MIDI COMPONENTS

In addition to the usual studio gear—monitors, computer DAW, cables, etc.—you can use the following equipment to record with MIDI:

- MIDI controller
- Sequencer
- Synthesizer
- Sampler and sample CDs
- Drum machine or soft synth playing drum-set samples
- MIDI computer interface, audio interface with MIDI I/O, or computer USB port
- MIDI cables
- Equipment stand

You learned about most of these in previous chapters, but a review might help at this point.

FIGURE 16.1
M-Audio Oxygen 49, an example of a MIDI keyboard controller. It communicates with a computer by USB as well as MIDI.

A **MIDI controller** is an instrument that generates MIDI data when you play on it. Examples are a piano-style keyboard (Figure 16.1), drum pads, MIDI guitar, or a MIDI breath controller. A synthesizer or drum machine can act as a controller.

Some keyboard controllers are compact and lightweight, and connect to your computer via a USB cable. Examples are the Novation 49 SL MkII series, M-Audio's Axiom and Oxygen series, CME M-Key v2, Akai MPK Mini Controller, and Roland A300Pro. Because they have fewer keys than a standard keyboard, they use octave up/down buttons.

Some MIDI controllers have an audio interface so you can connect microphones to your computer.

A **sequencer** is a device or program that lets you record, edit, and play back MIDI data. A recording done on a sequencer is called a sequence, which is a MIDI song. Unlike an audio recorder, a sequencer does not record audio. Instead, it records the key number of each note you play, note-on signals, note-off signals, and other parameters such as velocity, pitch-bend, and so on. A sequencer captures a performance, not the sound. Then any sound you want can be played by that performance. You can edit the recorded performance (sequence) to fix wrong notes, etc.

A sequencer records and plays MIDI files (.mid files). They are sequencer recordings of your own performances, or are MIDI files that can be downloaded from the Web.

The sequencer can be a standalone unit, a circuit built into a keyboard instrument, or a computer running a sequencer program (Figure 16.2). Like a multitrack audio recorder, a sequencer can record 8 or more tracks, with each track containing a performance of a different instrument.

A step up from a sequencer program is a **MIDI/audio recording program** (also called DAW software). It records both MIDI sequences and digital audio tracks on your hard

FIGURE 16.2
A screenshot of a piano-roll view, part of a MIDI sequencer program.

drive, and keeps them in sync. In other words, this program lets you add microphone signals such as a vocal, sax, etc., to MIDI sequences. This software and your computer form a digital audio workstation (DAW). Details are given in Chapter 14 on computer recording.

A **synthesizer** is a musical instrument that creates sounds electronically. It can play MIDI data, either from a built-in keyboard, a separate MIDI controller, a MIDI sequence, or a MIDI file downloaded from the Web. Synthesizers come in four forms: piano-style, sound module, software, or a synth chip on a sound card.

- A piano-style synth has a piano keyboard and built-in sound generators. You might want to use more than one synth to expand your palette of sounds.
- A sound module or tone module is a synthesizer without a keyboard. This stand-alone device is triggered by a sequencer or a MIDI controller.
- A soft synth or virtual instrument is a synthesizer that is simulated in software. It runs in your CPU. The GUI of the software looks like a hardware synthesizer. A wavetable soft synth plays samples of real instruments, which sound more realistic than an FM soft synth.
- Another option is a synth chip, which is built into many sound cards.

Some soft-synth companies are E-MU, Arturia, Quantum Leap, Waldorf, Spectrasonics, Rob Papen, FXpansion, MOTU, IK Multimedia, Native Instruments, Cakewalk, Vir2 Instruments, Vienna Instruments, Garritan, Toontrack, Wavemachine Labs, Inaudio, Submersible Music, Gen, Slate Digital, Avid, McDSP, and many more. Spectrasonics Stylus RMX sounds great for drums, and Trilian is popular for bass.

The musical sounds that synthesizers make—such as a fretless bass, grand piano, sax, or drum set—are called patches, programs, or presets. A multitimbral synthesizer can play two or more patches at once. A polyphonic synthesizer can play several notes at once (chords) with a single patch.

A **sampler** is a device that records sound events, or samples, into computer memory and plays them back when activated by a sequencer MIDI file or MIDI controller.

A sample is a digital recording of one note of a real sound source: a flute note, a bass pluck, a drum hit, etc. A sample also can be a digital recording of a short musical phrase. The sampling process is described in Chapter 13 under "Digital Recording."

A **soft sampler** is software that plays samples and lets you map them along your keyboard. In other words, you can tell the sampler which notes on the keyboard play which samples. A sampler lets you import any sound, such as a WAV file of a single note that you recorded, or a file from a sample CD. In contrast, a soft synth generates its own sounds—either the ones supplied with it, or your own customized versions.

Some software sample players are Korg Legacy Collection, IK Sampletank, IK Multimedia Miroslav Philharmonik, Native Instruments Kontakt 5 and Komplete 8, Quantum Leap, MOTU MachFive 3, EXS-24, Garritan Personal Orchestra, and Steinberg HALion. Synthogy Ivory is a great-sounding piano sample player. Also check out these websites: www.synthzone.com/softsyn.htm, www.sonicspot.com/softwaresynth.html, www.hitsquad.com/smm/, www.soundsonline.com, and www.kvraudio.com.

A **soundfont** is an audio sample (an instrument patch) in a special SF2 format. It's like a WAV file, but also includes a key range so that when you play a MIDI note number (keyboard key), it plays the sample pitch assigned to that note number. Soundfonts also include velocity switching, note envelope, looping, release sample, filter, and low-frequency oscillator (LFO) settings. A single soundfont can contain many WAV files of different pitches. You can import a soundfont into a sample player. To use soundfonts, you need a SoundFont player, a MIDI controller, MIDI interface, and MIDI sequencing software.

Often a sampler is built into a sample-playing keyboard, which resembles an electronic piano. It contains samples of several different musical instruments. When you play on the keyboard, you hear the sample notes. The higher the key you press, the higher the pitch of the reproduced sample.

Be sure to check out **sample libraries**. You can buy CDs with samples for use in your own projects. You copy the samples to your hard drive, then load them into a software sample player and trigger them with a sequencer or MIDI controller.

Several DAW recording programs have samples and loops included. You can also download samples from the Web. Do a Google search for piano sample freeware, Hammond B3 sample freeware, and so on.

A **drum machine** or beat box is a device that plays built-in samples of all the sounds of a drum set and percussion (Figure 16.3). It also is a sequencer that records and plays back drum patterns played or programmed with built-in keys or drum pads. Some units can sample sounds. Most recorder-mixers have a drum machine built in. In a DAW, you might use a soft sampler that plays drum samples—a virtual drummer.

Some drum plug-ins and drum machines for MIDI/audio recording programs include EastWest (Quantum Leap) Storm Drum 2; Native Instruments Battery 3; Toontrack

FIGURE 16.3
BOSS DR-880, an example of a beat box.

EZdrummer, Beatstation, and Superior Drummer 2; Monkey Machine freeware (http://rinki.net/pekka/monkey/); Gen 16 Digital Vault Z-Pack; Steven Slate Digital EX and Signature Drumkits; FXpansion BFD ECO; Sonic Reality Ocean Way Drums; Submersible DrumCore; and Cakewalk Session Drummer 2.

A **librarian program** organizes your collection of samples and synth patches.

A **voice editor program** lets you create your own synthesizer patches.

A **notation program** converts your performance to standard musical notation. You can edit the notes, add lyrics and chords, and print out a copy.

A **MIDI interface** plugs into a user port in your computer and converts MIDI signals into computer data so that it can be recorded, edited, and played back with sequencer software. Many sound cards and audio interfaces include a MIDI interface. Some keyboard controllers connect to a computer via USB, so no MIDI interface is needed.

The iConnectivity iConnectMIDI is a MIDI interface with two USB mini ports that can connect to an iOS device (iPad, iPod Touch) or to a Mac/PC.

FIGURE 16.4
The Kurzweil PC3K8, an example of a keyboard workstation.

If you have two or more hardware synthesizers, or a synth and a drum machine, you might need a **line mixer** to blend their audio outputs into a single stereo signal.

MIDI cables carry MIDI signals and are used to connect synths, drum machines, and computers together so that they can communicate with each other. A MIDI cable is a 2-conductor shielded cable with a 5-pin DIN plug on each end. Pins 4 and 5 are the MIDI signal, pin 2 is shield, and pins 1 and 3 are not connected.

An **equipment stand** is a system of tubes, rods, and platforms that supports all your equipment in a convenient arrangement. It lets you use short cables and has a small footprint.

A **keyboard workstation** (Figure 16.4) includes several MIDI components in one chassis: a keyboard, a sample player, a sequencer, and perhaps a synthesizer and disk drive. That's everything you need to compose, perform, and record instrumental music. Many workstations include drum sounds so that you can get by without a separate drum machine. A few include a mic input so you can record vocals or acoustic instruments.

The rest of this chapter describes recording procedures for some MIDI studio setups. You'll also need to read your instruction manuals thoroughly and simplify them into step-by-step procedures for various operations. Note that each piece of MIDI gear has its own idiosyncrasies, and the instructions may have errors or omissions. If you have questions, call or e-mail tech support for your equipment. Also looks for books and videos devoted to your specific musical instrument.

RECORDING MUSIC MADE BY SOFT SYNTHS

In this method, you will play a MIDI controller and record the performance on a MIDI track in a MIDI/audio recording program on your computer. Software synths, samples, and drums in your computer will play the sounds during recording and playback (Figure 16.5). They are included with most recording software, or you can buy them separately.

FIGURE 16.5
rgc:audio Pentagon, an example of a soft synth.

The MIDI Signal Chain

Figure 16.6 shows a MIDI signal chain that uses a soft synth. From start to finish (left to right), these are the components and their functions:

1. **MIDI controller.** A piano-style keyboard, drum pads, or breath controller that outputs MIDI code (note on, note off, key number, etc.) when you play it. The code is tagged with a channel number from 1 to 16, which you set in the controller.
2. A **MIDI cable** from your controller's MIDI OUT to your interface's MIDI IN.
3. A **MIDI interface** or **MIDI/audio interface.** It converts the MIDI signal into a data format that your computer can understand.
4. A **PCI, USB,** or **FireWire** connection from the interface to the computer. The MIDI data you generate with a controller travels on this cable to your computer.
5. A **MIDI driver** (not shown). This is a small program in your computer that allows your recording software to transfer data to and from an audio/MIDI interface.
6. **Sequencer** (MIDI/audio) software running in your computer. It records the MIDI signal to the computer's hard drive and plays it back.
7. **Soft synth/soft sampler.** A virtual instrument plug-in in your DAW program. As you record your performance, you are listening to the software synthesizer

FIGURE 16.6
A MIDI signal chain that uses a soft synth.

(or sample player) play music according to your keyboard presses. So in addition to being recorded, the MIDI signal goes to a soft synth or sample player of your choice.

8. A **PCI**, **USB**, or **FireWire** connection (same as number 4 above) carries the synth's digital audio to your audio interface. A single USB or FireWire cable is used to send MIDI data to your computer, and to send audio from your computer to your interface and monitor speakers.

9. The **MIDI/audio interface** converts the synth's digital audio to analog, and sends that musical signal to its stereo output channels. This is same interface as in number 3 above.

10. **Monitors**. The stereo output of your interface feeds amplified speakers or headphones so you can hear the sound of your performance.

During playback, the recorded MIDI sequence "plays" or triggers the soft-synth patch, and its audio is routed through your interface and monitors.

MIDI Recording Procedure

Now that you understand the MIDI signal chain, the steps below will make more sense. Here's how to create the audio and MIDI tracks in your sequencer (MIDI/audio) program.

1. Set Up the Audio and MIDI Tracks
   Insert a MIDI track to record your MIDI sequence, and insert an audio track that contains the soft synth as a plug-in. (Your sequencer might combine the MIDI and audio tracks into one track.)

Synth and sample patches are stored in groups called banks, and each bank contains several different patches. In the MIDI track, choose a synth bank and patch (bass, drums, piano, etc).

In the MIDI track input, specify the MIDI channel that the synth will respond to: either Omni mode (any channel) or the same channel number that your MIDI controller is set to.

In the MIDI track output, specify the synth or sample player that you want the MIDI track to play.

Set the audio track output to the stereo output of your interface.

2. Start Recording

In your interface control program, or in your recording program, set the latency or buffer size low (under 4 msec if possible) while tracking so that the synth responds without a noticeable delay. See Appendix B under the heading "Minimizing Latency."

Choose a tempo in your sequencer. Set the metronome to count off two measures.

Enable Record mode on the MIDI track. Using your mouse, click on the RECORD button on-screen or on your computer keyboard. You'll hear the sequencer's metronome ticking at the tempo you set.

3. Play Music on Your Keyboard

Listen to the metronome and play along with its beat. The sequencer keeps track of the measures, beats, and pulses. As you play, MIDI data from your keyboard goes from keyboard MIDI OUT to interface MIDI IN, and is recorded as a MIDI file or sequence. When the song is done, click on STOP.

Another way to record your performance is in step time, one note at a time. You also can set the sequencer tempo very slow, record while playing the synth at that tempo, and then play back the sequence at a faster tempo.

Suppose you are trying to simulate a real musical instrument. Stay within the limits of that instrument: its note range, the number of notes in a chord that it can play, and so on. Work on the sound of the synth in its GUI. Often, audio effects such as reverb and EQ can increase the realism of a synthesized instrument. Add volume automation to pads so they don't sound like a steady tone. Record a complete performance rather than using cut and paste.

4. Play Back the Sequencer Recording

Click PLAY. You'll hear the sequence playing your soft synth. If you change the patch and click PLAY, you'll hear the same performance played by a different instrument.

5. Layer the Drum Track

Using the recording software's sound-on-sound setting, you can record several passes of a MIDI drum track, adding another instrument each time. On the first pass tap-in the kick and snare, on the next pass add hi-hat, and so on.

6. Quantize the Track

Quantizing automatically corrects the timing of each note to the nearest note value (quarter note, eighth note, and so on). If you wish, quantize the performance by the desired amount. Caution: Quantizing can dehumanize the performance, making it too rhythmically perfect. It's better to adjust the timing of certain notes only, and to the smallest note value that works. Quantizing is essential if you want to use a notation program.

7. Punch In/Out to Correct Mistakes

To correct a musical error, you can punch in to Record mode before the mistake, record a new performance, and then punch out of Record mode. Another way to correct errors is to edit the MIDI notes in the piano roll view, described later in this chapter. Here's one way to punch in:

a. Go to a point in the song a few bars before the mistake.

b. Just before you get to the mistake, punch in to Record mode and play a new, correct performance.

c. As soon as you finish the correction, punch out of Record mode.

Alternatively, you can use autopunch. With this feature, the computer punches in and out automatically at preset measures; all you have to do is play the corrected musical part. Perform an autopunch as follows:

a. Using the computer keyboard or mouse, set the punch-out point (the measure, beat, and pulse where you want to go out of Record mode).

b. Set the punch-in point (just before the part you want to correct).

c. Set the cue or preroll point (where you want the track to start playing before the punch).

d. Click on PLAY.

e. When the screen indicates punch-in mode, or when the appropriate measure comes up, play the corrected part.

f. The sequencer punches out automatically at the specified point in the song.

These punch-in routines were done in real time. You can also punch in/out in step time:

a. Go to a point in the song just before the mistake.

b. Set the sequencer to step-time mode.

c. Step through the sequence note by note, and punch in to Record mode at the proper point.

d. Record the proper note in step time.

e. Punch out of Record mode.

8. Edit the Sequence Recording

You might find it easier to edit the MIDI performance. Go to the MIDI edit screen or piano-roll view (Figure 16.2). It's a grid showing pitch verses time. The pitch of each note is represented by its height on the grid, and the duration of each note is represented by it length. You can grab incorrectly played notes and put them at the correct pitch and timing, delete unwanted notes, copy and paste phrases of notes, stretch or shorten note durations, and so on.

When you're finished editing one MIDI performance, overdub other MIDI tracks, edit them, and set up a mix.

9. Arrange the Song by Combining Sequences

At this point you can put together your composition. Many songs have repeated sections: the verse and chorus are each repeated several times. If you wish, you can record the verse and chorus once. Then copy the verse section and paste it every place it occurs in the song. Do the same for the chorus.

10. Add Audio Tracks

Suppose you want to add vocals or live instruments to the mix. Here's a suggested procedure.

a. If you are using a sound card, plug a mic into a mic preamp or mic input of a mixer. Connect the preamp or mixer line output to your sound-card line input. If you are using an audio interface, plug a mic into one of its mic connectors. Connect the interface's PCI, USB, or FireWire port to your computer.

b. In your recording software, insert an audio track, and set its input source to the audio interface or sound-card channel that the mic signal is plugged into. For example if you plugged a mic into interface input 3, set the track's input to interface input 3. Enable Record mode on that track.

c. Set the recording level by adjusting the microphone input trim or gain on your preamp, mixer, or interface.

d. Go to the beginning of the tune and hit PLAY in your DAW. The MIDI sequences that you recorded earlier should start playing. (You may need to press the PLAY key on a drum machine first if it is an outboard device, rather than a part of software.)

e. While listening to the MIDI tracks playing through headphones, record the vocal on an audio track. Then overdub more vocals and non-MIDI instruments on other open tracks.

You might record a few takes and then cut and paste selected portions to create a perfect take. For example, record one good chorus and copy it in each chorus section in your song. You also can edit audio notes to correct their pitch or timing.

When you play back what you recorded, the recorded MIDI tracks play the soft synths, and the digital audio tracks play their audio signals. A mix of the soft synths and audio tracks plays from the audio interface line output, which is connected to your monitors.

11. Mix, Export the Mix, and Burn a CD

a. After all your tracks are recorded, use the on-screen mixer to set up a mix of the audio tracks and MIDI instruments. Adjust levels, panning, and effects (plug-ins).

b. Play the song several times to perfect the mix and to set up automation. See Chapter 11 under the sections on mixing procedures and automation.

c. Export the mix to a stereo WAV file on your hard drive. Then you can burn a CD of the song or convert it to an MP3 file.

RECORDING A HARDWARE SYNTH

Suppose you like the sound of a hardware synth. You want to record its sound and also record a MIDI sequence of your performance so you can edit it. Figure 16.7 shows the connections.

1. Monitor a mix of the synth audio and sequencer playback. In other words, monitor a mix of the interface input signal and output signal.

2. Insert a MIDI track in your sequencer program. Set its input to your MIDI/audio interface in Omni channel mode. Set the output of the MIDI track to your MIDI/audio interface. That way, during playback the recorded sequence will pass through the MIDI/audio interface to your hardware synth and will trigger it to play music.

3. Enable Record mode in the MIDI track. Record your performance on the MIDI track. (The sequencer will also record whatever MIDI channel number is set in the synth.)

4. If necessary, set your synthesizer to External Clock (MIDI clock) so it will follow the timing of the recorded sequence.

5. Play the sequence and edit it in the piano-roll view.

FIGURE 16.7
Setup for recording a hardware synth on a computer.

6. Insert an audio track. Set its input to the interface line input that is connected to the synth audio output. Disable Record mode on the MIDI track, and enable Record mode on the audio track.

7. Make sure that the synth MIDI channel is unchanged. Set levels and start recording. In other words, play the MIDI track while recording the synth audio on the audio track. The MIDI sequence will "play" or trigger the synthesizer as you record its output on an audio track.

8. When the song is done, you have an audio track produced by the hardware synth. If you wish, you can disable the MIDI track that drove the synth.

"NO SOUND" MIDI TROUBLESHOOTING

Suppose you load a MIDI file and hit PLAY on your sequencer, but you hear nothing. Or you play notes on your MIDI controller, but there's no sound. Not a peep.

That's not too surprising, considering that the MIDI signal chain is a long one. Every link in the chain has to work, whether it's hardware, software, or their settings. I'll give some MIDI troubleshooting advice that should coax your system to speak when you ask it to.

If any part of the chain is not turned on, or has the wrong settings, you won't hear a sound when you monitor the synth—no matter how hard you pound the keys! Listed below are some possible reasons why your synth can't sing, and what to do about it.

- MIDI OUT or MIDI THRU is not connected to MIDI IN somewhere in your system. Trace the cable connections from beginning to end, and see whether a MIDI cable is not plugged in where it should be. The cable itself might be broken.
- Your MIDI interface is not communicating with your computer. A missing or flaky PCI, USB, or FireWire connection can disrupt the data flow between interface and computer. Replug or replace the cable. You might need to restart your sequencer program too. Some recording software has an on-screen indicator that flashes when the software is receiving MIDI data, so check it out.
- When you are tracking, the MIDI sequencer track is not selected. Some sequencer software does not monitor the soft synth unless you select its MIDI track. In some sequencers, you can live-monitor only one synth at a time, but you can play back multiple synth tracks at once.
- The MIDI sequencer track has no MIDI channel assigned, or is set to a different channel than the MIDI controller. Just as you need to set your TV to a certain channel to see a station's programming, you need to set your MIDI sequencer's track to the same channel your controller is sending data on. It might be easiest just to set the MIDI track input to your interface in Omni mode, so the track will hear any channel that the controller is sending. You can set multiple MIDI tracks to Omni; just record one track at a time.
- The MIDI sequencer track output is not pointing at your soft synth. You have to tell the MIDI track which synth to play.

- The wrong sound bank was selected in a synth. You might have specified a bank that contains no sounds. Or the MIDI mapping on your controller is set so the notes aren't playing the patches you want them to play. Check the bank setting, patch settings, and MIDI mapping.
- In the sequencer's options menu, under MIDI devices, your MIDI interface is not selected. The sequencer does not know where to find the driver that communicates with your interface, so you don't hear anything. If your sequencer has a MIDI data indicator that is NOT flashing in response to your playing, maybe you forgot to select the MIDI driver.
- The synth's audio-track volume, or the MIDI track's key velocity, is turned down. Disable any mute buttons and turn up the faders for those tracks. Check that no other tracks are soloed.
- You started playing the MIDI track in the middle of a long note, rather than at its beginning. A synth needs to receive a note-on command in order to play a note. If you start playing the sequence from a point *after* the beginning of a note, the synth does not receive the note-on code, so it remains silent.
- Your hardware synthesizer lost the program-change command. Although you told the synth which patch to play, sometimes this information is lost. Change the patch on the synth and then set it back as it was. Or record the program change at the beginning of the song.
- The input to your sequencer track lost the setting for its MIDI input device. Sometimes when you close a sequencer program and reopen it, the MIDI track inputs might forget their settings. Temporarily set the MIDI track input to another device, then set it back as it was.
- The MIDI driver is buggy. Download the latest update from the interface manufacturer.
- The synth audio output is not assigned to your interface's stereo output channels. In other words, the synth is making music but not sending it anywhere.
- Can you hear any audio playing through your monitors? If not, the monitor power amp, or your mixer's monitor controls, might be off or turned down. Maybe the cables between your interface and monitors are not plugged in. Seems obvious, but it's happened to me.

I hope these suggestions make your speakers play notes when you press keys. The better you understand the MIDI/audio signal path, the better you can find errors and fix them.

RECORDING WITH A KEYBOARD WORKSTATION

This is another way to record in a MIDI studio. Basically you follow the steps for multi-track sequencer recording above, but do them in the keyboard workstation rather than in a computer. The workstation's manual explains how to do it.

You can access the keyboard's effects menus to set up the effects that are heard with a patch: hall reverb, chorus, flanging, echo, distortion, and so on. Press the correct number on the numeric keypad to get to the effects menus. (Note that these are built-in keyboard effects, not outboard studio effects.)

Save the completed song (MIDI sequence) in multitrack form to the keyboard's internal memory or to a plug-in RAM card. If you're satisfied with the final results, use a DAW to record the stereo output signals of the workstation. Then you can burn a CD or create an MP3 file from the WAV file you just recorded on your hard drive.

In addition to these basic operations, you can:

- Bounce tracks: copy one track's performance to another track so it will play another patch.
- Edit each note event.
- Create and copy patterns (for a drum or bass part, for example).
- Modify track and song parameters.
- Insert/delete/erase measures.
- Modify sounds and effects.
- Change the instrument (patch) that each track plays.

RECORDING WITH A DRUM MACHINE AND SYNTH

Suppose you want to record a drum pattern on a hardware drum machine, and then you want to add a synth part using your computer.

First record the drum pattern. Here's one suggested method:

1. On the drum machine, set the tempo, time signature, and pattern length in measures. In this example, the pattern will be 2 bars long.
2. Add a count-off (a few measures of clicks) at the beginning so that overdubs made later can start at the correct time.
3. Start recording, and play the hi-hat key in time with the metronome beat.
4. At the end of 2 bars, the hi-hat pattern you tapped repeats over and over (loops).
5. While this is happening, you can add a kick drum beat.
6. While the hi-hat and kick drum are looping, add a snare drum beat, and so on.
7. Mix the recording by adjusting the faders or keys on the drum machine for each instrument.

Next, you repeat the process for a different rhythmic pattern—say, a drum fill—and store this as Pattern 2. Then develop other patterns. Finally, you make a song by repeating patterns and chaining them together as described in the drum machine's instruction manual. A song is a list of patterns in order.

Some musicians like to compose by programming a simple repeating drum groove first. While listening to this, they improvise a synth part. After recording the synth part, they redo the drum part in detail, adding hand claps, tom-tom fills, accents, and so on.

Finally, copy the audio signal of the drum pattern to a stereo track in your computer recording program. Play back that track while overdubbing other parts.

USING EFFECTS

No matter how you record with MIDI, effects are an important part of the mix. To keep the sound lively, try to vary the effects throughout the song, or use several types of effects in the mix.

For example, suppose you have a multitimbral synth, and you want to add a different effect to each patch. Whether or not you can do this depends on your synth. If it has a separate output for each patch, you can use a different effect on each patch. But if your synth has only a single output (mono or stereo) and you run it through an effects device, the same effect is on all the patches.

If your song includes program changes (patch changes), you can have the effects change when the patch changes. Set up a MIDI-driven multieffects processor so that each synth program change corresponds to the desired effect. When the synth program changes, the effect also changes.

What if you want the effect, but not the synth patch, to change during a mix? Reserve a track and channel just for effects program changes. You don't hear these program changes in your synth, but you do hear the effects change. During a mixdown, it's usually easier to change effects automatically with your sequencer, rather than manually.

In a sample player, some samples might have reverberation or some other effect already on them: the effect is part of the sampled sound. Note that the sampled reverberation cuts off every time you play a new note. Although this sounds unnatural, you can use it for special effect.

Because effects are audio signals, audio recorders can record effects but MIDI sequencers can't. MIDI sequencers, however, can play synths or patches that have effects built in.

Some recording software lets you convert a MIDI track to an audio track, then apply audio effects to that track. Other software uses a separate MIDI track and audio track for the same instrument patch. You apply effects to the audio track.

MIDI effects (MFX) are nonaudio processes applied to MIDI signals, such as an arpeggiator, echo/delay, chord analyzer, quantize, transpose MIDI event filter, or velocity change. They are plug-ins in MIDI tracks.

LOOP-BASED RECORDING

Let's turn to a different aspect of computer MIDI recording. It's possible to compose, record, and perform music entirely in software. You might start by creating a variety of loops or grooves, which are constantly repeated rhythmic or musical patterns.

You can buy loops or make your own. You might create a loop by recording a 4-measure sample of a drum beat or bass riff that you play to a metronome or click track. Or use a handheld recorder to sample some interesting sounds in the field. The sounds might create a rhythm when repeated. If not, you can edit the timing of individual sounds until you get a rhythm you like. Soft synths are another source for loops, and some are bundled with ready-to-use patterns. Many recording programs include loop patterns. You can also create loops from parts of tracks that you recorded.

Making Your Own Loops

Let's make a 4-bar loop in 4/4 time using the editing function in a DAW. Start with a musical pattern longer than 4 measures that was made to a metronome or click track. Ideally the pattern has little or no reverb so that the reverb won't be cut off when you trim the loop.

Trim the start of the loop waveform just before beat 1, and trim the end of the loop after measure 4-beat 4, and just before the next beat 1. To avoid a click in the audio when the loop repeats, both trim points should be at zero crossings where the waveform crosses the 0-volt line (Figure 16.8)

Copy the trimmed 4-measure clip and paste it on the beat several times to repeat it. Or right-click the clip and select "Loop clip." If you hear no clicks and the rhythm is steady, export the 4-measure loop to your loop library. Store your loop collection by tempo, style and type (for example, 120 bpm, jazz, drum set with brushes).

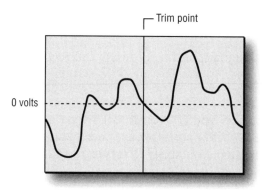

FIGURE 16.8
Trim loops at zero crossings in the waveform.

Once you have created several loops, you can drag and drop them (or import them) into your DAW software. There, you can copy and paste loops on the beat to repeat them. A SONAR groove clip (explained next) can be repeated by click-dragging the end of the clip to the right.

Types of Loops

There are five types of loop audio files based on their ability to have their tempo changed:

- A loop made from a standard digital audio file. It has a fixed tempo, so you must build your song around that tempo.
- A processed audio file. The tempo or pitch of the loop can be changed by a time-stretch or pitch-shift algorithm in your audio editing program. But if you need to change the tempo of a song, you need to adjust all the loops you stretched.
- A file with REX-based time stretching. The transients in the audio file are cut into slices, whose spacing depends on the tempo. REX files follow tempo changes in your composition.
- An acidized (RIFF) WAV file. Pitch and tempo information are in the file header, and the audio is sliced at transients as with REX files. Acidized files follow tempo and key changes. Slowing down the loop by more than 15 or 20 beats per minute can add artifacts, so it helps to start with a slow tempo of less than 120 bpm.
- Cakewalk SONAR recording software includes Groove Clips that have embedded tempo and root-note pitch. SONAR can stretch the clips to match changes in tempo, or transpose clips to match the project's key. You repeat a Groove Clip by dragging its end to the right in the track. Almost any audio clip can be converted to a SONAR Groove Clip; just select the clip and press Ctrl-L.

Working with Loops

Once you've made several repetitions of the groove, you can add effects, distortion, or pitch/tempo changes. You might put some echo or reverb on a drum loop and vary the dry/wet mix with automation as the loop plays. When the loops are complete, overdub audio and MIDI tracks such as soft-synth parts, percussion, vocals, and acoustic instruments. Import other loops and place them where you want them in the song.

You can also create loops externally, then import them into a DAW recording program. For example, compose a repetitive beat of drums, synth, and samples in a beat box. Launch your DAW software, set up a stereo track, and record the box output to that track as it plays the beat. On other DAW tracks you can overdub vocals, doubled vocals, rap sections, and harmony.

Repetitive loops can get boring, so edit or remove various notes in the piano-roll view to add some variety. You might insert some spaces here and there in the song to create breaks for solos or record scratches.

Some loop programs offer groove quantizing, which transfers the timing and dynamics from one groove to another. It allows human-like variations in timing and key velocity.

Loop Libraries

A loop library or sound library is a collection on CD or DVD of loops and sampled musical phrases that you can use to create a composition. The loops come as MIDI files and WAV files, and usually are organized by tempo and key. You drag them into DAW tracks and construct several song sections, each made of repetitive loops.

The major distributors of loop libraries are Big Fish Audio, Sony, and EastWest. Other libraries are available from Nine Volt Audio, PureMagnetik, www.producerloops.com, www.kellysmusicandcomputers.com/loop_sample_libraries.htm, www.loopmasters. com, www.ilio.com, and www.sonomic.com.

Some library examples are M-Audio ProSessions, Sample Logic, Audiobro LA Scoring Strings, EastWest Hollywood Strings, Hark Loops, Cakewalk loop libraries, Sony Creative Software Artist Integrated loop libraries, Sonic Foundry Acid loop libraries, Vienna Symphonic Library, GarageBand, Smart Loops, FL Studio, Beatboy, Keyfax Twiddly Bits, Groove Monkee, APO Multimedia Mix It, and Pocket Fuel RADS series. Some drum sample libraries include Wizoo VST Drum Sessions, FXpansion's BFD, Discrete Drums, DrumCore, Multiloops Naked Drums, and Toontrack EZdrummer.

Do a Google search for "Loop Libraries" or "ACID loops." Complete looped rhythm tracks are available online for purchase. Just add vocals.

Loop Creation Software (Construction Kits)

Many DAW recording programs include loop-creation software, which is also available separately as a loop construction kit. For example, Propellerheads Software (www. propellerheads.se) has a variety of programs to create and modify loops. They also offer soft synths, drum machines, and sequencers. Here are some of their products:

ReCycle is a standalone tool that starts with a loop and lets you change its tempo and pitch, and replace and process sounds within the loop. By detecting peaks in the waveform, ReCycle automatically breaks a loop into parts or slices. A slice might be a snare-drum hit, a kick-drum hit, or a kick-and-snare hit. When a loop's tempo is changed, the start point of each slice moves in time so that the beats occur at the right time. (You might need to touch this up manually.)

If the tempo is slowed down, ReCycle creates a decay after each drum hit to fill in the gaps. You can delete slices, change their length, attack, decay, and pitch; and add compression or EQ. Then you can import the improved loop as a REX2 file into an audio track in your sampler or sequencer program. There you can control all aspects of the loop.

Reason includes a group of synthesizers and samplers, a drum machine, a mixer, effects, a pattern sequencer, and more (Figure 16.9). It's all-in-one and easy to learn. Reason can import and play ReCycle's REX files.

ReWire is a useful feature found in some loop programs. It transfers or streams audio data between two computer applications in real time, almost like a cable. This allows programs to communicate with each other and synchronize together.

Balance is a 2-in/2-out USB audio interface designed to integrate into Reason with no setup or driver install.

Some other loop-based tools to compose and record music are listed below.

Ableton Live lets you compose with soft synths, record high-resolution multitrack audio, and play loops in a live performance. With Live you can make stuttering drums, create arrangements, do MIDI loops, and modify grooves, timing, pitch, volume, and effects.

FIGURE 16.9
The Thor synthesizer screen in Reason software.

FIGURE 16.10
Sonar's Matrix loop construction app.

Sony Creative Software's **ACID** products let you select audio loops from Windows Explorer, drag them into a recording program's track view, and arrange them into multi-track projects. The tempo and key of each loop are automatically matched to your project's music in real time, using the slicing technique described before. ACID uses time-stretching algorithms to lengthen sounds when the tempo is slowed down.

Other loop-intensive DAWs include FL Studio (PC) and Apple GarageBand (with Apple Loops included). Figure 16.10 shows Sonar's Matrix loop construction app, which comes with Cakewalk Sonar Producer X1. It's like a multitrack recording project for loops. You drag 2-bar and 4-bar instrument loops into the Matrix wherever you want them, stack several instruments, and play them. You can even trigger different loops in real time and record your performance.

A few loop construction kits are Native Instruments Discovery Series, Ueberschall Funk & Soul, Sony Charm, Freshtone Samples Lost Tapes Vol. 1, Best Service K-Size Electro Edition, and Big Fish Audio Pure Rock Hits.

Some excellent articles on loops were in the July 2004 and August 2002 issues of *EQ Magazine*, the January 2007 issue of *Recording Magazine*, and the October 2011 issue of *Electronic Musician* ("Refined Repetition" by Craig Anderton). Also be sure to see the article "Under Construction" by Len Sasso: go to http://www.emusician.com/techniques/0768/loops--under-construction/139936.

CHAPTER 17

On-Location Recording of Popular Music

Sooner or later you'll want to record a band—maybe your own—playing in a club or concert hall. Many bands want to be recorded in concert because they feel that's when they play best. Your job is to capture that performance on a recorder and bring it back alive.

You can record pop-music concerts in several ways. Listed below are a range of techniques from simple to elaborate. In general, sound quality improves as the recording setup becomes more complex.

- Record off the board (PA mixer) into a portable stereo recorder.
- Record with a portable stereo recorder.
- Record with a 4-tracker. Record with a stereo mic on tracks 1 and 2, while recording the PA mixer output on tracks 3 and 4. Mix the tracks later.
- Feed the PA mixer's insert-send signals to a multitrack recorder. Edit and mix the recording back in your studio.
- Use a mic splitter on stage to feed the PA mixer, monitor mixer, and recording mixer. Record to multitrack. Edit and mix the recording back in your studio.
- Do the multitrack recording in a van or truck.

We'll describe each method so you can decide what's best for you and try it out.

RECORD OFF THE BOARD

Let's start with the easiest technique: connect the PA mixing console (board) to a portable digital recorder.

Using cables with the appropriate mating connectors, connect the PA mixer's tape-out or 2-track-out connectors to your stereo recorder's line input(s). If your recorder's line input is a stereo mini phone jack, use an adapter cable: either two 1/4-inch phone plugs to a stereo mini phone, or two RCA connectors to a stereo mini phone. They are available at Radio Shack and other electronic suppliers. Outside the U.S., if your recorder's line input is a stereo mini socket, use an adapter cable: either two 6.35 mm jack plugs to a stereo mini jack, or two phono connectors to a stereo mini jack.

The recorded mix might be poor with this method. Here's why: The sound mixer hears a combination of the band's instruments, the stage monitors, and the house speakers. So the board mix is intended to *augment* the sound of the instruments and monitors on stage—not to sound good by itself. For example, if the bass-guitar amp is very loud on stage, it will be turned down in the mix that you record off the board. Vocals will be turned up high in the mix because they can't be heard otherwise. Board mixes can sound good if there is not much sound coming off the stage (as with acoustic groups), and the venue is large or outdoors. It helps to monitor the board mix with headphones or isolating earphones to hear what you're recording.

RECORD WITH A HANDHELD DIGITAL RECORDER

Here's another simple method. The process is easy and the required gear costs as low as $200. A stereo recording may not offer the sound quality of a professional multi-track recording. But if you record how the band sounds from a seat in the audience, that may be good enough—especially if the recording is just a demo, or is for the band members.

You can use a recorder with built-in microphones, or you can plug two mics (or a stereo mic) into the recorder. You'll pick up the group as a whole from several feet away. The mics will capture not only the musicians, but also the room reverb and background noise. You could call it a "documentary" or "audio snapshot" recording.

As you record, what you hear with headphones or earphones is what you get. There's no mixing back in the studio.

If you place the mics a few feet from a folk group or jazz group without a PA system, the sound can be quite good. But most bands use a PA. When you record the band you're also recording the sound of the PA speakers, so the mix or balance you get depends on the PA engineer's skill. The sound will be more distant and muddy than you get with several close-up mics.

Gear

You'll need a handheld digital recorder, described in Chapter 4. A mic stand is optional.

Headphones or earphones let you know whether the mics are working correctly, and let you hear what the mics are picking up. If the band and PA are loud, it's hard to hear what's being recorded unless you use *isolating headphones* (Remote Audio HN-7506) or *isolating earphones* (Etymotic Research models; Klipsch Image S4).

Preparing for the Session

Before going on the road, install fresh batteries and clean the connectors with isopropyl alcohol or DeoxIT from Caig Labs (www.caig.com). Do a trial recording to make sure everything works.

If possible, record the gig in a room where the audience is attentive and the background noise is low. You might visit some potential venues to check out the noise and acoustics. Avoid very live rooms because they can make the recording muddy.

Be sure you have enough free space on your flash-memory card before going on location. Listed below are approximate recording times for a 1 GB card. Double these times for a 2 GB card.

    24-bit/44.1 kHz stereo WAV..............1.0 hour
    16-bit/44.1 kHz stereo WAV..............1.5 hours
    256 kbps stereo MP3..........................9.0 hours
    128 kbps stereo MP3.........................16.5 hours

The recorder might have a record-level switch labeled "manual" and "auto." Set it to "manual" in order to retain the dynamics of the performance. If the switch is labeled "AGC" (automatic gain control), set it to "off."

At the Gig

Plug in some headphones and turn on Record Monitor mode. You'll hear the room acoustics and any background noise (audience, air conditioning, traffic). Room noises that you wouldn't otherwise notice become obvious when you listen on headphones. Also listen for buzzes, distortion, and crackles from bad cables or connections.

Where do you place the mics? The closer the mics are to the group, the clearer and cleaner the sound is. In other words, close placement captures more of the music and less of the room sound and background noise. Try to place the mics as close as possible to the stage where you still pick up the house PA speakers well, about a stage-width away from the stage (Figure 17.1). Another good placement is even with the right PA speaker's inner edge, 12-15 feet from the band, and at least 3 feet above the audience. Keep the mics away from a bar or other noise sources.

If there are dancers near the stage and the ceiling is low, you might try boundary mics (such as two PZM®s) gaffer-taped to the ceiling or mini mics hung from the ceiling.

To eliminate muddy room sound, you can record acoustic groups off the PA mixer into a line input. Or try placing the recorder/mics on stage (on a stool or mic stand) and record the sound of the monitor speakers plus the band. Some bluegrass or old-time bands sing into one microphone, so mount the recorder/mics next to that mic.

FIGURE 17.1
Typical mic placement for a stereo recording of a pop group.

If the group plays in a circle (as in an outdoor jam or rehearsal), try placing omnidirectional mics in the center. Or just walk around with the mics while monitoring the mics over headphones. Find a spot where you hear a good balance and put the mics there. Ask the musicians' permission to record them during a break in the music.

Set the recording level so the meter reads about −6 dB maximum. That allows some headroom for surprises. Peak levels above 0 dB result in distortion. Some portable recorders include a limiter which prevents recording levels above 0 dB. Others have a clip or peak LED that flashes when the recording level is too high.

Recorders have a mic gain switch or pad (attenuator) to prevent mic-preamp distortion. If you have to set the recording-level control less than 1/3 up to achieve a 0 dB recording level, use the low-gain setting or switch in the pad.

If you are using large-diaphragm condenser mics plugged into a phantom power supply, you might need to plug the output of the supply into the recorder line input—rather than the mic input—to prevent distortion.

A Recording Session with No PA

You might be able to record a small folk group or acoustic jazz group on location without a PA system or audience. This gives you the freedom to improve the sound.

1. If the room is too live (reverberant), put up some packing blankets, comforters, rugs, acoustic foam, or cushions.
2. Often a good-sounding spot for the musicians is near the center of a large room. Place the musicians around the stereo mic pair where you want them to appear in the recording. For example, you might place two singing guitarists on the left and right, with bass in the center.

FIGURE 17.2
A method of stereo miking a jazz group.

3. Experiment with microphone height to vary the vocal/guitar balance. Try different miking distances to vary the amount of ambience or room sound. Three to six feet distance is typical (Figure 17.2).
4. As the musicians are playing and during playback, monitor the mic signals with headphones. If some instruments or vocalists are too quiet, move them closer to the mics, and vice versa, until the balance sounds right.
5. While the band is rehearsing a loud song, make a trial recording. Set the recording level and mic gain as described earlier.
6. Record a tune. If someone makes a mistake, either record another take of the entire tune, or record starting from a few bars before the mistake, and edit the takes together later.

After the Gig

Back in the studio, connect the USB port in the recorder to the USB port in your computer. The recorder shows up as a storage device on your computer screen. Drag-and-drop the recorded sound files to the computer's hard drive for editing and CD burning. The files transfer in a few minutes. Then the flash-memory card is empty, free to make more recordings.

Now you can edit the recording and adjust its tonal balance (equalize it) with DAW recording software or with Harmonic Balancer software (www.har-bal.com). You might cut a few dB around 300 Hz to reduce boomy reverb and get a clearer recording. If your mics are weak in the bass, compensate by boosting a few dB at 50-100 Hz.

Try out these tips and enjoy your recordings!

RECORD WITH A 4-TRACKER

This method is fairly simple and provides good sound. Put a stereo mic or mic pair at the front-of-house (FOH) PA mixer position. Plug the mic connectors into mic

FIGURE 17.3
Recording two mics and a PA mix on a four-track recorder-mixer.

inputs 1 and 2 of a 4-track recorder-mixer (or any size recorder-mixer). Connect the PA mixer's tape outputs or 2-track outputs to line inputs 3 and 4 (Figure 17.3). Mix the recording to stereo back in the studio.

The FOH mics pick up the band as the audience hears it: lots of room reverb, lots of bass, but rather muddy or distant. The PA mixer output sounds tight and clear, but typically is thin in the bass. Luckily, a mix of all four tracks can sound surprisingly good. Tracks 1 and 2 provide ambience and bass; tracks 3 and 4 provide definition and clarity.

When you mix the four tracks, you might hear an echo because the FOH mics pick up the band with a delay (caused by the sound travel time from stage to mics). To remove the echo, import all the tracks into digital recording software, and delay the PA mixer tracks by sliding them a little to the right. Align the waveforms of the mic tracks and mixer tracks at big peaks.

CONNECT THE PA-MIXER INSERT SENDS TO A MULTITRACK RECORDER

Now we get into professional techniques. This is an easy way to record, and it offers very good sound quality with minimal equipment (Figure 17.4). Connect the mixer insert sends to the inputs of a multitrack hard-drive recorder. Use the mixer gain trims to set recording levels during the sound check. You don't have to mix while recording—instead, mix and monitor back in your studio. As for drawbacks, you might have to ask the PA operator to adjust the gain trims during the show to prevent recorder clipping.

What are some multitrack hard-disk recorders? The Alesis ADAT HD24XR is a reliable, great-sounding 24-track model with built-in swappable drives. The JoeCo BBR1B Recorder (www.joeco.co.uk) has 24 line-level inputs and an output for an external USB 2.0 hard drive.

*Plug tip-sleeve connector in halfway
or wire tip/ring together

FIGURE 17.4
Connecting insert sends to a multitrack recorder provides great sound and easy setup.

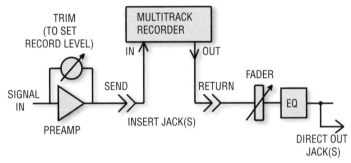

FIGURE 17.5
Simplified signal flow through part of a mixing console, showing insert and direct out.

Connections

This section describes how to connect a multitrack hard-drive recorder to a mixer's insert-send connectors.

In the console are several mic preamplifiers, one per mic, which amplify the mic-level signal up to line level. For each mic channel, this line-level signal typically appears at two connectors on the back of the mixer: *direct out* and *insert send*. That's where to connect to the recorder inputs.

Usually the *insert send* is the best connector to use. Here's why. Typically the direct-out signal is post-fader (Figure 17.5). This means the signal at the direct-out connector comes after the fader, so the signal is affected by the fader (volume) settings. Any fader movements will show up on your recording, which is undesirable. It's better to connect recorder tracks to insert sends, which are usually pre-fader, pre-EQ. However, any changes the PA operator makes in the trim settings during the show will affect your recording levels.

In some mixing consoles, the direct outs can be set to pre-fader. If so, use those, because insert sends are often in use to connect compressors or gates.

FIGURE 17.6
Three ways to wire cables based on the type of insert connector.

First, find out what kind of insert connectors the PA mixer has, and what kind of input connectors your multitrack recorder has. Buy or make some shielded cables (or a snake) that mate with those connectors. Figure 17.6 shows three ways to wire cables based on the type of insert connector.

Some boards have a single TRS (tip-ring-sleeve) insert connector per channel, instead of separate insert-send and insert-return connectors. Usually the tip is send and the ring is return. In the TRS connector that you plug into the TRS insert, wire tip and ring together, and also to the cable hot conductor (Figure 17.6, top). That way, the insert-send goes directly to the insert return. If nothing is connected to the insert return, no mic signal goes through the mixer.

On some PA mixers with a tip-ring-sleeve insert connector, you can use a TS (tip-sleeve) connector (Figure 17.6, middle). Plug it in halfway to the first click so you don't break the signal path—the mic signal still goes through the PA mixer. If you plug in all the way to the second click, the signal does not go through the PA mixer, just to the recorder. Cover the connections with a mixer case cover or board because someone could bump into the mixer and dislodge a cable.

If the PA mixer has separate send and return insert connectors, connect the send to the recorder track input, and connect the return to the recorder track output (Figure 17.6, bottom). If necessary, set your multitrack recorder to monitor the input analog signal so that the PA mixer will receive a signal. Another option: Use an insert snake with TRS connectors at the mixer end. Carry some TRS-to-dual-TS adapters to handle mixers that have separate connectors for insert send and return.

The insert sends are balanced or unbalanced, and the same is true of the recorder inputs. To connect balanced and unbalanced equipment correctly, see the article "Sound System Interconnection" on the Rane website, www.rane.com/note110.html.

You often encounter PA consoles where some insert sends are tied up with signal processors. You must use those channels' direct-out connectors instead, which usually are post-fader (unless they can be switched to pre-fader). Another option is to "Y" the insert send to your recorder and to the processor input. Or, assign those mic channels to unused groups (buses) and get your recording signals from there.

Some rack-mounted mic preamps have insert connectors. You could plug the mic-snake XLRs into these preamps, then connect the preamp's inserts to the PA mixer line inputs. That way, any gain changes made on the PA console will not affect your recording. Caution: Any gain changes you make on the mic preamps will affect the PA levels.

If the PA mixer has a FireWire or USB port, simply connect the port to a laptop that is running recording software. Set up the software to recognize the mixer as its input/output device. The signal from each mic channel goes to a separate track in the software.

What if you want to record several instruments on one track, such as a drum mix? Assign all the drum mics to one or two output buses in the PA mixer. Plug the bus-out insert connector to the recorder track input. Use two buses for stereo. Adjust the faders to get a good drum mix.

Monitoring

To monitor the quality of the signals you're recording, you generally let the PA system be your monitor system. But you may want to set up a monitor mix over isolating headphones or earphones so you can hear more clearly.

Here is a suggested procedure. Connect all the recorder outputs to unused line inputs in the PA mixer, or to a separate mixer. Use those faders to set up a monitor mix. Assign them to an unused bus, and monitor that bus with headphones/earphones. If you can spare only a few inputs, plug in just one track at a time to check its sound quality. Listen closely for any hum, noise, or distortion.

Setting Levels

Set recording levels with the PA mixer's gain-trim or input-attenuator knobs. This affects the levels in the PA mix, so be sure to discuss your trim adjustment in advance with the PA mixer operator. If you turn down an input trim, the PA operator must compensate by turning up that channel's fader and monitor send.

As we said, if the PA operator changes the input trim during the show, these changes will show up on your recording.

Set recording levels before the concert during the sound check (if any!). It is better to set the levels a little too low than too high because during mixdown you can reduce noise but not distortion. A suggested starting level is −10 dB, which allows for surprises. Do not exceed 0 dB because the signal will distort. Also, if you set the recording level conservatively, you are less likely to change the gain trims during the performance. You don't want to hassle the PA operator.

Some recording engineers run each insert-send signal through a potentiometer to set the recording level. By adjusting levels on a rack-mounted panel of potentiometers, you don't have to ask the PA operator to change the gain trims for you. Of course, if you are operating the PA console, you can set the gain trims yourself.

If you have a spare recorder, record a safety copy on it at the same time. This provides a backup in case one recorder fails.

SPLITTING THE MIC SIGNALS

Now let's consider a different way to make a multitrack recording. Plug each mic into a mic splitter, which sends the mic signal to two destinations: the PA mixer and recording mixer. This gives you independent control of each microphone. The splitter has one XLR input and two or three XLR outputs per mic. The third output on some splitters goes to a monitor mixer. You can use transformer-isolated splitters or Y-cable splitters.

Splitters are made by such companies as Pro Co, Radial, Whirlwind, Rapco, Klark Technik, and Wireworks.

As shown in Figure 17.7, connect the outputs from all the splitter channels to the PA snake and to your recording snake. Connect the recording snake to your recording mixer (or audio interface) mic inputs. Connect the recording mixer's insert sends to a multitrack recorder, or connect the interface USB/FireWire cable to a laptop computer. When the recording is done, you can edit and mix the recording back in your studio.

Let's explain the advantages of splitting the mic signals. You use your own mic preamps, so you are not dependent on the quality of the PA console mic preamps. Also, you are not hassling the operator about adjusting gain trims. Each mix engineer can work without interfering with the others. The FOH engineer can change trims, level, or EQ, and it will have no effect on the signals going to the recording engineer. Another

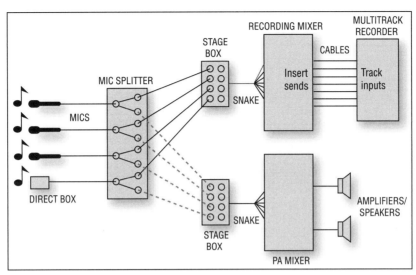

FIGURE 17.7
Splitting the mic signals to the recording mixer and PA mixer.

plus: A splitter provides consistent, unprocessed recordings of the mic signals. This consistency makes it easy to edit between different performances. What's more, splitters let you use mic preamps on stage if you wish. That way, the cable from each mic to its preamp is short, which reduces hum and radio-frequency interference.

Using Splitters

To use a splitter or multichannel splitter, plug each mic into a splitter input. Decide which mixer you want to supply phantom power (usually the PA mixer). Connect the splitter's direct outputs to that mixer's snake. Connect one set of isolated outputs to the recording snake, and another set to the monitor snake (if used). Plug the snakes into the mixers. Figure 17.8 shows splitter connections to two mixers.

Splitters have a ground-lift switch on each output channel. This switch connects or disconnects (floats) the cable shield from pin 1 of the XLR connector. When the ground-lift switches are set correctly, you should get no ground loops and their resulting buzzes.

How do you set the ground-lift switches?

1. First, turn *off* phantom power in each console. Turn down all the faders.
2. Make sure the direct feed's ground-lift switches (if any) are set to *ground*, not *lift*. Otherwise phantom power won't work.
3. Go to the mixer connected to the direct feed. Turn on the mixer, switch on phantom power, and bring up each fader to listen for a signal.

FIGURE 17.8
A transformer-isolated microphone splitter.

4. On the splitter, find the ground-lift switches for the recording-mixer feed. Set them to the position where you monitor the least hum and buzz at the recording mixer. In XLR Y-cable splitters, solder a ground-lift switch between the shield and pin 1 in the nondirect feeds.

5. Repeat step 4 for the monitor mixer.

MULTITRACK RECORDING IN A TRUCK

Here's the ultimate setup. Each mic is split three ways to feed the stage boxes for the recording, reinforcement, and monitor consoles. A long snake is run to a recording truck or van parked outside the concert hall or club. In the truck, the snake connects to a mixing console plugged into a multitrack recorder. The interior of the truck is acoustically treated, and powered Nearfield monitor speakers allow accurate monitoring of the signal. Designing a recording truck is a subject in itself, and is beyond the scope of this book.

The truck contains the console, monitors, and recorders, so you don't need to cart them into the venue. This saves setup time. Also, a truck provides a quiet, controlled monitoring environment.

There you have a full choice of methods for recording live. Check out all the options, and you'll find a system that works well for your style of recording.

PREPARING FOR THE SESSION

Ready to record a live gig? The recording will go a lot smoother if you plan what you're going to do. So sit down, grab a pen, and make some lists and diagrams as described here. We'll go over the steps to plan a recording.

Preproduction Meeting

Call or meet with the sound-reinforcement company and the production company putting on the event. Find out the date of the event, location, phone numbers, and e-mails of everyone involved; when the job starts; when you can get into the hall; when the second set starts; and other pertinent information. Decide who will provide the split, which system will be plugged in first, second, and so on. Draw block diagrams for the audio system and communications (comm) system. Determine who will provide the comm headphones.

If you're using a mic splitter, work out the splitter feeds. The mixer getting the direct side of the split provides phantom power for condenser mics that are not powered on stage. If the house system has been in use for a long time, give them the direct side of the split.

Overloud stage monitors can ruin a recording, so work with the sound-reinforcement people toward a workable compromise. Ask them to start with the monitors quiet because the musicians always want them turned up louder.

Make copies of the meeting notes for all participants. Don't leave things unresolved. Know who is responsible for supplying what equipment.

Site Survey

If possible, visit the recording site in advance and go through the following checklist:

- Check the AC power to make sure the voltage is adequate, the third pin is grounded, and the waveform is clean.
- Listen for ambient noises: ice machines, coolers, 400-Hz generators, heating pipes, air conditioning, nearby clubs, etc. Try to have these noise sources under control by the day of the concert.
- Sketch dimensions of all rooms related to the job. Estimate distances for cable runs.
- Turn on the sound-reinforcement system to see if it functions okay by itself (no hum, and so on). Turn the lighting on at various levels with the sound system on. Listen for buzzes. Try to correct any problem so that you don't document bad PA sound on your recording.
- Determine locations for any audience/ambience mics. Keep them away from air-conditioning ducts and noisy machinery.

- Plan your cable runs from stage to recording mixer.
- If you plan to hang mic cables, feel the supports for vibration. You may need microphone shock mounts. If there's a breeze in the room, plan on taking windscreens.
- Make a file on each recording venue including the dimensions and the location of the circuit breakers.
- Determine where the control room will be. Find out what surrounds it—any noisy machinery?
- Visit the site when a crowd is there to see where there may be traffic problems.

Mic List

Now write down all the instruments and vocals in the band. If you want to put several mics on the drum kit, list each drum that you want to mike. As for keyboards, decide whether you want to record off each keyboard's output, or off the keyboard mixer (if any).

Next, write down the mic or direct box you want to use on each instrument (see Table 17.1).

Make copies of this mic list. At the gig, you'll place one list by the stage box, and the other by each mixing console. The list will act as a guide to keep things organized.

Table 17.1	Mic List	
Input	**Instrument**	**Mic**
1	Bass	DI
2	Kick	AKG D112
3	Snare	Shure Beta 57
4	Hi-hat	Crown CM-700
5	Small rack tom	Shure SM57
6	Large rack tom	Shure SM57
7	Small floor tom	Senn. MD421
8	Large floor tom	Senn. MD421
9	Overhead L	AT 4051A
10	Overhead R	AT 4051A
11	Lead guitar	Shure SM57
12	Rhythm guitar	Shure SM57
13	Keyboard mixer	DI
14	Lead vocal	Beyer M88
15	Harmony vocal	Crown CM-311A

Table 17.2	Track Sheet
Track	**Instrument**
1	Drum mix L
2	Drum mix R
3	Bass
4	Lead guitar
5	Rhythm guitar
6	Keys mix
7	Lead vocal
8	Harmony vocal

Track Sheet

Next, decide what will go on each track of your multitrack recorder. If you have enough tracks, your job is easy. Just assign each instrument or vocal to its own track: bass to track 1, kick to track 2, and so on. If you need to assign several instruments or vocals to the same track, set up a submix as described earlier. Table 17.2 shows a sample track sheet.

Block Diagram

Now that your track assignments are planned, you can figure out what equipment you'll need. Draw a block diagram of your recording setup from input to output (such as Figure 17.9). Include mics, mic cables, mic stands and booms, DI boxes, insert cables, multitrack recorder(s), outlet strip and extension cord, and recording media. You might bring your own mixer and snake, or use those from the house system. On your diagram, label the cable connectors on each end so you'll know what kinds of cables to bring. It's a good idea to keep a file of system block diagrams for various recording venues.

In Figure 17.9, the block diagram shows a typical recording method: feeding FOH console insert jacks to a multitrack recorder. We'll use this example in the rest of this chapter.

Equipment List

Generate a list of recording equipment from your block diagram. Based on Figure 17.9, you'd need the following recording gear:

- Direct boxes, mics, mic cables, mic stands, and booms (unless these are part of the house system)
- Insert cables (for example, 8 stereo phone plugs to 8 RCA connectors)
- Multitrack recorder

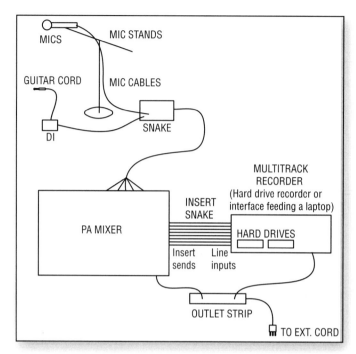

FIGURE 17.9
Sample block diagram.

- Outlet strip and extension cord
- Recording medium (bring enough for the duration of the gig)

Don't forget the incidentals: a cleaning tape, pen, notebook, flashlight, guitar picks, heavy-duty guitar strings, drum keys, spare tape, mic pop filters, gaffer's tape, guitar tuner, ear plugs, audio-connector adapters, audio cable ground-lift adapters, in-line pads, in-line polarity reversers, spare cables, gooseneck lights for the console, spare batteries, water, food—and aspirin!

Bring a tool kit with screwdrivers, pliers, soldering iron and solder, AC-outlet checkers, fuses, a pocket radio to listen for interference, ferrite beads of various sizes for RFI suppression, canned air to shoot out dirt, cotton swabs and pipe cleaners, and DeoxIT from Caig Labs to remove oxide from connectors.

Check off each item on the list as you pack it. After the gig, you can check the list to see whether you reclaimed all your gear.

PREPARING FOR EASIER SETUP

You want to make your setup as fast and easy as possible. Here are some tips to help this process.

Protective Cases

Mount your console and recorders in protective carrying cases. Install casters or swivel wheels under racks and carrying cases so you can roll them in. Rolling is so much easier than lifting and carrying. You might permanently install the multitrack recorders in carrying cases that act as racks. When a remote job comes up, just grab them and go.

A very helpful item is a dolly, wheeled cart, or hand truck to transport heavy equipment into the venue. Consider getting some lightweight tubular carts. Being collapsible, they store easily in your car or truck. One maker of equipment carts is www.kart-a-bag.com.

Pack mics, headphones, and other small pieces in cloth bags, trunks, or milk crates.

CAUTION: Keep hard drives separated from magnets, such as in headphones, monitor speakers, and dynamic mics.

You might want to build a mic container (a big box full of foam rubber with cutouts for all the mics). Or construct a wheeled cabinet with drawers for mics, DIs, and speaker cables.

In an article in the September-October 1985 issue of *db* magazine, remote recording engineer Ron Streicher offered these suggestions:

> "Especially for international travel, make sure your documentation is up to date and matches the equipment you're carrying. Make a list of everything you take: all the details, such as each pencil, razor blade, connector, etc. Also make sure your insurance is up to date. You need insurance for en route as well as at the destination.

> "I organize my cases so I know where every item is. They're ready to go anytime and make setup much faster. The cables are packed with their associated equipment, not in a cables case. I check everything coming and going, and try to have 100% redundancy, such as a small mixer to substitute for the large console."

Mic Mounts

If you'll be recording a singer/guitarist, take a short mic mount that clamps onto the singer's mic stand. Put the guitar mic in the mount. Also bring some short mounts or mini mics to clamp onto drum rims. By using these mounts, you eliminate the weight and clutter of several mic stands.

Some examples of short mounts are the Mic-Eze units by Ac-cetera (www.ac-cetera.com). They have standard 5/8–27 threads and mic clamps that either spring shut or are screw-tightened. Flex-Eze is two clamps joined by a short gooseneck and Min-Eze is two clamps joined by a swivel.

Snakes and Cables

You can store mic cables on a cable spool, available in the electrical department of a hardware store. Wrap one mic cable around the spool, plug it into the next cable and wrap it, etc.

A snake can be wrapped around a large reel or can be coiled in a trunk. Commercial snake reels are made by such companies as Whirlwind (www.whirlwindusa.com), ProCo (www.procosound.com), and Hannay (www.hannay.com).

Use wire ties to join cables that you normally run together, such as PA sends and returns.

Snake hookup is quicker if the snake has a multipin twist-lock connector (such as the Whirlwind W1 or W2). This connector plugs into a mating connector that divides into several male XLRs. Those XLRs plug into the mixing console. Leave the XLRs in the console carrying case. You'll find that the snake is easier to handle without the XLR pigtails on it.

For a clean, rapid hookup of drum mics, put a small snake (subsnake) near the drum kit, and run it to the main stage box. Or, put the main stage box by the drum kit. Snakes are made by such companies as Whirlwind (www.whirlwindusa.com), ProCo (www.procosound.com), and Horizon (www.horizonmusic.com).

Check that all your mic cables are wired in the same polarity (pin 2 hot on both ends).

You might want to use 3-conductor shielded mic cables. Connect the shield to ground only at the male XLR end. Also use cables with 100% shielding. Those measures enhance the shielding capability of the shield and reduce pickup of lighting buzzes.

In XLR-type cable connectors, do not wire pin 1 to the shell, or you may get ground loops when the shell contacts a metallic surface.

Label all your cables on both ends according to what they plug into; for example, DSP-9 effects in, track 12 out, power amp in, snake aux 2 out. Or you might prefer to number the cables near their connectors. Cover these labels with clear heat-shrink tubing.

Label both ends of each mic cable with the cable length. Put a drop of glue on each connector screw to temporarily lock it in place.

Rack Wiring

You can speed the console wiring by using a small snake between the rack and the console, and between the multitrack recorders and the console. When packing, plug the snakes into the rack gear and multitracks, and coil the snake inside the rack and the multitrack carrying case. In other words, have all your equipment prewired. At the gig, pull out a bundled harness and plug it into the console jacks.

You might be feeding your multitrack from the insert jacks of the house console. If so, use a snake with TRS stereo phone plugs at the console end. Carry some TRS-to-dual-mono phone adapters to handle consoles that have separate jacks for insert send and return.

Some engineers prefer to make a clearly marked interface panel on the rear of the racks and plug into the panels. This is easier than trying to find the right connectors on each piece of equipment.

Small bands might get by with all their equipment in a single, tall rack. Mount a small mixer on top, wired to effects in the middle, with a coiled snake and power cord on the bottom.

Small snakes for rack and multitrack connections are made by Hosa (www.hosatech.com), Horizon (www.horizonmusic.com), and ProCo (www.procosound.com), among others.

Other Tips

Here are some more helpful hints for successful on-location recordings.

- Record several gigs and pick the best performances to use in an album.
- Plan to use a talkback mic from the board to the stage monitor speakers during sound checks. You might bring a small instrument amp for talkback so that you can always be heard.
- Hook up and use unfamiliar equipment before going on the road. Don't experiment on the job!
- Consider recording with redundant (double) systems so you have a backup if one fails.
- Walkie-talkies are okay for preshow use, but don't use them during the performance because they cause RF interference. Use hardwired communications headsets. Assistants can relay messages to and from the stage crew while you're mixing.
- During short set changes, use a laptop computer local area network to show what set changes and mic-layout changes are coming up next; transmit this information to the monitor mixer and sound-reinforcement mixer.
- Don't put tapes or hard drives through airport X-ray machines because the transformer in these machines is not always well shielded. Have the tapes or hard drives inspected by hand.
- Hand-carry your mics on airplanes. Arrange to load and unload your own freight containers, rather than trusting them to airline freight loaders. Expect delays here and at security checkpoints.
- Get a public-liability insurance policy to protect yourself against lawsuits.
- Call the venue and ask directions to the load-in door. Make sure that someone will be there at setup time to let you in. Ask the custodian not to lock the circuit-breaker box the day of the recording.

- A few days before the session, check out the parking situation.
- Just before you go, check out all your equipment to make sure it's working.
- Arrive several hours ahead of time for parking and setup. Expect failures—there's always something going wrong, something unexpected. Allow 50% more time for troubleshooting than you think you'll need. Have backup plans if equipment fails.
- In general, plan everything in advance so you can relax at the gig and have fun!

AT THE SESSION: SETUP

Okay! You've arrived at the venue. After parking, offload your gear to a holding area, rather than onstage, because gear on stage will likely need to be moved.

Learn the names of the house sound-crew members, and be friendly. These people can be your assets or your enemies. Think before you comment to them! Try to remain in the background and do not interfere with their normal way of doing things (for example, take the secondary side of the split).

Power Distribution System

At the job, you need to take special precautions with power distribution, interconnecting multiple sound systems, and electric guitar grounding.

Consider buying, renting, or making your own single-phase power distribution system (distro). It will greatly reduce ground loops and increase reliability. One source of AC power distribution equipment is www.furmansound.com. Figure 17.10 shows a suggested AC power distribution system.

The amp rating of the distro's main breaker box should exceed the current drain of all the equipment that will be plugged into the distro system.

Power Source

If you're using a remote truck, find a source of power that can handle the truck's power requirements, usually at a breaker panel. Some newer clubs have separate breaker boxes for sound, lights, and a remote truck. Find out whether you'll need a union electrician to make those connections. Label your breakers.

Check that your AC power source is not shared with lighting dimmers or heavy machinery; these devices can cause noises or buzzes in the audio.

The industry-standard power connector for high-current applications is the Cam-Lok, a large cylindrical connector. Male and female Cam-Loks join together and lock when you twist the connector ring. Distro systems and power cables with Cam-Lok connectors can be rented from rental houses for film, lighting, electrical equipment, or entertainment equipment. One such rental house is Mole-Richardson at www.mole.com.

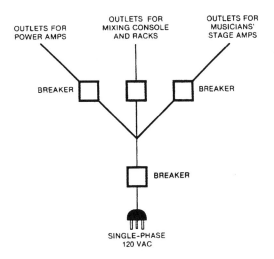

FIGURE 17.10
An AC power distribution system for a touring sound system.

Use an adapter from Cam-Lok to bare wires. Pull the panel off the breaker box, insert the bare wires, and connect the Cam-Lok to your truck's power.

CAUTION: Have an electrician do the wiring if you don't know what you're doing. A union electrician might be required anyway. Some breaker boxes have Cam-Loks already built in.

To reduce ground-loop problems, get on the same power that the house sound system is using. From that point, run your distribution system, or at least run one or two thick (14- or 16-gauge) extension cords to your recording system. These cords may need to be 100 to 200 feet long. Plug AC outlet strips into the extension cord, and then plug all your equipment into the outlet strips.

If your recording system is one or two multitrack recorders that will connect to the FOH console, simply plug into the same outlet strip that the FOH console is using.

Measure the AC line voltage. If the AC voltage varies widely, use a line voltage regulator (power conditioner) for your recording equipment. If the AC power is noisy, you might need a power isolation transformer.

Check AC power on stage with a circuit checker. Are grounded outlets actually grounded? Is there low resistance to ground? Are the outlets the correct polarity? There should be a substantial voltage between hot and ground, and no voltage between neutral and ground.

Some recording companies have a gasoline-powered generator ready to use if the house power fails. If there are a lot of lighting and dimmer racks at the gig, you might want to put the truck on a generator to keep it isolated from the lighting power.

Interconnecting Multiple Sound Systems

If you hear hum or buzz when the systems are connected, first make sure that the signal source is clean. You might be hearing a broken snake shield or an unused bass-guitar input.

If hum persists, experiment with flipping the ground-lift switches on the splitter and on the direct boxes. On some jobs you need to lift almost every ground. On others you need to tie all the grounds. The correct ground-lift setting can change from day to day because of a change in the lighting. Expect to do some trial-and-error adjustments.

If the house system has serious hum and buzz problems, offer help. You can hear buzzes in your quiet truck that they can't hear over the main system with noise in the background.

Often, a radio station or video crew will take an audio feed from your mixing console. In this case, you can prevent a hum problem by using a console with transformer-isolated inputs and outputs. Or you can use a 1:1 audio isolation transformer between the console and the feeds. For best isolation, use a distribution amp with several transformer-isolated feeds. Lift the cable shield at the input of the system you're feeding. Some excellent isolation transformers are made by Jensen (www.jensen-transformers.com).

Connections

After unpacking, place one mic list by the stage box so you know what to plug in where. Place a duplicate list by each console. Attach a strip of white tape just below the mixer faders. Use this strip to write down the instrument that each fader affects.

Based on the mic list you wrote, you might plug the bass DI into snake input 1, plug the kick mic into snake input 2, and so on. Label fader 1 "BASS," label fader 2 "KICK," etc. Also plug in equipment cables according to your block diagram.

Have an extra microphone and cable offstage ready to use if a mic fails.

Don't unplug mics plugged into phantom power because this will make a popping noise in the sound-reinforcement system.

Running Cables

To reduce hum pickup and ground-loop problems associated with cable connectors, try to use a single mic cable between each mic and its snake-box connector.

Avoid bundling mic cables, line-level cables, and power cables together. If you must cross mic cables and power cables, do so at right angles and space them vertically.

Plug each mic cable into the stage box; then run the cable out to the each mic and plug it in. This leaves less of a mess at the stage box. Leave the excess cable at each mic stand so you can move the mics. Don't tape the mic cables down until the musicians are settled.

It's important that audience members do not trip over your cables. In high-traffic areas, cover cables with rubber floor mats or cable crossovers (metal ramps). At least tape them down with gaffer's tape pressed lengthwise onto the cables.

It helps to set up a closed-circuit TV camera and TV monitor to see what's happening on stage. You need to know when mics get moved accidentally, or when singers use the wrong mic, etc.

Recording-Console Setup

Here's a suggested procedure for setting up the recording system efficiently:

1. If the console is set up in a dressing room or locker room, add some acoustic absorption to deaden the room reflections. You might bring a carpet for the floor, plus acoustic foam or moving blankets for the walls.
2. Turn up the recording monitor system and verify that it is clean.
3. Plug in one mic at a time and monitor it to check for hums and buzzes. Troubleshooting is easier if you listen to each mic as you connect it, rather than plugging them all in, and trying to find a hum or buzz.
4. Check and clean up one system at a time: first the sound-reinforcement system, then the stage-monitor system, then the recording system. Again, this makes troubleshooting easier because you have only one system to troubleshoot.
5. Use as many designation strips as you need for complex consoles. Label the input faders bottom and top. Also label the monitor-mix knobs and the meters.
6. Monitor the reverb returns (if any) and check for a clean signal.
7. Make a short test recording and listen to the playback.
8. Verify that the left and right channels are correct, and that the pan-pot action is not reversed audibly.
9. If you are setting up a separate recording monitor mix, do a preliminary pan-pot setup. Panning similar instruments to different locations helps you identify them.

MIC TECHNIQUES

Usually the miking is left up to the sound-reinforcement company. But there are some mic-related problems you should know about, such as feedback, leakage, room acoustics, and noise. Here are some ways to control these problems:

- Use directional microphones, such as cardioid, supercardioid, or hypercardioid. These mics pick up less feedback, leakage, and noise than omnidirectional mics at the same miking distance.

- With vocal mics, aim the null of the polar pattern at the floor monitors. The null (area of least pickup) of a cardioid is at the rear of the mic—180 degrees off-axis. The null of a supercardioid is 125 degrees off-axis and it is 110 degrees off-axis for a hypercardioid.
- To reduce breath pops with vocal mics, be sure to use foam pop filters. Allow a little spacing between the pop filter and the mic grille. It also helps to switch in a low-cut filter (100-Hz highpass filter).
- Mike close. Place each mic within a few inches of its instrument. Ask vocalists to sing with lips touching the mic's foam pop filter.
- Use direct boxes. Bass guitar and electric guitar can be recorded direct to eliminate leakage and noise in their signals. However, you might prefer the sound of a miked guitar amp. You could record the guitar direct from its effects boxes. Then use a guitar-amp emulator during mixdown. Note that sequencers and some keyboards have high-level outputs, so their DI boxes need transformers that can handle line level.
- Use contact pickups. On acoustic guitar, acoustic bass, and violin, you can avoid leakage by using a contact pickup. Such a pickup is sensitive only to the instrument's vibration, not so much to sound waves. The sound of a pickup is not as natural as a microphone, but a pickup may be your only choice. Consider using both a pickup and a microphone on the instrument. Feed the pickup to the house and monitor speakers; feed the mic and pickup to the recording mixer.

When you're recording a band that has been on tour, should you use their PA mics or your own mics? In general, go with their mics. The artists and PA company have been using their mics for a while and may not want to change anything. Most mics currently used in PA are good quality anyway, unless they are dirty or defective.

If you're not happy with their choice, you could add your own instrument mics. Let the PA people listen to the sound in the recording truck, or in headphones. If it sounds bad because of their mic choice, ask "Would it be okay if we tried a different mic (or mic placement)?" Usually it's all right with them—it's a team effort.

Electric Guitar Grounding

While setting up mics, you need to be aware of a safety issue with the electric guitar. Electric-guitar players can receive a shock when they touch their guitar and a mic simultaneously. This occurs when the guitar amp is plugged into an electrical outlet on stage, and the mixing console (to which the mics are grounded) is plugged into a separate outlet across the room. If you're not using a power distro, these two power points may be at widely different ground voltages. So a current can flow between the grounded mic housing and the grounded guitar strings.

CAUTION: Electric guitar shock is especially dangerous when the guitar amp and the console are on different phases of the AC mains.

It helps to power all instrument amps and audio gear from the same AC distribution outlets. If you lack a power distro, run a heavy extension cord from a stage outlet back to the mixing console (or vice versa). Plug all the power-cord ground pins into grounded outlets. That way, you prevent shocks and hum at the same time.

If you're picking up the electric guitar direct, use a transformer-isolated direct box and set the ground-lift switch to the minimum-hum position.

Using a neon tester or voltmeter, measure the voltage between the electric-guitar strings and the metal grille of the microphones. If there is a voltage, flip the AC polarity switch on the amp (if any). Use foam windscreens for additional protection against shocks.

Audience Mics

If you have enough mic inputs, you can use two audience mics to pick up the room acoustics and audience sounds. This helps the recording to sound "live." Without audience mics, the recording may sound too dry, as if it were done in a studio.

One easy method is to aim two cardioid mics at the audience. Put them on tall mic stands, on the stage floor, on either side of the stage. If those mic stands must not be seen, try hanging mics (Figure 17.11).

Some engineers pick up the audience and hall with two mics placed at the FOH mix position. That's an easy, effective way to capture the crowd. But the signal from these mics is delayed relative to the on-stage mics. If the FOH mix position is far from the stage (say, 50 feet or more), this delay will cause an echo when the audience mics are mixed with the stage mics. Prevent this by placing the audience mics fairly near the stage. Or, during computer mixdown, move the FOH mic tracks to the left, earlier in time so that large peaks in the FOH tracks' waveforms coincide with large peaks in on-stage tracks. If you're using the audience mics only to capture applause, the delay is not an issue.

FIGURE 17.11
Some audience miking techniques.

What if you don't have enough tracks for the audience mics? Record them on a 2-track recorder. Load this recording into your DAW along with the stage-mic mix. Align the two recordings in time as described above.

If the audience mics are run through the FOH mixer, leave the audience mics unassigned in that mixer to prevent feedback.

To get more isolation from the house speakers in the audience mics, use several mics hung close to the audience. Some engineers put up four audience mics maximum; some use eight to ten. Use directional mics and aim the rear null at the house speakers.

Another option is not miking the audience, or not using the audience tracks. Instead, during mixdown, you could simulate an audience with audience-reaction CDs. Simulate room reverb with an effects unit.

SOUND CHECK AND RECORDING

Now that the mics are set up, you might have time for a sound check. That's when you set recording levels as described earlier in this chapter under "Setting Levels."

A few minutes before the band starts playing, start recording. Keep a close eye on recording levels. If a track is going into the red, slowly turn down its input trim and note the recorder-counter time where this change occurred.

CAUTION: If you are recording off the FOH mixer, turning down its input trim will affect the house levels. The FOH mixer operator will need to turn up the corresponding monitor send and channel fader.

This is a touchy situation that demands cooperation. Ideally, you set enough headroom during the sound check so you won't have to change levels. But be sure the house mixer operator knows in advance that you might need to make changes. Ask the operator whether she wants to adjust the gain trims for you, so she can adjust corresponding levels at the same time. Thank the operator for helping you get a good recording.

If you are recording with a splitter and mic preamps on stage, assign someone to watch the levels and adjust them during the concert. Preamps with meters allow more precise level setting than preamps with clip LEDs.

Keep a track sheet and log as you record. For each song in the set list, note the recorder-counter time when the song starts, or press the Set Locate button on your multitrack recorder. Later, during mixdown, you can go to those counter times or locate points to find songs you want to mix. Also note where any level changes occurred so you can compensate during mixdown. It helps to note a counter time when the signal level was very high. When you mix the recording you can start at that point in setting your overall mix levels.

TEARDOWN

After the gig, pack your mics away first because they may be stolen or damaged. Refer to your equipment list as you repack everything. Note equipment failures and fix broken equipment as soon as possible.

After you haul your gear back to the studio, it's time for mixing and editing.

1. If necessary, copy all the tracks into your recording software's multitrack template.
2. Select all the tracks, and split all the tracks at the same point—several seconds before and after each song.
3. Export only the tracks in each split section to separate song folders. That way you can mix each song separately by importing the tracks from each folder into a multitrack template. Another option: Copy the tracks in each split section, then paste them into new projects.

Alternatively, use volume envelopes or automation to mix all the songs, then export one long concert mix. If you want, you can import that stereo mix and split it into individual songs, saving only the best ones. Then overlap one song's ending applause with the beginning of the next song.

SAME-DAY RECORDING/DISTRIBUTION

You can record a concert and duplicate the recording immediately so the concertgoers can buy that recording soon after the gig. Use the 2-track or 4-track recording methods described earlier, then copy the stereo WAV file to microSDs, USB thumb drives, MP3 files for download, or even CDs.

Want to try same-day recording? Check out USB duplicators and microSD duplicators on Google. At concerts, collect email addresses on a sign-up sheet, and send them a link to the concert recording's MP3 file.

Some of the information in this chapter was derived from two workshops presented at the 79th convention of the Audio Engineering Society in October 1985. These workshops were titled "On the Repeal of Murphy's Law—Interfacing Problem Solving, Planning, and General Efficiency On-Location," given by Paul Blakemore, Neil Muncy, and Skip Pizzi and "Popular Music Recording Techniques," given by Paul Blakemore, Dave Moulton, Neil Muncy, Skip Pizzi, and Curt Wittig.

CHAPTER 18

On-Location Recording of Classical Music

This chapter explains how to do an on-location stereo recording of a classical-music ensemble. We cover both equipment and procedures.

EQUIPMENT

- Microphones (low-noise condenser or ribbon type, omni or directional, free field or boundary, stereo or matched pair)
- Multitrack recorder or stereo recorder
- Low-noise mic preamps or low-noise mini mixer (necessary if the mic preamps in your recorder are low quality)
- Phantom power supply (necessary if your mixer or mic preamp lacks phantom power)
- Stereo bar
- Mic stands and booms, or fishing line to hang mics
- Shock mount (optional)
- Mixer (necessary if you use more than two mics)
- MS matrix box (if you are recording with the mid-side technique)
- Headphones and/or speakers
- Power amplifier for speakers (optional) or powered Nearfield monitors
- Recording medium: hard drive or flash memory
- Power strip, extension cords
- Notebook and pen
- Talkback mic and powered speaker (optional)
- Tool kit
- Fresh batteries

First on the list are microphones. You'll need at least two or three of the same model number or one or two stereo microphones. Good microphones are essential, for the microphones—and their placement—determine the sound of your recording.

For classical-music recording, the preferred microphones are condenser or ribbon types with a wide, flat frequency response and very low self-noise (explained in Chapter 6). A self-noise spec of less than 20 dB equivalent SPL, A-weighted, is recommended.

If you want to do spaced-pair recording, you can use either omnidirectional or directional microphones. Omnis are preferred because they generally have a deeper low-frequency response. If you want to do coincident or near-coincident recording for sharper imaging, use directional microphones (cardioid, supercardioid, hypercardioid, or bidirectional). Mics with those polar patterns tend to roll off in the low frequencies, but you can compensate for this with equalization.

You need a power supply for condenser microphones; you can use an external phantom power supply, a mixer or mic preamp with phantom power, or internal batteries. A low-noise stereo mic preamp (or low-noise portable mixer) lets you make recordings free of electronic hiss.

You can mount the microphones on stands or hang them from the ceiling with nylon fishing line. Make sure that the fishing line's tensile strength exceeds the weight of the mics. Check legal safety issues with hanging mics; different rules apply in different places. Stands are much easier to set up but more visually distracting at live concerts. Stands are more suitable for recording rehearsals or sessions with no audience present. Neumann makes tiltable "auditorium hangers" that suspend mic capsules of the KM 100 system from their cables.

The mic stands should have a tripod folding base and should extend at least 13 feet high. Some suitable products are telescoping stands, which are are lightweight and compact. Some examples are the Shure S15A (www.shure.com), Quik Lok A85 (www.quiklok.com), AEA-13MDV (www.wesdooley.com), and K&M 21411B (various vendors). You can use baby booms or stand extenders to increase the height of regular mic stands.

A useful accessory is a stereo bar (stereo microphone adapter, stereo mic mount). This device mounts two microphones on a single stand for coincident or near-coincident stereo recording. A long stereo bar can accommodate spaced-pair miking. Check out the AES, Grace, and DPA stereo bars.

Another needed accessory in most cases is a shock mount to prevent pickup of floor vibrations. Some models are the On-Stage MY410 (various vendors), Sabra Som SSM-1 (www.sabra-som.com), Shure A55M (www.shure.com), and AKG H85 (www.akg.com).

In difficult mounting situations, boundary microphones may come in handy. They can lie flat on the stage floor to pick up small ensembles or can be mounted on the ceiling or on the front edge of a balcony. They also can be attached to clear Plexiglas panels that are hung or mounted on mic stands. Boundary mics are made by most microphone companies.

To monitor in the same room as the musicians, you need some closed-cup, circumaural (around the ear) headphones or isolating earphones to block out the sound of the musicians. You want to hear only what is being recorded. Of course, the headphones should be wide range and smooth for accurate monitoring. A better monitoring arrangement might be to set up powered Nearfield loudspeakers in a separate room.

If you're in the same room as the musicians, you have to sit far from them to clearly monitor what you're recording. To do that, you need a pair of microphone extension cables. Longer extensions are needed if the mics are hung from the ceiling or if you monitor in a separate room.

A mixer is necessary when you want to record more than one source, for example, an orchestra and a choir, an orchestra with spot mics, or a band and a soloist. You might put a pair of microphones on the orchestra and another pair on the choir. Connect the insert sends from the mixer to a multitrack recorder, then mix the tracks to stereo back in the studio. If your budget doesn't allow a multitrack recorder you could mix the mic signals live to a stereo recorder.

For monitoring a mid-side recording, bring an MS matrix box that converts the MS signals to left-right signals, which you monitor.

SELECTING A VENUE

If possible, plan to record in a venue with good acoustics. It should have adequate reverberation time for the music being performed (about 2 seconds for orchestral music). This is very important, because it can make the difference between an amateur-sounding recording and a commercial-sounding one. Try to record in an auditorium, concert hall, or house of worship rather than in a band room or gymnasium. Halls with wooden surfaces and a shoebox shape tend to sound best.

You may be forced to record in a hall that is too dead, that is, the reverberation time is too short. In this case, you may want to add artificial reverberation from a digital reverb unit or plug-in, or cover the seats with plywood sheets or 4-millimeter polyethylene plastic sheeting. Strong echoes can be controlled with carpets, RPG diffusors, or drapes.

If the venue is surrounded by noisy traffic, consider recording after midnight. Turn off noisy air conditioning if possible while recording.

Call the venue manager and ask that the circuit breakers for the stage outlets be turned on the day of the session. Ask where you can load in your equipment. Also make sure that the load-in door will be unlocked when you plan to load in.

SESSION SETUP

Be sure to test all your equipment for correct operation before going on the job. If you are using battery-powered devices, install fresh batteries in them just before the concert. Keep your equipment inside your home or studio until you're ready to leave. A recorder left outside in a cold car may become sluggish if the lubricant stiffens, and batteries may lose some voltage.

Make sure your hard drive or flash card has enough capacity for the session. Table 14.1 in Chapter 14 shows the storage required for a 1-hour recording.

Arrive at the venue a few hours early to allow for setup and for fixing problems.

First, power up the equipment. You can use batteries or an AC extension cord plugged into an outlet near the stage. Make sure this outlet is live. Gaffer-tape the extension cord lengthwise or cover it with mats to prevent tripping. Tie your outlet strip's AC cord to the extension cord so they can't separate if someone pulls on the extension cord.

If this is a session, listen to the ensemble playing on the stage. Is the sound bad? You might want to move the musicians out onto the floor of the hall.

Mounting the Mics

Place your microphones in the desired stereo miking arrangement. As an example, suppose you're recording an orchestra rehearsal with two crossed cardioids on a stereo bar (the near-coincident method). Screw the stereo bar onto a mic stand and mount two cardioid microphones on the stereo bar. For starters, angle them 110 degrees apart and space them 7 inches apart horizontally. Aim them downward so that they'll point at the orchestra when raised. You may want to mount the microphones in shock mounts or put the stands on sponges to isolate the mics from floor vibration.

Basically, place two or three mics (or a stereo mic) several feet in front of the group, raised up high (as in Figure 18.1). The microphone placement controls the acoustic perspective or sense of distance to the ensemble, the balance among instruments, and the stereo imaging.

As a starting position, place the mic stand behind the conductor's podium, about 12 feet from the front-row musicians. Connect mic cables and gaffer-tape them to the top of each mic stand for strain relief. Raise the microphones about 14 feet off the floor. This prevents overly loud pickup of the front row relative to the back row of the orchestra. It also reduces audience noise by mounting the mics farther from the audience. Gaffer-tape the mic cables to the bottom of the stand to keep it from being pulled over.

Leave some extra turns of mic cable at the base of each stand so you can reposition the stands. This slack also allows for people accidentally pulling on the cables. Try to route the mic cables where they won't be stepped on, or cover them with mats.

FIGURE 18.1
Typical microphone placement for on-location recording of a classical music ensemble. A: top view. B: side view.

Live, broadcast, or filmed concerts require an inconspicuous mic placement, which may not be sonically ideal. In these cases, or for permanent installations, you probably want to hang the microphones from the ceiling rather than using stands. You can position a stereo bar with three nylon fishing lines spaced apart. Make sure that the tensile strength of the fishing line exceeds the weight of the mics and stereo bar. Hang individual mics by their cables and attach two fishing lines to the front of each mic to aim it. Another inconspicuous mic placement is on mic-stand booms projecting forward of a balcony. For drama or musicals, directional boundary mics can be placed on the stage floor near the footlights.

If you are uncertain which mic technique to use, you could set up two or more mic arrays and record them to multitrack. When you master the program, choose the best-sounding technique. You might even mix the signals of two or more mic pairs.

Connections

Now you're ready to make connections. Here are some connection methods for just two mics:

- If your stereo recorder has high-quality mic preamps and phantom power, plug the mics directly into the recorder mic inputs.
- If your recorder lacks phantom power (or the phantom voltage is lower than the mics require), plug the mics into a phantom supply, and from there into your recorder mic inputs.
- If your recorder lacks high-quality mic preamps, plug the mics into a low-noise mic preamp or mixer with phantom power. Connect cables from there into your recorder line inputs.

Here are some connection methods for more than two mics:

- Plug the mics into a stage box (described in Chapter 4) and run the snake back to your mixer. Plug the snake mic connectors into your mixer mic inputs. If you are recording to 2-track, plug the mixer stereo outputs into the recorder line inputs. If you are recording to multitrack, plug the mixer insert-send jacks into the recorder line inputs.
- If you have a multichannel audio interface with mic preamps, plug the mics into the interface. Connect the interface USB or FireWire port to a laptop computer.
- If you want to feed your mic signals to several mixers—recording, broadcast, PA—plug your mic cables into a mic splitter or distribution amp (described in Chapter 17 under the heading "Splitting the Mic Signals"). Connect the splitter outputs to the snakes for each mixer. Supply phantom from one mixer only, on the direct side of the split. Each split will have a ground-lift switch on the splitter. Set it to *ground* for the mixer supplying phantom. Set it to *lift* for the other mixers (or to *ground* if that results in the least hum). This prevents hum caused by ground loops between the different mixers.
- If you're using directional microphones and want to make their response flat at low frequencies, you can run them through a mixer with equalization for bass boost. You might prefer to equalize the recording during mastering instead. Boost around 50-100 Hz until the bass sounds natural or until it matches the bass response of omni condenser mics. You won't need this EQ if the microphones have been factory-equalized for flat response at a distance.

Monitoring

Put on your headphones or listen over loudspeakers in a separate room. Sit equidistant from the speakers, as far from them as they are spaced apart. You'll probably need to use a Nearfield arrangement (speakers about 3 feet apart and 3 feet from you) to reduce coloration of the speakers' sound from the room acoustics.

Turn up the recording-level controls and monitor the signal. When the orchestra starts to tune up, set the recording levels to peak around −20 dBFS so you have a clean signal to monitor. You can set levels more carefully later on.

MICROPHONE PLACEMENT

Nothing has more effect on the production style of a classical-music recording than microphone placement. Miking distance, polar patterns, angling, spacing, and spot miking all influence the recorded sound character.

Miking Distance

The microphones must be placed closer to the musicians than a good live listening position. If you place the mics out in the audience where the live sound is good, the recording probably will sound muddy and distant when played over speakers. That is because all the recorded reverberation is reproduced up front along a line between the monitor speakers, along with the direct sound of the orchestra. Close miking (4 to 20 ft from the front row) compensates for this effect by increasing the ratio of direct sound to reverberant sound.

The closer the mics are to the orchestra, the closer it sounds in the recording. If the instruments sound too close, too edgy, or too detailed, or if the recording lacks hall ambience, the mics are too close to the ensemble. Move the mic stand a foot or two farther from the orchestra and listen again.

If the orchestra sounds too distant, muddy, or reverberant, the mics are too far from the ensemble. Move the mic stand a little closer to the musicians and listen again.

Eventually you'll find a sweet spot, where the direct sound of the orchestra is in a pleasing balance with the ambience of the concert hall. Then the reproduced orchestra will sound neither too close nor too far.

Here's why miking distance affects the perceived closeness (perspective) of the musical ensemble: the level of reverberation is fairly constant throughout a room, but the level of the direct sound from the ensemble increases as you get closer to it. Close miking picks up a high ratio of direct-to-reverberant sound; distant miking picks up a low ratio. The higher the direct-to-reverb ratio, the closer the sound source is perceived to be.

If the recording venue is "live" because of hard surfaces—brick, glass, stone—chances are you will need to mike closely. On the other hand, if the venue is "dead" because of soft surfaces—carpet, drapes, stuffed seats—expect to mike farther away.

An alternative to finding the sweet spot is to place a stereo pair close to the ensemble (for clarity) and another stereo pair distant from the ensemble (for ambience). According to Delos recording director John Eargle, the distant pair should be no more than 30 feet from the main pair, otherwise its signal might simulate an echo. You record the two pairs to a multitrack recorder and mix them back in the studio. The advantages of this method are

- It avoids pickup of bad-sounding early reflections.
- Close miking reduces pickup of background noise.
- After the recording is finished, you can adjust the direct/reverb ratio or the perceived distance to the ensemble.
- Comb filtering due to phase cancellations between the two pairs is not severe because the delay between them is great and their levels and spectra are different.

Similarly, Skip Pizzi recommends a "double MS" technique, which uses a close MS microphone mixed with a distant MS microphone (as shown in Figure 18.2). One MS microphone is close to the performing ensemble for clarity and sharp imaging, and the

FIGURE 18.2
Double MS technique using a close main pair and a distant pair for ambience. Spot mics also are shown.

other is out in the hall for ambience and depth. The distant mic could be replaced by an XY pair for lower cost.

If the ensemble is being amplified through a sound-reinforcement system, you might be forced to mike very close to avoid picking up amplified sound and feedback from the reinforcement speakers. In that case you will need to add high-quality artificial reverberation or convolution reverb.

For broadcast or communications, consider miking the conductor with a wireless lavalier mic or stand-mounted mic.

Stereo-Spread Control

Concentrate on the stereo spread. If the monitored spread is too narrow, it means that the mics are angled or spaced too close together. Increase the angle or spacing between mics until localization is accurate. Note: Increasing the angle between mics will make the instruments sound farther away; increasing the spacing will not.

If off-center instruments are heard far-left or far-right, that indicates your mics are angled or spaced too far apart. Move them closer together until localization is accurate.

If you record with a mid-side microphone, you can change the monitored stereo spread either during the recording or after.

To change the spread during the recording, connect the stereo-mic outputs to the matrix box and connect the matrix-box L–R output to the recorder. Use the stereo-spread control (M/S ratio) in the matrix box to adjust the stereo spread.

To alter the spread after the recording using a matrix box: Record the mid signal on one track and the side signal on another track. Monitor the output of the recorder with a matrix box. Back in the studio, run the mid and side tracks through the matrix box, adjust the stereo spread as desired, and record the left and right outputs.

To alter the spread after the recording using a DAW:

1. Record the mid mic on track 1; record the side mic on track 2.
2. Copy or clone track 2 to track 3. Be sure the waveforms are aligned.
3. Pan track 2 hard left; pan track 3 hard right.
4. Reverse the polarity of track 3 or use an "invert polarity" plug-in.
5. Group tracks 2 and 3 so their faders move together.
6. To change the stereo spread, vary the levels of tracks 2 and 3 relative to track 1.

If you are set up before the musicians arrive, check the localization by recording yourself speaking from various positions in front of the microphone pair while announcing your position (e.g., "left side," "mid-left," "center"). Play back the recording to judge the localization accuracy provided by your chosen stereo array. Recording this localization test is an excellent practice.

MONITORING STEREO SPREAD

Full stereo spread on speakers is a spread of images all the way between speakers, from the left speaker to the right speaker. Full stereo spread on headphones can be described as stereo spread from ear to ear. The stereo spread heard on headphones may or may not match the stereo spread heard over speakers, depending on the microphone technique used.

Due to psychoacoustic phenomena, coincident-pair recordings have less stereo spread over headphones than over loudspeakers. Take this into account when monitoring with headphones or use only loudspeakers for monitoring.

If you are monitoring your recording over headphones or anticipate headphone listening to the playback, you may want to use near-coincident miking techniques, which have similar stereo spread on headphones and loudspeakers.

Ideally, monitor speakers should be set up in a Nearfield arrangement (say, 3 feet from you and 3 feet apart) to reduce the influence of room acoustics and to improve stereo imaging. On the wall behind the monitors, attach a panel of acoustic foam that extends a few feet beyond the speaker spacing.

If you want to use large monitor speakers placed farther away from you, deaden the control-room acoustics with acoustic foam or thick fiberglass insulation (covered with muslin). Place the acoustic treatment on the walls behind and to the sides of the loudspeakers. This smoothes the frequency response and sharpens stereo imaging.

You might include a stereo/mono switch in your monitoring system to check for mono compatibility.

Soloist Pickup and Spot Microphones

Sometimes a soloist plays in front of the orchestra. You have to capture a tasteful balance between the soloist and the ensemble. That is, the main stereo pair should be placed so that the relative loudness of the soloist and the accompaniment is musically appropriate. If the soloist is too loud relative to the orchestra (as heard on headphones or loudspeakers), raise the mics. If the soloist is too quiet, lower the mics. You may want to add a spot mic (accent mic) about 3 feet from the soloist and mix it with the other microphones. Take care that the soloist appears at the proper depth relative to the orchestra.

If a sound-reinforcement system is in use, place the soloist mic 8 to 12 inches away to prevent feedback. Use a foam windscreen on a vocal mic. To make the soloist mic inconspicuous, you could place a small-diaphragm mic at about chest height aiming at the mouth, and use a slender mic stand such as the Schoeps Active Tube RC.

Many record companies prefer to use multiple microphones and multitrack techniques when recording classical music. Such methods provide extra control of balance and definition and are necessary in difficult situations. Often you must add spot or accent mics on various instruments or instrumental sections to improve the balance or enhance clarity (as shown in Figure 18.2). In fact, classical recording expert John Eargle contended that a single stereo pair of mics rarely works well.

As pkautzsch posted on Gearslutz.com, "Recording is not about the real picture, but about the picture we want. I like to have it similar to the real picture—that is, choir behind orchestra, soloists in front, violins left, etc.—but I'd never try to make a recording that is real. At the best, it will be credible. What spot mics do in recording for home stereo use is nothing else than what our eyes and brains do when attending a concert."

A choir that sings behind the orchestra can be miked separately with two to four cardioids. You might place the choir in the audience area facing the orchestra, and mike the choir.

If the recording mics are also used for sound reinforcement, place the choir mics about 3 feet out front and 3 feet above the head height of the back row. Add piano mics and windscreened soloist mics about 8 inches from their sound sources. Record all the mics to multitrack, and mix the tracks back in your studio. Add reverb and EQ as needed.

Pan each spot mic so that its image position coincides with that of the main microphone pair. Using the mute switches on your mixing console, alternately monitor the main pair and each spot mic to compare image positions.

When you use spot mics, mix them at a low level relative to the main pair—just loud enough to add definition but not loud enough to destroy depth. Operate the spot-mic faders subtly or leave them untouched. Otherwise the close-miked instruments may seem to jump forward when the fader is brought up, then fall back in when the fader is brought down. If you bring up a spot-mic fader for a solo, drop it only 6 dB when the solo is over, not all the way off.

Often the timbre of the instrument(s) picked up by the spot mic is too bright. You can fix it with a high-frequency rolloff or by using a mic with that characteristic. Adding artificial reverb to the spot mic can help too.

To further integrate the sound of the spots with the main pair, you might want to delay each spot's signal to coincide with those of the main pair. That way, the main and spot signals are heard at the same time. For each spot mic, the formula for the required delay is

$$T = D/C$$

where

T = delay time in seconds
D = distance between each spot mic and the main pair in feet
C = speed of sound, 1130 ft per second

For example, if a spot mic is 20 feet in front of the main pair, the required delay is 20/1130 or 17.7 msec. Some engineers add even more delay (10–15 msec) to the spot mics to make them less noticeable. As a rule of thumb, 1 foot corresponds to about 1 millisecond of delay.

A suggested track assignment for multitrack concert recordings is shown below:

Tracks 1–2: main pair
Tracks 2–3: distant pair
Tracks 4–5: "outriggers" or widely spaced pair
Tracks 6 and up: spot mics and conductor's mic (for announcing takes).

Once the microphones are positioned properly, gaffer-tape the mic stand legs to the floor (or add sandbags) so that the stands can't be knocked over.

SETTING LEVELS

Now you're ready to set recording levels. Ask the conductor to have the orchestra play the loudest part of the composition, and set the recording level for the desired meter reading. A typical recording level is −6 dB maximum on a peak-reading meter for a digital recorder. The level can go up to 0 dB maximum without distortion, but aiming for −6 dB allows for surprises. Bass-drum and tympani hits produce the highest peaks.

If you plan to record a concert with no sound check, you have to set the record-level knobs to a nearly correct position ahead of time. Do this during a preconcert trial recording, or just go by experience: set the knobs where you did at previous sessions (assuming you are using the same microphones in the same venue). Another way to preset the recording level is to aim for a peak meter level of −20 dBFS when the orchestra tunes up.

Before going on location, you could play loud orchestral music over your studio monitors or home stereo, set up your mics and recorder, and set approximate recording levels. If you need to set the level less than one-third up to get a 0 dB meter level, you probably need to insert a pad or set the mic-gain switch to low gain.

RECORDING A CONCERT

Before the concert obtain a printed program (and ideally a score) of the musical selections. On this program, next to each piece, you will write down the recorder counter times of the beginning and end of that piece. On the score, you will mark any edits that might need to be made such as a troublesome cough or a note patched from a rehearsal take. That will help you locate and identify the pieces correctly when you edit the recording later.

Start recording when the conductor walks out (or sooner). Record the concert nonstop if your recording medium allows. Document your mic placement and recording level for future jobs.

EDITING

At this point, the recording is finished and you brought it back to your studio. The concert recording has long applause after each piece. Suppose you want to edit the applause and make it shorter and insert a few seconds of silence between the compositions. Here is a suggested procedure for editing a stereo recording using a digital audio workstation:

1. Connect a USB cable between the computer and recorder. Then the computer recognizes the recorder as a storage device. Click-drag the recorder's WAV file of the concert to your hard drive.
2. In the audio editing software, start a new project and set up an audio track.
3. Import the concert file from the hard drive into the audio track.
4. Play the track and locate the first piece. Refer to the counter times you wrote on your session notes or concert program.
5. Delete the part of the recording before the first piece but do not close up the space yet.
6. Find the applause at the end of the first piece. About 10 seconds into the applause, split the clip or region. Also split the clip a few seconds before the start of the next piece. Cut out the audio between the split points, but do not close up the space.
7. Label the clip of the composition with its title.
8. Repeat step 6 for the rest of the pieces.

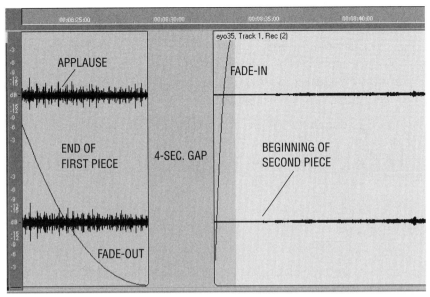

FIGURE 18.3
A DAW screen showing an applause fade-out, a 4-second gap, and a fade-in before the next piece.

Now each musical piece is in a separate, labeled clip or region on your screen. Next you will add fades and adjust the spacing between the pieces. Please refer to Figure 18.3.

1. Slide the first piece back to time zero, then slide it to the right 10 frames (1/3 second).
2. At the end of each piece, let the applause play for 3 seconds then fade it out over about 8 seconds. Use a fade that starts quickly and ends slowly.
3. If there is background noise such as air-conditioning rumble, insert a fade-in about 2 seconds before the beginning of each piece (Figure 18.3). Or you may want the track to start right when the music starts, that is, with no ambience before it starts.
4. Time-slide the clips to create a 4-second gap between them (or whatever interval sounds right).

If you don't want total silence between pieces, do not fade out the applause. End each piece's clip where the applause stops, and leave about 4 seconds of recorded "room tone" between clips.

To edit a session that has no applause, set the beginning of each clip just before the first note. Set the end of each clip just after the reverberant tail fades to silence. Leave about a 4-second silent gap between clips (or whatever sounds right).

When you have finished editing, note the start time and duration of each clip. You will use this information to write a cue sheet that determines the start IDs when burning a CD. Export the edited program to a 24-bit stereo file.

Sometimes, despite your best intentions and the finest microphones, the spectral balance of the recording might be poor. The bass might be weak, the strings strident, and so on. Room acoustics play a strong part in this. Fortunately, a recording having a skewed tonal balance often can be salvaged with equalization. An effective tool for this purpose is Harmonic Balancer (www.har-bal.com), which shows the spectrum of the recording and lets you equalize it as needed.

Once the program is equalized (if necessary), import the equalized file and normalize it so that the highest peak reaches −0.3 dBFS. Finally, enable dither and export the normalized mix to a 16-bit stereo file. You will burn a CD from this file. There's your finished recording!

CHAPTER 19

Web Audio and Online Collaboration

You've recorded a new song and you can't wait for people to hear it. You might want to distribute it on the Web, in addition to (or instead of) CDs.

Why put your music online? Web excerpts of your songs can entice listeners to buy your CDs or download your songs. Songs on the Web cost less than CDs. Also, distributing online is much quicker than distributing CDs. Keep in mind, though, that it's hard to get your songs noticed among the millions of other titles that are online. Many books and websites offer suggestions on marketing your music.

STREAMING VERSUS DOWNLOADING

There are two basic ways to transmit music on the Internet: streaming files and downloading files. A streaming file plays as soon as you click on its title. A downloaded file doesn't play until you copy the entire file to your hard disk. Streaming audio is heard almost instantly (after the buffer memory fills), but often sounds muffled and can be interrupted by net congestion.

The sound quality of streaming audio depends on the speed of the modem and ranges from funky AM radio (with a 56 K modem) to near-CD quality (with a cable modem). With downloaded audio you must wait a few seconds to several minutes to download the song. But when it plays, the sound is high-fidelity and continuous.

DATA COMPRESSION

Audio files that you record on your computer are usually WAV or AIFF files. They have no data reduction (data compression), so they take up lots of memory. A 3-minute song recorded at 16 bits, 44.1 kHz consumes about 32 MB. Downloading a 32 MB WAV file on the Web would take several hours using a 56 K modem. So, audio WAV files intended for Web download are usually data-reduced or data-compressed by encoding

Table 19.1	Data Compression Ratio and Sound Quality of Various MP3 Bitrates	
Bitrate (CBR)	**Compression**	**Sound quality**
64 kbps	20:1	AM radio quality
128 kbps	10:1	Cassette quality
192 kbps	7:1	Near-CD quality
256 kbps	5:1	CD quality
320 kbps	4.4:1	CD quality

them as MP3, WMA, or other formats. For example, if you compressed a 3-minute song by 10:1 it would consume about 3.2 MB and would download 10 times faster than the equivalent WAV file.

Most types of data compression tend to degrade sound quality. The sound quality of a compressed-data format depends on its bitrate, measured in kilobits per second (kbps). The higher the bitrate, the better the sound, but the greater the file size. A bitrate of 128 kbps for stereo MP3 files is considered to be a good compromise between sound quality and file size. At low bitrates (at and below 128 kbps), you can hear artifacts such as "swirly" cymbals, smeared transients such as drum hits, a general "phasey" effect, and less treble. At higher bitrates above 192 kbps, the artifacts start to disappear, and to most listeners the sound is CD quality.

In short, data compression reduces the file size and the download time at the expense of sound quality.

A stereo MP3 file encoded at 128 kbps is 64 kbps per channel. A mono MP3 file encoded at 64 kbps is equal in quality to a stereo MP3 file encoded at 128 kbps. Mono files consume half the file space of stereo files if both are the same bitrate.

Table 19.1 relates MP3 stereo bitrate to sound quality.

WEB-RELATED AUDIO FILES

You can put audio on the Web in several file formats. Let's look at some of the file types in current use.

WAV (.wav): A standard PC format for audio files. It encodes sound without any data reduction by using pulse code modulation. WAV files that are used to make audio CDs are 16-bit, 44.1 kHz. Because WAV files consume a lot of memory (about 32 MB for a 3-minute song in stereo), they are seldom used on the Internet for songs because they take so long to download.

However, an Internet connection by fiber-optic cable (such as Verizon's FiOS) is much faster than a standard copper-wire connection. FiOS is claimed to offer up to 50 or 30 Mbps downloads and up to 20 or 5 Mbps uploads. With FiOS, it's possible to transfer wave files—even hi-res ones—in a short time. For example, www. itrax.com offers songs for sale online that are surround files or 24-bit/96 kHz stereo files.

AIFF (.aif): Audio Interchange File Format, a standard Mac format for audio files. Use this format to transfer audio files between a PC and a Mac. Like WAV files, AIFF files are not data-compressed.

MID (MIDI, or Musical Instrument Digital Interface): This is a MIDI sequence, not an audio file. A .mid file is a string of numbers that represent performance gestures, such as note on/off, which note is played, key velocity, and so on. Chapter 16 on MIDI describes how to create MIDI files. Because MIDI files do not include audio, they are very compact. The user downloads the MIDI file and imports it into a MIDI sequencer or MIDI/audio DAW. The sequence generates instrumental music in a synth, sound module, or soft synth when played. To ensure that the proper instrument sounds play, most people use a General MIDI file format.

The following formats use data compression.

MP3 (MPEG Level-1 Layer-3): The most popular format. You can choose a low bitrate to make a small file with lower sound quality, or choose a high bitrate to make a larger file with higher quality.

When encoding an MP3 file, you have a choice of constant bitrate (CBR) or variable bitrate (VBR). VBR uses more bits on complex musical passages and fewer bits on simpler passages. CBR encodes every frame at the same bitrate, so it is less efficient. Use VBR to get a fixed level of quality at the lowest possible bitrate and file size. CBR is recommended only for streaming where the bitrate must be fixed. Average bitrate (ABR) is a compromise between CBR and VBR modes. Use ABR when you need to know the size of the encoded file but still allow some variation in the bitrate.

MP3PRO: An improvement over MP3. Songs encoded at 64 kbps with MP3PRO are said to sound as good as songs encoded at 128 kbps with MP3. MP3PRO offers faster downloads and nearly doubles the amount of music you can put on a flash memory player. MP3 and MP3PRO files are compatible with each others' players, but a free MP3PRO player is needed to hear MP3PRO's improvement in sound quality. See www. mp3prozone.com.

WMA (Windows Media Audio): This popular Microsoft format provides digital rights management (DRM), which is a technology that provides copy protection. It's useful for online music stores. For more information see www.microsoft.com/windows/ windowsmedia. WMA Pro also supports multichannel and high-resolution audio.

Suppose you want to download a 3-minute song. We'll assume that the download speed is 90 kilobytes per second (typical of DSL).

- The original WAV file of the song is 30.3 MB, CD quality or better, and downloads in about 5.6 minutes.
- An MP3 or WMA file at 128 kbps is 3.0 MB, cassette quality, and downloads in about 33 seconds.
- An MP3 or WMA file at 320 kbps is 6.9 MB, CD quality, and downloads in about 1.3 minutes.

RealAudio: This format is used for streaming audio and for music downloads from RealAudio's music store. The streaming fidelity depends on modem speed and the current Internet bandwidth. RealAudio files (.ra or. rm) are often used as short excerpts or previews of songs.

OGG (Ogg Vorbis): License-free open software. For a given file size, Vorbis sounds better than MP3. Vorbis takes up less file size than MP3 files of equal quality. For more information see www.xiph.org and www.vorbis.com.

AAC (MPEG Advanced Audio Coding): AAC offers better sound quality than MP3 at the same bitrate. Many listeners claim that AAC files made at 128 kbps sound like the original uncompressed audio source. What's more, AAC supports multichannel audio and a wide range of sample rates and bit depths. Digital rights management technology is also used with this format to help control the copying and distributing of music.

HD-AAC combines MPEG-4 AAC lossy compression with MPEG-4 SLS lossless coding. It provides a single file which can be scaled anywhere from audiophile quality to low-bitrate quality. From a single HD-AAC encoded file, you can play a lossless, high-quality version on your home stereo, and play a lossy, small file size version on an iPod.

FLAC (Free Lossless Audio Codec): A codec that keeps the sound quality of your mix, but reduces file size by about half. It's lossless: the audio is compressed (data-reduced) without any loss in sound quality. Unlike WAV files, FLAC lets you embed metadata into the file, which is displayed by a music player—artist, album, song, ISRC, and so on. See http://flac.sourceforge.net/.

Currently, WMA and MP3 are the most popular types. *Audio clip 41 at www.elsevierdirect. com/companions/9780240811444 demonstrates MP3 and WMA data compression.*

WHAT YOU NEED

To put your music on the Web, you need to download a few pieces of software, which are low cost or free:

Ripper: Also called a grabber, this program converts audio from a CD or CD-R to a WAV file. Examples are Exact Audio Copy (www.exactaudiocopy.de), CDex (http://cdexos.sourceforge.net), AltoMP3 from www.yuansoft.com, and FreeRip3 (www.mgshareware.com). Windows Media Player 12 rips to MP3 and other formats. Not all CD-ROM drives support ripping. Most DAWs can export MP3 files.

MP3 encoder: A program that converts a WAV or AIFF file to an MP3 file. Examples of freeware are RazorLame 1.1.5a (www.free-codecs.com/download/RazorLame.htm) and Lame MP3 Encoder (www.free-codecs.com/download/Lame_Encoder.htm). You need both. Get Lame version 3.98 or later. Another option is Wave to mp3 from www.softpedia.com. In Google, search for WAV to MP3 converter. Some recording software includes an MP3 encoder, and so does Winamp, FreeRip3, CDex, dBpowerAmp, foobar2000, Audio MP3 Editor, and Exact Audio Copy.

ID3 tag editor: This program lets you add song title, artist, genre, and other metadata information to an MP3 file. This information displays on screen when the MP3 file plays. You can add studio contact information in a comments field. Two freeware programs are Stamp ID3 Tag Editor and Mp3tag. Several MP3 playback programs include a tag editor. If you create Broadcast WAV files for download, use this utility to create the appropriate metadata: http://sourceforge.net/projects/bwfmetaedit/.

WMA encoder: A program that converts a WAV or AIFF file to a WMA file. Several are at www.team-mp3.com/wma/wma_encoder.htm. Windows Media Technologies has digital rights management, which limits copies or format conversions.

Ogg Vorbis encoder (optional) from www.vorbis.com: Converts CD tracks or WAV files to Vorbis format. For a given file size, Vorbis sounds better than MP3. Vorbis takes up less file size than MP3 files of equal quality.

To listen to your music that you put on the Web, you need:

MP3 player: A program or a device that plays MP3 files. Some software player examples are Winamp from www.winamp.com, Windows Media Player from http://www.microsoft.com/windows/windowsmedia, RealPlayer from www.real.com, and QuickTime in Apple's OS or www.apple.com/quicktime/. Portable MP3 players play MP3 files downloaded from the Web.

Windows Media Player: Although intended to play mainly Windows Media files, this free program plays many file types:.avi,.asf,.asc,.rmf,.wav,.wma,.wax,.mpg,.mpeg,.m1v, .mp2,.mp3,.mpa,.mpe,.ra,.mid,.rmi,.qt,.aif,.aifc,.aiff,.mov,.au,.snd, and .vod. This player is a free download from www.microsoft.com/windows/windowsmedia/.

A source for players, rippers, and encoders is www.team-mp3.com. Some digital audio editing programs can output MP3, WMA, and RealAudio files.

HOW TO UPLOAD COMPRESSED AUDIO FILES

Got everything you need? Let's go. Basically, here's what you will be doing:

1. Start with a WAV file of a song on your computer. Or convert your song from a CD to a WAV file by using a ripper.
2. Edit and process the WAV file to optimize it for the Internet.
3. Then use an encoder to convert the WAV file to an MP3 or WMA file.
4. Finally, send (upload) the converted file to a website that features music in the MP3 or WMA format.

Let's go over the procedure in more detail. You can substitute WMA for MP3 in the procedure below.

1. Start with a WAV file of the song. Edit the start and stop points of the song. You might want to edit a 30-second excerpt of a song to use as a preview of your music. If so, add a fade-in and fade-out to the preview.
2. Next, you might want to process the audio so that it will sound louder and clearer when played on the Web. To do this, reduce the bandwidth: apply low cut below 40 Hz and high cut above 15 kHz. You might want to try a higher-frequency low cut and a lower-frequency high cut. Apply a little compression or multiband compression, apply peak limiting, and then normalize the song to maximize its level.
Save the edited, processed song as a WAV file (PC) or AIFF file (Mac) on your hard drive.
3. Use an MP3 encoder to convert the song's WAV or AIFF file to an MP3 file on your hard drive. You might want to use a bitrate of 128 kbps, which many websites require. It's a good compromise between file size and sound quality.
4. If you want, add the song's title, artist, and comments using an ID3 tag editor.
5. Next you will upload your MP3 files to an MP3 server—a website that accepts MP3 files for distribution on the Web. Examples include www.soundcloud.com, www.bandcamp.com, www.reverbnation.com, and www.digstation.com (for independents only). The CD duplicator Disc Makers will upload your songs quickly to a variety of services for a small fee. Also see the MP3 Sites page at www.team-mp3.com. Some sites offer free downloads of their music files; others charge the listener so that you make some income.

As an alternative, you can send your physical CD to CD Baby (www.cdbaby.com), and they will convert it to data-compressed files for you (Figure 19.1). They are a popular, affordable way for independent artists to sell CDs.

FIGURE 19.1
CD Baby website.

CD Baby's MusicStore on Facebook is a free, easy way to stream and sell your music from the Facebook page for your band. You can also upload photos and videos, customize your store's design, and more. Create a CD Baby account at www.members. cdbaby.com. Then you can sell your music on Facebook, as well as iTunes, Amazon, Spotify, and many other online music stores. CD Baby will handle all the transactions and accounting for you.

ReverbNation is another digital distributor worth checking out. With it you can build an online store for your music and integrate it with Facebook. Bandcamp, Nimbit, and Topspin also provide that service.

eMusic (www.emusic.com) does not work with unsigned artists, but they do accept music from independent labels. Usually this is done through a service such as The Orchard (www.theorchard.com), INgrooves (www.ingrooves.com), or TuneCore (www.tunecore.com) (Figure 19.2). They can compress and deliver a label's CD catalog

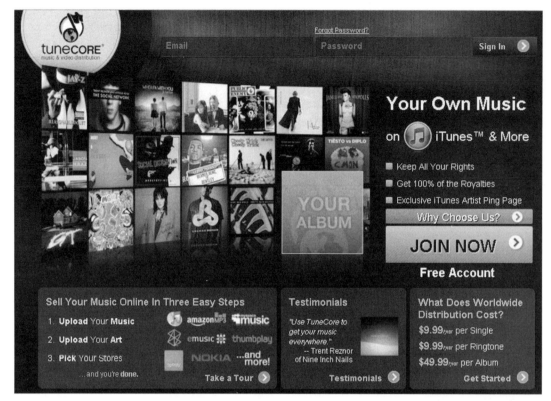

FIGURE 19.2
TuneCore website.

to all digital music stores. All income from digital sales is combined and reported to you.

Once you have chosen a site, go there and click on the button labeled "Artists only," "Submit your music," "Sell your music," or something similar. This is where you upload your MP3 files.

Once you sign up and fill out some forms, click on "Upload" (or whatever). You can also upload scanned photos of your band, your album cover, and text describing your band and its music. Some sites take a few days or weeks to approve your songs.

Congratulations—you're on the Web!

PUTTING YOUR MUSIC ON YOUR WEBSITE

So far I've covered how to upload your songs to an MP3 server. Now I'll explain how to put your songs on your own website. People visiting your site can click on a song title to hear a streaming song preview, or to download an entire song.

You might upload your MP3 files to SoundCloud, and use a SoundCloud player on your website or Facebook page. Or upload them directly to your website's server as described below.

Let's start by creating a webpage link to each MP3 file. It's easy. You can create the link with webpage design software or with one line of HTML code.

Again, you can substitute WMA for MP3 in the procedure below.

1. Suppose you have a song called "Blues Bash" on a CD or on your hard drive as a wave file. If the song is on a CD, use ripper software to transfer that song to your hard drive as a WAV file (PC) or AIFF file (Mac). In this example, we'll call the file **blues.wav**.
2. Use an MP3 encoder to convert the WAV file to an MP3 file. I recommend a setting of 160 kbps bitrate, VBR, stereo. This setting gives pretty good sound with a relatively small file, so it's fairly quick to download. You now have an MP3 file called **blues.mp3**.
3. You can add song-title and artist information with an ID3 tag editor.
4. In your computer, move **blues.mp3** to the same directory on your hard drive that your webpage is in.
5. Use webpage design software to open your band's webpage. Here we'll call it **bandpage.htm**.
6. On that page, where you want the song title to appear, type in the title of the song. In this example it's "Blues Bash." Link that title to **blues.mp3**.

When you left-click on the song title, **blues.mp3** should load and play. When you right-click on the song title, you can select "Save Target As" to copy the MP3 file to a directory on your hard drive.

If you want to use HTML code:

1. Open your webpage with a browser such as Windows Explorer.
2. Select View > Source. You'll see the HTML code for that page in plain text format.
3. Find a spot on the page where you want the song title to appear. Type in the title of the song and the link to its MP3 file. In HTML it would look something like this:

<a href="blues.mp3">Blues Bash</a>

For a WMA file it would be

<a href="blues.wma">Blues Bash</a>

4. Save and close the text file, go to Windows Explorer, and select View > Refresh. You should see the link to the MP3 or WMA song. When you left-click on the link, you should hear the song. When you right click on it, you can select "Save Target As" to copy the MP3 file to a directory on your hard drive.

Now that your song links are working correctly on your computer, it's time to upload them.

1. Using an FTP app, upload (send) **bandpage.html** and **blues.mp3** to your Web server. Make sure that both files go to the same directory. Some Web servers have an upload page for this purpose. Note: Some Web servers do not allow MP3 files. Find the page on the Web server site that details what files they permit.
2. Now start your Web browser and go to **bandpage.html** on your site. Click "Refresh" in your browser so that you see the **bandpage.html** that you just uploaded.
3. Left-click the song title (in this case, "Blues Bash"), and it should play after the buffer fills. Or right-click the song title and select "Save Target As." The song should download over a few minutes. Then you can play it with your MP3 player or WMA player in glorious hi-fi stereo.

Other good information on marketing your music online is in the article "Music's New Messiahs: You!" by Gary Mraz in the April 2008 issue of *EQ* magazine. Also see http://newmusicstrategies.com/ebook/. CD Baby offers marketing advice after you create an account. Disc Makers and Distrophonix publish many articles about this topic online.

The future of music consumption seems to be streaming audio. Facebook is partnering with streaming sites—such as Spotify, MOG, turntable.fm, and Pandora—to stream music for free. Spotify and others also offer a subscription services for a monthly fee. It doesn't generate much income for artists—fractions of a cent per stream—but the exposure can be huge, especially on Facebook.

COLLABORATING BY SHARING FILES

Recording engineers and musicians in different cities can work together on a common project by sharing files—sending digital audio files to others in WAV, MP3, or WMA format.

For example, you can send a song mix to a studio musician in another city, who overdubs their part and sends it back to you as a WAV file to import back into your multitrack project. Here's the procedure:

1. Export a WAV file of the song's stereo mix.
2. Copy that WAV file to a data CD-R.
3. Mail or FedEx the CD-R to the musician.

Then the musician follows these steps:

1. Import the mix into their recording program.
2. Overdub their part.
3. Export their part as a WAV file.

4. Copy that WAV file to a data CD-R.

5. Send the CD-R back to the recording engineer.

6. The engineer imports the WAV file on that CD-R into the multitrack project for mixing and editing.

You could e-mail MP3 or WMA files instead. If you want to use those compressed-data formats, encode the MP3 or WMA files at 256 kbps or higher, VBR, for best sound quality.

Sending WAV, MP3, or WMA files by e-mail can be cumbersome because some e-mail programs limit attachments to 10 MB. Also, the person receiving the file attachment will have to wait a long time for the e-mail to download. A better arrangement is to upload the audio files to a file-sharing website such as www.yousendit.com. When the recipient is ready to download the file, they simply click on a link.

Rogue Amoeba makes Nicecast software, which broadcasts audio in real time over the Internet. That lets a musician or producer in another location hear your mix with high quality and comment back to you via phone or iChat.

Record at Home, Mix in a Commercial Studio

This can be a great way to make an album. You track all the instruments and vocals in your home studio. Later, when all the tracks are recorded, you take them to a commercial studio for mixing and mastering.

What are the benefits of working this way? You can take all the time you need to create music and record it without spending money. You're more relaxed because you're not watching the clock. When your tracks are ready, a competent mixing engineer can make your tracks sound really good, and probably will take much less time to create a good mix than you could.

What are the drawbacks of recording at home? Compared to a commercial studio, your home studio might not have the best mics, studio acoustics, or recording techniques. You are likely to get a better sound with a pro recording.

Some musicians thrive on the discipline of recording efficiently in a commercial studio. Others need the lack of time pressure when recording at home.

HOW TO RECORD AND PREPARE YOUR TRACKS

Let's say you want to record in your own studio and have a pro do the mixes. Of course, you want your tracks to sound as clean as possible before handing them over to an outside recording engineer. Here are some tips:

During recording:

- Avoid clipping your recorded tracks—distortion usually can't be fixed during mixdown. Record each track at −6 dBFS maximum in peak meter mode.

- Avoid recording room reverb—it can't be removed during mixdown. Mount acoustic material on the walls to reduce reverb. Use acoustic foam or dense fiberglass insulation from atsacoustics.com. Place dense fiberglass panels across room corners to reduce boomy-sounding standing waves. Put a 3-foot square piece of acoustic foam on the ceiling over singers and their mics. Mike close and record direct when possible.
- Prevent noise—it's difficult or impossible to remove during mixdown. Close doors and windows, and turn off air conditioning or the furnace while tracking.
- Delete unused tracks and delete unused audio files from the session. Some recording software has a tool called "Consolidate audio files" that puts all the audio files used in a project in a common folder.

When exporting your tracks:

- Turn off pitch correction; don't use time correction. Let the mixing engineer do it with better tools.
- Turn off compression, EQ, reverb, and other effects. Again, the mixing engineer is likely to have better-sounding effects, and probably knows how to use them more effectively.
- Set each track's fader to 0 dB. Save (export) each track as a soloed WAV file, 24-bit, 44.1 kHz or higher. Each exported file should be the mix with one track soloed, played from time zero to the end of the song. The exported tracks will be continuous but will include your edits, and will have automation moves embedded in the audio. Name each track's file according to the song and instrument. For example, "On the Edge-bass.wav." You might also export each track with its effects turned on if you really like them and they are hard to duplicate. If so, add -fx to the filename of each track exported with effects.
- Bounce any MIDI soft-synth tracks into audio tracks, or export the MIDI tracks in Standard MIDI File format.
- Do NOT burn an audio CD of your tracks. Instead copy the tracks' WAV files to a USB thumb drive, USB hard drive, or data DVD-R.
- Many DAWs can read OMF (Open Media Format). If you want to export your entire session via OMF, first follow the steps above. Then export your audio mix file using File > Export > OMF or something similar. In Pro Tools check the "Include audio" option. Recent alternatives to OMF are Advanced Authoring Format (AAF) and Material eXchange Format (MXF). See www.aafassociation.org.
- Cakewalk SONAR saves its projects as a .bun (bundle) file, which includes all the track recordings, edits, balances, and plug-ins. Another studio with SONAR Producer can import and reproduce that project except for plug-ins that they don't have.

Take or send the track files to the mixing studio. There, the engineer will import your WAV or AIFF files, put each file on its own track, and start mixing. Bring notes on the plug-ins you used, along with any presets.

At the mixing session, you might bring some commercial CDs whose sound you admire so that the mixing engineer can get an idea for the sound you want. Make sure the genre of the commercial CDs is the same as your music. Also bring some rough mixes that you made at home.

You might be delighted to hear how good your mixes can sound when handled by a pro. And your album can have a relaxed vibe because you tracked it at home, on your own time.

For more information, read the "NARAS DAW Guidelines for Music Production" at http://www.grammy.org/recording-academy/producers-and-engineers/guidelines. Also read the "Pro Tools Guidelines for Music Production" at www.charlesdye.com/ptguidelines2.0.pdf.

You might want to read AES31-3-2008: AES standard for network and file transfer of audio - Audio-file transfer and exchange - Part 3: Simple project interchange. This standard provides a way to express audio-edit data in readable text form. Edit precision is sample-accurate and SMPTE time code is displayed. All AES standards documents are available at www.aes.org/publications/standards. Enter "aes31-3" into the Quick Search window and click "Search."

When sending out your mixes for mastering for CD release, a good format is one stereo WAV file per song recorded at 24 bits/44.1 kHz. Many studios can handle higher sample rates. Ideally your stereo mixes are recorded without overall compression or limiting, and with a recorded level of −6 dBFS to −3 dBFS maximum in peak meter mode.

FINDING STUDIO MUSICIANS, PRODUCERS, AND ENGINEERS

At many websites you can find online session work, or find online session musicians to enhance your projects. Using their services, you can work with people all over the world from the comfort of your own studio. Some sites accept only musicians who are with major labels, or they divide their services into major-label musicians and minor-label or indie musicians.

For example, sessionplayers.com seeks major-label musicians, singers, and engineers to join their group of available studio professionals. In their Player's Network, they also offer directories of professional but more affordable session musicians for recording work and musical collaborations.

Here are some websites where session players and engineers can do online collaboration and find musical jobs: sessionplayers.com, ejamming.com, esession.com, kompoz.com, riffworld.com, musiciansjunction.com, bandmix.com, studiopros.com, and shenandoahmusic.com.

Also, search Google for these terms: music collaboration technology, musician finder, musician's forum, online recording session, and online session musicians.

dB or Not dB

In the studio, you need to know how to set and measure signal levels and match equipment levels. You also need to evaluate microphones by their sensitivity specs. To learn these skills, you must understand the decibel, the unit of measurement of audio level.

DEFINITIONS

In a recording studio, level originally meant power, and amplitude referred to voltage. Nowadays, many audio people also define "level" in terms of voltage or sound pressure, even though this terminology is not strictly correct. You should know both definitions in order to communicate.

Audio level is measured in decibels (dB). One dB is the smallest change in level that most people can hear—the just-noticeable difference. Actually, the just-noticeable difference varies from 0.1 dB to about 5 dB, depending on bandwidth, frequency, program material, and the individual. But 1 dB is generally accepted as the smallest change in level that most people can detect. A 6- to 10-dB increase in level is considered by most listeners to be "twice as loud." Sound pressure level, signal level, and change in signal level all are measured in dB. *Play audio clip 42 at www.elsevierdirect.com/companions/9780240811444 to hear the effects of various decibel changes on loudness. Audio clip 43 wraps up the demonstration.*

SOUND PRESSURE LEVEL

Sound pressure level (SPL) is the pressure of sound vibration measured at a point. It's usually measured with a sound level meter in dBSPL (decibels of sound pressure level).

The higher the sound pressure level, the louder the sound (Figure A.1). The quietest sound you can hear, the threshold of hearing, is 0 dBSPL. Average conversation at 1 foot is 70 dB SPL. Average home-stereo listening level is around 85 dBSPL. The threshold of pain—so loud that the ears hurt and can be damaged—is 125 to 130 dBSPL.

FIGURE A.1
A chart of sound pressure levels.

Sound pressure level in decibels is 20 times the logarithm of the ratio of two sound pressures:

$$\text{dB SPL} = 20 \log P/P_{ref}$$

where P is the measured sound pressure in dynes/cm², and P_{ref} is a reference sound pressure: 0.0002 dyne/cm² (the threshold of hearing).

SIGNAL LEVEL

Signal level also is measured in dB. The level in decibels is 10 times the logarithm of the ratio of two power levels:

$$\text{dB} = 10 \log P/P_{ref}$$

where P is the measured power in watts, and P_{ref} is a reference power in watts.

Recently it's become common to use the decibel to refer to voltage ratios as well:

$$\text{dB} = 20 \log V/V_{ref}$$

where V is the measured voltage, and V_{ref} is a reference voltage.

This expression is mathematically equivalent to the previous one, because power equals the square of the voltage divided by the circuit resistance:

$$dB = 10 \log P_1/P_2 = 10 \log ((V_1^2/R)/(V_2^2/R))$$
$$= 10 \log (V_1^2/V_2^2) = 20 \log (V_1/V_2)$$

The resistance R (or impedance) in this equation is assumed to be the same for both measurements, and thus divides out.

Signal level in decibels can be expressed in various ways, using various units of measurement:

- dBm: decibels referenced to 1 milliwatt, with no defined load impedance
- dBu: decibels referenced to 0.775 volt
- dBv: decibels referenced to 0.7746 volt
- dBV: decibels referenced to 1 volt

dBm

If you're measuring signal power, the decibel unit to use is dBm, expressed in the equation

$$dBm = 10 \log P/P_{ref}$$

where P is the measured power, and P_{ref} is the reference power (1 milliwatt).

For an example of signal power, use the following equation to convert 0.01 watt to dBm:

$$dBm = 10 \log (P/P_{ref})$$
$$= 10 \log (0.01/0.001)$$
$$= 10$$

So, 0.01 watt is 10 dBm (10 decibels above 1 milliwatt).

Now convert 0.001 watt (1 milliwatt) into dBm:

$$dBm = 10 \log (P/P_{ref})$$
$$= 10 \log (0.001/0.001)$$
$$= 0$$

So, 0 dBm = 1 milliwatt. This has a bearing on voltage measurement as well. Any voltage across any resistance that results in 1 milliwatt is 0 dBm. This relationship can be expressed in the equation

$$0 dBm = V^2/R = 1 \text{ milliwatt}$$

where V is the voltage in volts, and R is the circuit resistance in ohms.

For example, 0.7746 volt across 600 ohms is 0 dBm. One volt across 1000 ohms is 0 dBm. Each results in 1 milliwatt.

Some voltmeters are calibrated in dBm. The meter reading in dBm is accurate only when you're measuring across 600 ohms. For an accurate dBm measurement, measure the voltage and circuit resistance, then calculate with the following equation:

$$dBm = 10 \log ((V^2/R)/0.001)$$

dBu

Another unit of measurement expressing the relationship of decibels to voltage is dBu. This means decibels referenced to 0.775 volt:

$$dBu = 20 \log V/V_{ref}$$

where V_{ref} is 0.775 volt.

dBv

The unit dBv with a small "v" means decibels referenced to 0.7746 volt. This figure comes from 0 dBm, which equals 0.7746 volt across 600 ohms (600 ohms used to be a standard impedance for audio connections):

$$dBv = 20 \log V/V_{ref}$$

where V_{ref} is 0.7746 volt.

dBV

Signal level also is measured in dBV (with a capital V), or decibels referenced to 1 volt:

$$dBV = 20 \log V/V_{ref}$$

where V_{ref} is 1 volt. For example, use the following equation to convert 1 millivolt (0.001 volt) to dBV:

$$\begin{aligned} dBV &= 20 \log (V/V_{ref}) \\ &= 20 \log (0.001/1) \\ &= -60 \end{aligned}$$

So, 1 millivolt $= -60$ dBV (60 decibels below 1 volt). Now convert 1 volt to dBV:

$$\begin{aligned} dBV &= 20 \log (1/1) \\ &= 0 \end{aligned}$$

So, 1 volt $= 0$ dBV.

To convert dBV to voltage, use the formula

$$Volts = 10^{(dBV/20)}$$

Change in Signal Level

Decibels also are used to measure the change in power or voltage across a fixed resistance. The formula is

$$dB = 10 \log (P_1/P_2)$$

or

$$dB = 20 \log (V_1/V_2)$$

where P_1 is the new power level, P_2 is the old power level, V_1 is the new voltage level, and V_2 is the old voltage level.

For example, if the voltage across a resistor is 0.01 volt, and it changes to 1 volt, the change in dB is

$$\begin{aligned} dB &= 20 \log (V_1/V_2) \\ &= 20 \log (1/0.01) \\ &= 40 \text{ dB} \end{aligned}$$

Doubling the power results in an increase of 3 dB; doubling the voltage results in an increase of 6 dB.

THE VU METER, ZERO VU, AND PEAK INDICATORS

A VU meter is a voltmeter of specified transient response, calibrated in volume units or VU. It shows approximately the relative volume or loudness of the measured audio signal. VU meters are used on analog tape recorders, broadcast consoles, some live-sound mixing consoles, and older recording consoles.

The VU-meter scale is divided into volume units, which are not necessarily the same as dB. The volume unit corresponds to the decibel only when measuring a steady sine-wave tone. In other words, a change of 1 VU is the same as a change of 1 dB only when a steady tone is applied.

Most recording engineers use 0 VU to define a convenient "zero reference level" on the VU meter. When the meter on your mixer or recorder reads "0" on a steady tone, your equipment is producing a certain level at its output. Different types of equipment produce different levels when the meter reads 0 (Figure A.2). Zero VU corresponds to:

- +8 dBm in older broadcast and telephone equipment
- +4 dBm in balanced recording equipment
- −10 dBV in unbalanced recording equipment
- −20 dBFS on a peak-reading meter in a digital recorder or recording software

FIGURE A.2
VU-meter scale.

A 0 VU recording level (0 on the record level meter) is the normal operating level of an analog tape recorder; it produces the desired recorded flux on tape. A "0 VU recording level" does not mean a "0 VU signal level."

With a VU meter, 0 VU corresponds to a recording level 8 dB below the level that produces 3% third-harmonic distortion on tape at 400 Hz. Distortion at 0 VU typically is below 1%.

The response of a VU meter is not fast enough to track rapid transients accurately. In addition, when a complex waveform is applied to a VU meter, the meter reads less than the peak voltage of the waveform. (This means you must allow for undisplayed peaks above 0 VU that use up headroom.)

In contrast, a peak indicator responds quickly to peak program levels, making it a more accurate indicator of recording levels. One type of peak indicator is an LED that flashes on peak overloads. Another is the LED bar graph meter commonly seen on digital recorders. Yet another is the PPM (peak program meter). It is calibrated in dB, rather than VU. Unlike the VU meter reading, the PPM reading does not correlate with perceived volume.

In a digital recorder, the meter is an LED or LCD bar graph meter that reads up to 0 dBFS (FS means full scale). In a 16-bit digital recorder, 0 dBFS means all 16 bits are ON (1) at the peak of the waveform. In a 24-bit digital recorder, 0 dBFS means all 24 bits are on. The OVER indication means that the input level exceeded the voltage needed to produce 0 dBFS, and there is some short-duration clipping of the output analog waveform. Some manufacturers calibrate their meters so that 0 dBFS is less than 16 bits or 24 bits ON; this allows a little headroom.

BALANCED VERSUS UNBALANCED EQUIPMENT LEVELS

Generally, audio equipment with balanced (3-pin) connectors works at a higher nominal line level than equipment with unbalanced (phono) connectors. There's nothing inherent in balanced or unbalanced connections that makes them operate at different levels; they're just standardized at different levels.

These are the nominal (normal) input and output levels for the two types of equipment:

- Balanced: +4 dBu (1.23 volts)
- Unbalanced: −10 dBV (0.316 volt)

In other words, when a balanced-output recorder reads 0 VU on its meter with a steady tone, it is producing 1.23 volts at its output connector. This voltage is called +4 dBu when referenced to 1 milliwatt. When an unbalanced-output recorder reads 0 on its meter with a steady tone, it is usually producing 0.316 volt at its output connector. This voltage is called −10 dBV when referenced to 1 volt.

INTERFACING BALANCED AND UNBALANCED EQUIPMENT

There's a difference of 11.8 dB between +4 dBu and −10 dBV. To find this, convert both levels to voltages:

$$dB = 20 \log (1.23/0.316)$$
$$= 11.8$$

So, +4 dBu is 11.8 dB higher in voltage than −10 dBV (assuming the resistances are the same).

A cable carrying a nominal +4 dBu signal has a signal-to-noise ratio (S/N) 11.8 dB better than the same cable carrying a −10 dBV signal. This is an advantage in environments with strong radio-frequency or hum fields, such as in a computer. But in most studios with short cables, the difference is negligible.

Connecting a +4 dBu output to a −10 dBV input might cause distortion if the signal peaks of the +4 equipment exceed the headroom of the −10 equipment. If this happens, use a pad to attenuate the level 12 dB (Figure A.3, top). The wiring shown converts from balanced to unbalanced as well as reducing the level 12 dB. If you hear hum with this connection, add an isolation transformer as shown in Figure A.3, bottom. All the leads should be twisted.

You don't always need that pad. Many pieces of equipment have a +4/−10 level switch. Set the switch to the nominal level of the connected equipment. If there is no such switch on either device, connect between them a +4/−10 converter box such as

FIGURE A.3

Top: Wiring balanced out to unbalanced in. The resistors form a 12 dB pad to match the balanced +4 dBu output to the unbalanced −10 dBV input. Bottom: Same, with an isolation transformer added to reduce hum.

the Ebtech Line Level Shifter (www.ebtechaudio.com/llsdes.html) or use the wiring shown here.

To connect an unbalanced −10 dBV output to a balanced +4 dBu input, use the wiring shown in Figure A.4. All the exposed leads should be twisted.

MICROPHONE SENSITIVITY

Decibels are an important concern in another area: microphone sensitivity. A high-sensitivity mic puts out a stronger signal (higher voltage) than a low-sensitivity mic when both are exposed to the same sound pressure level.

FIGURE A.4
Top: Wiring unbalanced out to balanced in. Bottom: Same, with an isolation transformer added to reduce hum.

A **microphone-sensitivity specification** tells how much output (in volts) a microphone produces for a certain input (in SPL). The standard is millivolts per pascal, where one pascal (Pa) is 94 dBSPL.

A typical "open-circuit sensitivity" spec is 5.5 mV/Pa for a condenser mic and 1.8 mV/Pa for a dynamic mic. "Open-circuit" means that the mic is unloaded (not connected to a load, or connected to a mic preamp with a very high input impedance). If the spec is 5.5 mV/Pa, that means the mic produces 5.5 mV when the SPL at the mic is 94 dBSPL.

If you put a microphone in a 20 dB louder sound field, it produces 20 dB more signal voltage. For example, if 74 dBSPL in gives 0.18 mV out (−75 dBV), then 94 dB SPL in gives 1.8 mV out (−55 dBV); 150 dBSPL in gives 1.1 volt out (+1 dBV), which is approximately line level! That's why you need so much input padding when you record a kick drum or other loud source.

Appendix B

Optimizing Your Computer for Multitrack Recording

Once you've chosen some recording software and installed it, you'll want to make your computer run as fast and glitch-free as possible. Speed becomes critical as you use more tracks, more effects, high bit rates, and high sampling rates.

The data flow of multitrack digital audio places high demands on computer speed. Recording and playback of digital audio requires long, continuous periods of streaming audio data. The more tracks in use, and the higher the bit depth and sampling rate, the faster the data flow has to be. And the more soft synths and effects plug-ins that are in use, the greater the load on the CPU.

Clearly, you need a fast computer for multitrack recording. But you also need to optimize the computer's settings for best results. I will cover some ways to speed up data flow and to reduce CPU usage. If you follow these tips, you will have a faster system that handles more plug-ins and more tracks at once. Also, when you play tracks or burn CDs, clicks and drop-outs in the audio will be less likely.

Most of these tips apply to a PC computer. Mac advice is offered at the end of this appendix in the section "Optimizing Mac for Multitrack Recording."

Disclaimer: Back up your system and data files first, then proceed at your own risk. Neither I nor Focal Press will be responsible for damage to your computer system or files from making these changes. If you are not comfortable doing a particular tweak, don't do it. However, I have done all these adjustments with no problems, and they are reversible.

SPEEDING UP YOUR HARD DRIVE

The hard drive should have a fast average access time (under 9 msec) and a high internal sustained transfer rate. Some recommended hard drives are the Seagate Barracuda, Maxtor DiamondMax by Seagate, and Western Digital models.

What transfer rate is needed? The data rate of 24 tracks at 24 bit/96 kHz is 6.6 MB/sec. That is a practical minimum.

The best current hard drives can transfer about 40 MB/sec continuously with a SATA interface. SSD (solid state drive) units can reach 70–250 MB/sec but tend to have lower capacity than conventional hard drives.

Hard drives with high rotational speed or spindle rate (7200 rpm or greater) tend to have faster transfer rate. They are recommended for 24-bit multitrack productions. The typical sustained transfer rate of a 7200 rpm drive is 30 to 40 MB/sec. This provides up to 160 16 bit/44.1 kHz tracks or 50 24/96 tracks, depending on file fragmentation.

For fastest speed, use one hard drive for applications and Windows files, and another for audio data. That way the audio drive head does not waste time looking for system files. If you don't want to open your computer, or you are using a laptop, connect an external drive (for audio files) to a USB or FireWire port.

If your Windows and audio drives are the SATA type, you can't put them on the same cable. Put the Windows drive on SATA channel 1 and the audio drive on SATA channel 2. If channel 2 is already in use, connect an external drive (for audio files) to a USB or FireWire port.

Then, in your recording software under Options or Preferences, indicate the drive letter that contains the audio data.

If you have only one hard drive, you might use one disk partition for audio files and another partition for programs. That way you can defragment or reformat the audio data partition frequently.

It's a good idea to have a dedicated drive just for musical instrument samples because they generate high bandwidth streams when you play them. For best speed, leave a third of any hard drive empty.

Defragment the audio file drive monthly. Defragging reorganizes files into contiguous areas on the disk so that each audio file can be accessed with minimum head movements. Before you defrag, disable programs in the System Tray. In Windows XP double-click My Computer. In Windows 7 or Vista, select Start > Computer. Right-click your audio drive, and select Properties > Tools > Defragmentation.

You can get faster, more thorough defragging with Raxco PerfectDisk 12 Professional (www.raxco.com). Another way to defrag: copy the files off the audio drive, reformat the audio drive, and copy the files back. Note that SSD (solid state drive) units do not need defragging.

When you are finished using all the files on the audio data drive, reformat it to zero-out the clusters. This is necessary even if you deleted all the files.

Install lots of RAM, at least 2 GB for audio, and at least 4 to 8 GB for soft synths. How does more RAM help? When your DAW application is streaming audio, it continuously feeds data into a RAM buffer for each track. If the RAM buffer is big, it needs to be filled less often, so the disk-drive heads need to search less often. A larger RAM buffer lets the disk heads work more efficiently by getting bigger continuous chunks of data into RAM. Also, having lots of RAM prevents frequent access of the swap file (described later). The goal is to have your programs running in RAM, rather than swapping data to your hard drive.

Disable write caching on your hard drives: write caching waits until the system is idle before writing data to the hard drive, causing a delay. Here's how to turn it off: select Start > Settings (XP only) > Control Panel > System or System and Maintenance or System and Security > Hardware or System > Device Manager > Disk drives > Double-click on your audio data drive > Policies > Uncheck "Enable Write Caching" > OK.

Disable read-ahead caching on your hard drive. If it is enabled, data is read out of the cache instead of directly from the hard drive—not good for streaming audio. Your recording program might have a setting for read-ahead caching.

INCREASING PROCESSING SPEED

A computer has a central processing unit (CPU) that does most of the calculations needed to run software. For example, in some DAWs the CPU performs all the signal processing for real-time effects. Get a computer or motherboard with the fastest CPU that you can afford—a clock speed of 2 GHz or greater.

Consider getting a computer or motherboard with a dual-core or quad-core CPU, an x64 processor, a 64-bit operating system, and 64-bit drivers for your hardware. The AMD FX, AMD Phenom II X2 or X4, and Intel Core i5 and i7 are good choices for audio work. When using a multi-core CPU, enable multiprocessing in your recording software. An x64 CPU lets the computer carry more RAM, which is helpful for soft synths.

The following tips reduce the amount of data that the CPU has to handle:

Do not enable hyperthreading in the BIOS unless your software requires it.

If you hear drop-outs or glitches, maybe too many real-time effects are running. Select a track that has real-time effects, and bounce (export) that track with effects to an open track. That way the effects are recorded or embedded in the bounced track, rather than running in real time. Then archive (save) the original track and delete it from the project. That reduces CPU loading.

You can temporarily disable plug-ins when doing MIDI keyboard overdubs. That lets you set the latency to a lower value without getting drop-outs. The result is a faster audio response when you press keys on the MIDI keyboard. There is more on this under the heading "Minimizing Latency" later in this appendix.

Archive any tracks that are muted, then delete those tracks from the project to further reduce the load on the processor.

It also helps to convert MIDI soft-synth tracks to audio tracks. Some DAWS have a "freeze" or "bounce MIDI to track" function for this purpose. Freezing deactivates the synth plug-in, replacing it with an audio track of the synth output. When you delete or archive the original MIDI and synth tracks, it frees up processor power for the mix.

Limit polyphony in a soft synth if you don't need it.

Do not insert the same delay effect (echo, reverb, chorus, flanging) in several tracks. Instead, set up an aux bus that has the desired effect. Use aux send controls on tracks to adjust the amount of effect on each track. This reduces the number of effects processes that are running and reduces CPU loading.

In extreme cases, consider installing a digital signal processing (DSP) card, such as the TC Works PowerCore. It handles the processing for the plug-ins. The plug-ins access the card rather than the CPU. You can install multiple cards to expand the system. Pro Tools includes its own DSP cards.

Before recording, turn off the waveform drawing function of your sound clips or regions.

Reduce the number of colors: right-click the Desktop > Properties > Settings > set Colors to 16 bit or 256 colors. A higher setting reduces audio performance when the level meters redraw in your audio application.

Reduce video acceleration in Windows XP: select Start > Settings > Control Panel > System > Performance > Graphics > Advanced. Reduce Hardware Acceleration as much as you can without degrading the display of your audio program.

In XP, adjust visual effects for best performance: select Start > Right-click My Computer > Properties > Advanced > Performance > Settings > Visual Effects Tab > Adjust for best performance.

Important: Change the Performance mode for the operating system. In XP right-click My Computer > Properties > Advanced > Performance > Settings > Advanced > Processor scheduling > select Background Services. This setting allocates more processor time to background activities, such as streaming audio and ASIO drivers. However, in Windows Vista or 7 you'll select Programs instead of Background Services. Select Start > Control Panel > System and Security > System > Advanced system settings (on left) > Advanced tab > Performance: Settings > Advanced tab > Adjust for best performance of Programs > Apply > OK.

PREVENTING INTERRUPTIONS

Unnecessary background programs can rob CPU cycles, interrupt the audio program, and interrupt the hard-drive head from playing audio. The following tweaks to your PC can increase the number of tracks and effects you can play without drop-outs or clicks by disabling various background programs or increasing free memory.

Don't multitask (run other programs) while the audio is streaming. Each additional task slows down the system. Press Ctrl-Alt-Del to see what's running, highlight anything you don't need, and select End Task. Leave Explorer and Systray running. Don't scroll while recording or playing back.

Important: Disable system sounds such as the chimes and fanfares in Windows: select Start > Settings > Control Panel or Start > Control Panel. In XP select Sounds and audio devices > Sounds > set to No Sounds. In Windows 7 select Sound > Sounds tab > Sound Scheme: No Sounds > Apply > OK. In Vista select Hardware and Sound > Sound > Sounds tab > Sound Scheme: No Sounds.

You might use a PS/2 mouse instead of a USB or wireless mouse, which can cause audio instability.

Switch off your anti-virus program when using your programs. Do this by right-clicking on its icon on the bottom right of your screen, then click "Disable." You need anti-virus for downloading files from outside sources but not for DAW work. Better yet, do not connect your recording computer to the Internet.

In the preferences menu of your programs, disable auto update to prevent interruptions. Also select Start > Settings > Control Panel > System or System and Security > Automatic Updates or Windows Update > Turn off automatic updates.

Important: Disable the computer's internal sound card when you are using your own sound card or audio interface. In Windows XP select Start > Settings > Control Panel > System > Hardware > Device Manager > Expand "Sound Video and Game Controllers" > Find and right-click the computer's internal sound card > Properties > Disable. Also disable the Microsoft device mapper. In Windows 7 and Vista select Start > Control Panel > Hardware and Sound > Device Manager > Expand "Sound Video and Game Controllers" > Find and right-click the computer's internal sound card > Disable. Optional: While you are in sound controllers, remove the legacy audio drivers except the ones you are using.

Disable or disconnect devices that you are not using; this frees up interrupts.

Disable the background image: in XP right-click Desktop > Properties > Background > None. In Windows 7 or Vista right-click Desktop > Personalize > Desktop background > choose a solid color.

Disable the screen saver: in XP right-click Desktop > Properties > Screen Saver: None. In Windows 7 or Vista right-click Desktop > Personalize > Screen Saver: None.

Important: Switch off power schemes: in Windows XP select Start > Settings > Control Panel > Power Options > Set "Turn off hard discs" to "Never." Switch off hibernation: Start > Settings > Control Panel > Power Options > Hibernate > Uncheck "Enable hibernate support." In Vista select Start > Control Panel > Power Options > High Performance. In Windows 7 select Start > Control Panel > System and Security > Power Options > High Performance (not Power Saver).

Remove programs and services that load on startup. Some are necessary, but many are not. They consume memory and resources. In XP, select Start > Run. In Windows 7 or Vista, select Start. Type in MSCONFIG and press OK. Select the Startup tab and disable everything but the system tray and the speaker volume control. Keep any programs that control hardware. Select the Services tab. At the bottom of the window, check "Hide all Microsoft services." Then press the "Disable All" button. Press OK then Restart. When Windows reboots, check "Don't show this message…" and click OK.

Some experts say that you should NOT disable Multimedia Class Scheduler, Plug and Play, Superfetch (in Vista), Task Scheduler, Windows Audio, and Windows Driver Foundation. If you use Pro Tools, recheck "MMERefresh" in the Startup tab and "Digidesign MME Refresh Service" in the Services tab.

SETTING THE BUFFER SIZE

Audio data streaming from your hard drive goes into a buffer memory, then is read out at a constant clock rate. The buffer must be filled faster than the clock empties it; otherwise there is nothing to read. This requires the CPU reading from the buffer to pause its processing while the buffer refills. Then you hear a short silence or audio drop-out. So if the hard drive is too slow, or the buffer is too small, this causes buffer underrun and results in drop-outs.

If the hard drive streams data faster than the buffer can be emptied, this causes buffer overflow or overrun. The extra data from the hard drive might be lost, causing drop-outs, clicks, or stopped playback. Or the extra data might overwrite adjacent memory locations, causing a system crash.

A small buffer forces the CPU to fill and empty the buffer much more often, giving the CPU less time for processing. In other words, the smaller the buffer, the higher the percentage of CPU usage (the harder the CPU has to work).

So, if you experience drop-outs or crashes and the CPU load indicator reads near the red, increase the buffer size in your recording software or audio interface software. You might start with a setting of 256 kB and go from there. Don't make the buffer bigger than necessary, because then playback will start with a delay.

Another buffer is the I/O buffer (hard-drive buffer), used for reading and writing data to the hard drive. If your disk activity indicator reads greater than 20% or so, set the disk I/O buffer size in your recording software to 256, 512, or whatever is needed to prevent drop-outs.

MINIMIZING LATENCY

If you are using soft synths, you want them to respond as fast as possible to MIDI-controller key presses. In other words, you want to minimize latency (monitoring delay). Latency is directly related to buffer size. Latency (msec) = buffer size (samples) × 1000 divided by the sample rate. For example, if the buffer size is 256 samples and the sample rate is 44,100 Hz, the latency is 5.8 msec.

A small buffer setting reduces latency—the application responds faster—but tends to create drop-outs, or reduces the number of tracks that play without dropouts. A large buffer setting increases latency but prevents drop-outs. Which setting is best?

To solve this conflict, use two different buffer settings. Open the Preferences menu in your recording software, or open the audio interface software. Set the buffer/latency small (under 4 msec if possible) while tracking MIDI synths or while processing the input signal with effects (Echo Input Monitor setting). During playback and mixdown, set it large (maybe 25 msec). The bigger buffer will reduce the load on your CPU and let you use more plug-ins.

If you can't set the latency under 4 msec without drop-outs or crashes, freeze or bounce tracks that have effects, or temporarily disable any plug-ins. You might also convert MIDI tracks to audio, then disable soft synths.

Note that the audio-interface buffer size affects latency, but the disk I/O buffer does not.

Monitoring with headphones has less perceived latency than monitoring with speakers. That's because sound takes time to travel from the speakers to your ears. For example, if your speakers are 6 feet away, that adds 5.3 msec of perceived delay to the monitored signal.

Drivers affect latency, so download the latest drivers for your audio interface. However, note that earlier versions of drivers are sometimes faster than later versions. Also check out third-party drivers such as CEntrance Ideal Driver (www.centrance.com) and ASIO-4ALL, a universal ASIO driver for WDM audio. You have to set up your audio software to use ASIO in order to notice any change in latency.

As we said before, set processor scheduling to "Background Services." If you use ASIO drivers, this tweak ensures low latency without drop-outs.

The round-trip latency of a MIDI keyboard playing a soft synth is less than the round-trip latency of audio in to audio out. When you trigger a soft synth with a MIDI keyboard, the round-trip latency is the sum of these latencies:

Driver (about 4 to 6 msec depending on the driver)
Output buffer (whatever you set it to)
D/A conversion (about 1 to 1.5 msec).

With the buffer set to 3 msec, the calculated latency when you play a MIDI soft synth is about 8 msec. That's not bad, considering that 9 msec is just noticeable.

OTHER TIPS

- Check out www.tweakXP.com, www.Tweak3D.net, and www.extremetech.com.
- To learn about buying or building a PC for recording digital audio, see http://www.cakewalk.com/PCResource/.
- Obedia.com offers 24/7 help for computer recording: tech support, consulting, and troubleshooting.
- More information on computer recording can be found at www.pcrecording.com.
- If you are using your computer professionally as part of a commercial recording studio, do NOT have a connection to the Internet on that computer. Have it be a dedicated digital audio workstation so that viruses, etc., won't destroy a customer's session. Use another computer for the Internet.
- Six vendors of audio-optimized computers are www.raincomputers.com, www.sweetwater.com (Creation Station), Vision DAW, ADK Pro Audio, PCAudioLabs, and www.liquiddaw.com.
- Consider using a RAID hard drive system which stores audio on two identical drives at the same time. If one drive fails, none of your work is lost.
- Go to the manufacturer websites of your sound card (or audio interface) and audio editing software. Check out the support sections for advice on optimizing your computer and on troubleshooting sound problems.
- Keep your sound card(s) away from the AGP graphics card to prevent hum.
- Don't install much software (it can corrupt the registry).
- On the Web, download the latest drivers for your motherboard, video card, IDE or SATA controller, CD/DVD burner, recording software, and sound card or audio interface. Also download the Windows updates.
- Consider using an AGP video card rather than a PCI video card to reduce the load on the PCI bus.
- Disable Disc Indexing Service on NTFS formatted drives if you do NOT do frequent searches for file names. In XP, double-click My Computer > Right-click each hard drive > Properties > Uncheck "Allow Indexing Service to index this disc for fast file searching" > Apply. Choose all files and subfolders within the drive. This action takes a while as all the files are scanned. If you get a message that says Access Denied..., press "Ignore All."

In Windows 7 and Vista, select Start, and type "services" into the search box. Select Services (Local). Scroll down to find Windows Search. Right-click, and choose Properties. Choose Startup Type > Disabled.

- Use nLite before installing Windows XP to get a DAW-optimized installation (www.nliteos.com/nlite.html).
- If you have a utility that measures disk throughput (sustained disk transfer rate) you might want to measure your system's speed before and after each hard-drive change mentioned above. That way you will know which changes were effective. A good utility is HD Tune, available from hdtune.com.
- If you hear lots of drop-outs, stuttering, or a buzzy sound during audio playback, try a different audio driver. In your recording software preferences, change the driver from WDM to ASIO or vice-versa.
- While recording, disable Windows Firewall, anti-virus, and spyware detection software. Better yet, do not connect your recording computer to the Internet.
- Press Ctrl-Alt-Del to open the task manager. Select the Processes Tab. Go to www.answersthatwork.com and go to the task list page. Check the processes running on your computer against their task list to see what they are. Disable any tasks that you feel are unnecessary.

WINDOWS VISTA AND WINDOWS 7

Windows Vista and Windows 7 offer several benefits for recording: improved stability, more processing in the graphics card and less in the CPU, WaveRT low-latency driver support, and MMCSS (Multimedia Class Scheduler Service) to allow the audio engine higher priority to the CPU. Here are some additional tips:

- Vista Home Premium or Windows 7 in 64-bit versions are recommended for audio.
- Be sure to download the latest Vista drivers for your software, hardware, and plug-ins.
- Go to www.update.microsoft.com for performance-increasing patches.
- If your motherboard allows, disable HPET in BIOS.
- Use Vista Lite (www.vlite.net) before installing Vista to get a DAW-optimized installation.
- Don't change the page file (swap file or virtual memory) because Vista handles it well.
- Disable startup items that you don't need. Select Start > Type Defender and press Enter. Select Tools > Software Explorer > All Users. Highlight an item you don't need and click Disable. Repeat for other items. Reboot when done.
- Turn off User Account Control (if you don't need this security feature). Select Start > Control Panel > User Accounts and Family Safety > User Accounts > Turn User Account Control on or off > Uncheck Use User Account Control (UAC) > OK > reboot.
- Disable Windows components that you don't need. Select Start > Type appwiz.cpl. Select Turn Windows features on or off. Unselect any components that you are not using. Click OK.

- Right-click on the desktop and select Personalize. Select Windows Color and Appearance. In the Color Scheme list box, select Windows Vista Basic, Windows Standard, or Windows Classic. Click OK
- Turn off Windows Aero. Select Start > Control Panel > System and Security > System > Advanced system settings (on left) > Advanced tab > Performance: Settings > Visual Effects tab > Adjust for best performance > Apply > OK.
- Turn off Windows Sidebar. Locate the Sidebar icon in your system tray. Right-click on the icon and select Exit. Uncheck Start Sidebar when Windows starts. Click OK.
- Check www.extremetech.com for Vista hot fixes.

OPTIMIZING MAC FOR MULTITRACK RECORDING

- Many of the PC tips apply to Mac: Use a second, fast hard drive running at 7200 rpm or higher for audio files. Use a minimum of nonaudio software. Disable energy-saver or sleep software. Update drivers. Reduce monitor resolution. Disable video acceleration software and cards. Check online user groups for advice.
- Check the Memory panel and turn off virtual memory.
- Check for enough application memory. Select the program's icon and press Command-I. In the info window that appears, try increasing the amount of allotted memory.
- Minimize the number of Preference Panes installed and the number of User Login items.
- Upgrade to the most recent version of OS X.
- For maximum speed, use computers with PCI Express (PCIe).
- Equip your Mac with the most powerful CPU possible. For instance, a Mac Pro with an 8-core processor.
- Buy more RAM: 2 GB or more is recommended.
- Delete outtakes and unused files to prevent a "too many files open" message.
- If you can't play enough tracks, increase the buffer size.
- If you hear pops or clicks, check your sound card's clock settings. When recording an analog source, set it to Internal Clock or Sync. If your source is digital, set it to external S/PDIF sync or the format you are using. Be sure sample rates match. Move audio and digital cables away from USB cables and power lines. Try turning off USB hubs, modems, and printers.
- If your system crashes frequently, try removing downloaded software. Delete your application's preferences file if it has become corrupt.
- If the application stops in the middle of a recording, check to see whether you have set a maximum recording time in advance.

Impedance

Impedance is one of audio's more confusing concepts. To clarify this topic, I'll present a few questions and answers about impedance.

WHAT IS IMPEDANCE?

Impedance (Z) is the resistance of a circuit to alternating current, such as an audio signal. Technically, impedance is the total opposition (including resistance and reactance) that a circuit has to passing alternating current.

A high-impedance circuit tends to have high voltage and low current. A low-impedance circuit tends to have relatively low voltage and high current.

I'M CONNECTING TWO AUDIO DEVICES. IS IT IMPORTANT TO MATCH THEIR IMPEDANCES? WHAT IF I DON'T?

First, let's review some definitions. When you connect two devices, one is the source and one is the load. The source is the device that puts out a signal. The load is the device you are feeding the signal into. The source has a certain output impedance, and the load has a certain input impedance.

A few decades ago in the vacuum-tube era, it was important to match the output impedance of the source to the input impedance of the load. Usually the source and load impedances were both 600 ohms. If the source impedance equals the load impedance, this is called "matching" impedances. It results in maximum power transfer from the source to the load.

In contrast, suppose the source is low Z and the load is high Z. If the load impedance is 10 times or more the source impedance, it is called a "bridging" impedance. Bridging results in maximum voltage transfer from the source to the load. Today, nearly all devices use bridging—low-Z out to high-Z in—because we want the most voltage transferred between components.

For best sound quality and highest voltage transfer, the input impedance of the connector you are plugging into should be at least 7 to 10 times higher than the output impedance of the source (such as an electric guitar or microphone).

For example, if a mic's impedance is 200 ohms, the input impedance of a mic input (that you plug the mic into) should be 7 to 10 times higher, or 1400 to 2000 ohms. If you look at the input impedance spec for mixer mic inputs, it is typically around 1500 ohms.

The typical impedance of a guitar pickup is 10–25 kilohms (or higher—depending on if it's single coil or humbucking coil). But impedance can rise at the pickup's resonance around 3–5 kHz (depending on pickup) to 100 kilohms or more. So guitar-amp input impedances are typically 1 megohms. Anything less loads the pickup and changes the sound.

If you connect a low-Z source to a high-Z load, there is no distortion or frequency-response change caused by this connection. But if you connect a high-Z source to a low-Z load, you might get distortion or altered response. For example, suppose you connect an electric bass guitar (a high-Z device) into an XLR-type mic input (a low-Z load). The low frequencies in the signal will roll off, so the bass will sound thin. And the highs might roll off, making the sound dull.

We want the bass guitar to be loaded by a high impedance, and we want the mic input to be fed by a low-impedance signal. A direct box or impedance-matching adapter does this (Figure C.1). Such an adapter is available from Radio Shack, part no. 274-017.

The adapter is a tube with a phone jack on one end and a male XLR connector on the other. Inside the tube is a transformer. Its primary winding is high Z, wired to the phone jack. The transformer's secondary winding is low Z, wired to the XLR. You plug the guitar cord into the phone jack, and plug the XLR into a mic input in a snake or mixer. Use it with a bass guitar, electric guitar, or synth.

This impedance-matching adapter works, but is not ideal. The load it presents to the bass guitar might be 12 kilohms, which will slightly load down the high-Z guitar

HI-TO-LOW IMPEDANCE MATCHING ADAPTER

FIGURE C.1
High-to-low impedance-matching adapter.

pickup, causing thin bass. An active direct box solves this problem. In place of a transformer, the active DI usually has a field effect transistor (FET). The FET has a very high input impedance that does not load down the bass guitar.

In general, the input impedance of a direct box (or an instrument jack) is higher than the input impedance of a line input.

WHAT ABOUT MICROPHONE IMPEDANCE?

Recording-quality mics have XLR (3-pin) connectors and are low Z (150 to 300 ohms). A low-Z mic can be used with hundreds of feet of cable without picking up hum or losing high frequencies.

I'M CONNECTING A MIC TO A MIXER. IS IMPEDANCE A CONSIDERATION?

Yes. If your mixer has phone-jack inputs, they are probably high Z. But most mics are low Z. When you plug a low-Z mic into a high-Z input you get a weak signal. That's because a high-Z mic input is designed to receive a relatively high voltage from a high-Z mic, and so the input is designed to have low gain. So you don't get much signal amplification.

If you can't get enough level when you plug a mic into a phone-jack input, here's a solution: between the mic cable and the input jack, connect an impedance-matching adapter (Figure C.2) such as Radio Shack 274-016. It steps up the voltage of the mic, giving it a stronger signal.

The adapter is a tube with a female XLR input and a phone-plug output. Inside the tube is a transformer. Its primary winding is low Z, wired to the XLR. It secondary winding is high Z, wired to the phone plug. Connect the mic to the XLR; connect the phone plug to the mixer's phone jack. Then the mixer will receive a strong signal from the mic.

If you're using a phantom-powered condenser mic, the connections are different. First, turn off any phantom power in your recorder-mixer. Connect your mic to a standalone

FIGURE C.2
Low-to-high impedance-matching adapter.

phantom power supply, and connect the supply output to the impedance-matching adapter.

If your mixer has XLR inputs, they are low-Z balanced. In this case, simply connect the mic to the mixer using a mic cable with a female XLR on the mic end and a male XLR on the mixer end. A low-Z mic input is typically about 1500 ohms, so it provides a bridging load to a mic that is 150 to 300 ohms.

SHOULD I CONSIDER IMPEDANCE WHEN I CONNECT TWO LINE-LEVEL DEVICES?

This is seldom a problem. In most audio devices, the impedance of the line output is low—about 100 to 1000 ohms. The impedance of the line input is high—about 10 kilohms to 1 megohm. So every connection is bridging, and you get maximum voltage transfer. Some audio devices, such as passive equalizers, require a terminating resistor at the input or output for best performance.

CAN I CONNECT ONE SOURCE TO TWO OR MORE LOADS?

Usually yes. You can connect several devices in parallel across one line output. Suppose you connect a mixer output simultaneously to a recorder input, an amplifier input, and another mixer's input in parallel. The combined input impedance of those three loads might be 4000 ohms, which still presents a bridging load to the mixer's 100-ohm output impedance.

Mics are a different story. Suppose you're recording a concert. If you connect one mic to the PA mixer and also to your recording mixer with a Y cable, the combined input impedance will be about 700 ohms or less. This can load down some microphones, reducing the bass in dynamic mics or causing distortion in condenser mics. To prevent the loading, put a 270 ohm 1% resistor in series with pin 2 and pin 3 in each leg of the Y cable.

CAN I CONNECT TWO OR MORE SOURCES TO ONE INPUT?

Not recommended. The low output impedance of one source will load down the output of the other source, and vice versa. This can cause level loss and distortion. The same goes for two mics into one mic input. It's better to use a small external mixer.

As another option, put a series resistor in line with each device before combining them. That prevents each device from loading down the other. A minimum resistor value might be 470 ohms per source. If the source is balanced, use one resistor on pin 2 and one on pin 3—two 1% resistors per source. This is more thoroughly examined in the Rane note "Why Not Wye" (http://www.rane.com/note109.html).

SUMMARY

- Impedance (Z) is the opposition to alternating current, measured in ohms.
- Microphones and line outputs are usually low Z.
- Electric guitars, acoustic-guitar pickups, instrument inputs, and line inputs are usually high Z.
- XLR mic inputs are low impedance; phone jack mic inputs are high impedance.
- Speakers are usually 4 to 8 ohms.
- Equal impedances in parallel result in half the impedance.
- Equal impedances in series result in twice the impedance.
- Connect low-Z sources to low-Z inputs. (A low-Z input is usually 7 to 10 times the source impedance, but it's still called a low-Z input.)
- Connect high-Z sources to high-Z inputs.
- Connect a low-Z source to a high-Z input through a step-up transformer (impedance-matching adapter).
- Connect a high-Z source to a low-Z input through a step-down transformer (impedance-matching adapter, or direct box).

Phantom Power Explained

Most condenser microphones need phantom power to operate their internal circuitry. Let's explain what phantom power is and how to use it.

DEFINITION

Phantom power is electrical power that is sent to a condenser mic through its mic cable. The cable must have XLR connectors to pass phantom power.

Each of these devices can provide phantom powering:

- A mic preamp with phantom power built in
- A mixing console with phantom power built in
- A standalone phantom power supply

In each device above, a female XLR mic connector supplies the phantom power to operate the mic. The microphone receives power from, and sends audio to, the device along the same cable. It's called "phantom" because the power does not need a separate wire; it's "hiding" in the mic signal wires inside the mic cable.

Technically, phantom power is a positive voltage (12–48 V DC) on XLR pins 2 and 3 with respect to pin 1. The cable shield is connected to the phantom-power ground.

Figure D.1 shows a microphone plugged into a standalone phantom power supply.

After you plug a condenser mic into a mixer, turn on phantom power in the mixer. The phantom on/off switch in many consoles is labeled "P48" to mean "phantom power 48 volts."

USING A STANDALONE SUPPLY

What if your mixer or mic preamp does not have phantom power? You can buy a phantom power supply (Figure D.2) from a dealer or online.

Connect the supply in series with the mic cable. The supply has XLR input and output connectors, one pair per channel. As shown in Figure D.1, connect a mic cable between

FIGURE D.1
Phantom power circuit.

FIGURE D.2
An example of a single-channel, standalone phantom power supply.

your microphone and the supply's input connector. Plug another mic cable between the supply's output connector and your mixer or preamp mic input.

Some supplies are AC powered, some are battery powered, and some are both. Some can power a single microphone; others can power several at once.

CAUTIONS FOR USE

Don't plug a mic into an input with phantom already switched on, or you'll hear a loud pop. If you have no choice (as during a live concert), mute the mic channel when you plug the mic in.

Make sure your phantom voltage is adequate for your microphones. Check the mic's datasheet under the "Powering" spec. Some mics start to distort or lose level if the phantom voltage drops much below 48 volts.

DC BIAS

Some microphones or mic capsules work on "DC bias" rather than phantom power. A separate wire supplies a voltage to the mic capsule. This system is used in the cable between a mic capsule and an XLR connector in lavalier mics, choir mics, and clip-on instrument mics. The mic capsule itself runs off DC bias, while the XLR connector houses a circuit that runs off phantom power. That circuit converts phantom power to DC bias for the mic capsule.

Where To Learn More

We recommend the following books, magazines, videos, and literature to anyone who wants more education in recording technology.

BOOKS AND VIDEOS

The Library

Go to your local library and search their database for books on recording, home recording, music recording, audio recording, MIDI, or sound production. Read everything you can on the subject.

Pro Audio Books Plus

This company offers an online catalog of all sorts of pro audio and recording books, videos, and courses at www.proaudiobooksplus.com.

Music Books Plus

This is a catalog with a huge number of audio and recording books. The website is www.musicbooksplus.com. They also carry tutorial videos on DVD.

Mix Books

All sorts of recording topics are covered in digital format with embedded videos. The website is www.mixbooks.com.

howaudio.com

This site offers over 1000 tutorial videos on home recording and related topics.

Focal Press

This is another major audio/recording books catalog and it is found at www.focal-press.com. Focal Press also offers video DVDs about DAW operation, recording, mic techniques, mixing, and setting up a home studio. If you want to record concerts or gigs, check out the Focal Press book "Recording Music On Location."

Amazon.com

This online bookstore has a great search engine. In the search field, type in whatever audio subject you are interested in: DVD, MIDI, digital audio, recording, mixing, microphones, etc.

One Amazon book that I recommend is "Hot Tips for the Home Recording Studio" by Hank Linderman (1994). It offers valuable advice on song writing, arranging, and music production.

RECORDING MAGAZINES

Recording (home and project studio recording), www.recordingmag.com

Mix (pro recording and concert sound), www.mixonline.com

Electronic Musician (home and project studio recording), www.emusician.com

Tape Op (home and project studio creative recording), www.tapeop.com

Future Music (home, project, and pro recording), www.musicradar.com/futuremusic

Sound On Sound (Britain's premier recording publication), www.soundonsound.com

The preceding magazines also include ads related to recording products and services.

PRO AUDIO MAGAZINES

Journal of the Audio Engineering Society (JAES); pro audio engineering. Scholarly. With its journal, conventions, workshops, and local chapters, the Audio Engineering Society is a tremendous resource; www.aes.org.

Pro Audio Review (reviews of pro audio equipment), www.prosoundnetwork.com

Live Sound (concert sound reinforcement), www.livesoundint.com

Church Production (audio for houses of worship), www.churchproduction.com

Technologies for Worship (audio for houses of worship), www.tfwm.com

CONSUMER AUDIO MAGAZINES

Sound and Vision (consumer audio and video), www.soundandvisionmag.com

Stereophile (high-end audio), www.stereophile.com

The Absolute Sound (high-end audio), www.avguide.com/channel/the-absolute-sound

GUIDES, BROCHURES, AND OTHER LITERATURE

Microphone application guides are available from

Crown International, www.crownaudio.com

Shure Inc., www.shure.com

Countryman Associates Inc., www.countryman.com

AKG Acoustics Inc., www.akg.com

Sennheiser Electronic Corp., www.sennheiserusa.com

Audio-Technica U.S. Inc., www.audio-technica.com

Neumann/USA, www.neumannusa.com

Schoeps Mikrofone, www.schoeps.de

Danish Pro Audio, www.dpamicrophones.com (be sure to check out their Microphone University)

Bartlett Microphones, www.bartlettmics.com

You can find valuable information in user manuals and free sales literature provided by manufacturers of recording equipment. Ask your equipment dealer for manufacturers' phone numbers and website URLs.

GUIDES TO RECORDING SCHOOLS

The Audio Engineering Society offers a directory of educational programs at www.aes.org/education/directory/.

Each July issue of *Mix* magazine contains a comprehensive directory of recording schools, seminars, and programs. Universities and colleges in most major cities have recording-engineering courses.

An index of recording schools is at www.modrec.com/schools/. Two recent programs are www.recordingconnection.com, and the Bachelor of Science in Music Technology at the University of Saint Francis, School of Creative Arts at www.sf.edu/sf/art/undergrad/music-tech.

The Sound Schools Audio Career and Education Handbook offers a list of recording schools. The website is www.soundschools.com.

THE INTERNET

A great place to ask questions, besides magazines, is on the Internet. In Google, select Groups, then type in rec.audio.pro. You can ask questions and get answers from pro engineers. You may get conflicting answers, because often there are many ways to do the same thing. Also, some who reply are more expert than others. But you'll often find stimulating debates.

Some other valuable websites (all starting with www.) are homerecording.com, soundprolivesocial.com, prorec.com, plug-inhomerecording.com, hometracked.com, kvraudio.com, gearslutz.com, recordbetteraudio.com, recording.org, recordproduction.com, vintagesynth.com, moultonlabs.com, microphones.com, digido.com, prosoundweb.com, digifreq.com, cdrfaq.org, harmony-central.com, and musicplayer.com (forums).

Also see pcrecording.com, thestereobus.com, musicradar.com, audiogrid.com, digitalprosound.com, coutant.org, microphone-data.com, themicrophonevault.com, audiodirectory.nl/, binaural.com, sweetwater.com (check out publications and forums), cakewalk.com (forums), and www.solorb.com/dat-heads/.

If you lack good multitrack recordings with which to practice mixing, go to www.raw-tracks.com. There you can download individual tracks in WAV or MP3 format, or purchase a CD of raw tracks. Another site is telefunken-elektroakustik.com/download/multi-track-session.php.

Some online MIDI resources are midi.org, midiworld.com, and harmony-central.com.

The websites of audio equipment manufacturers have support sections, online discussion groups, and FAQs with lots of information. Search for the name and model number of equipment you have, or are interested in.

Do a Google search with the keywords audio, recording, recording vocals, audio recording links, MIDI, digital audio, and so on. Also search for FAQs on various audio topics. You'll discover hundreds of audio-related websites and links. In Google, you can search Groups as well as websites.

"Webcasts on Demand" offers free web tutorials on many recording topics: www.mix-online.com/webcasts

RECORDING EQUIPMENT CATALOGS

Here are but a few catalogs from which to order recording gear:

ZZounds, www.zzounds.com

All Pro Sound, www.allprosound.com

American Musical Supply, www.americanmusical.com

BSW, www.bswusa.com

Guitar Center, www.guitarcenter.com

Musician's Friend, www.musiciansfriend.com

Sam Ash, www.samash.com

Prodigy Pro, www.prodigy-pro.com

B&H Pro Audio, www.bhproaudio.com

Sweetwater Sound, www.sweetwater.com

The Woodwind and the Brasswind, www.wwbw.com

EXPERIENCE

It's the best teacher. Record all you can with any equipment you have. You may be able to buy a 4-track digital recorder on eBay, or download free multitrack recording software. Then buy a cheap mic, and practice mic techniques and overdubbing.

STARTING A CAREER IN RECORDING

After you read the material in this book and apply it in your own work, you might consider a career in recording engineering. See the section "Guides to Recording Schools" earlier in this appendix. Some schools offer job placement upon graduation.

Currently there is more work for recording engineers in film, computer games, and TV than in music recording. One reason is that many more musicians are recording themselves at home.

Rather than trying to find a job in a music recording studio, you might start your own. The gear costs so little these days. After you acquire some high-quality equipment, offer to record local bands for free while you gain experience. You might specialize in one genre of music that matches your market and your recording skills.

You'll need to promote your studio. Post flyers in music stores. You can advertise your recording service for free on Craigslist. Create a website where you list your studio's musical instruments, recording equipment, and customers. Find potential customers at clubs or festivals, and give them demo CDs and business cards with your website's URL.

If you can't find much recording business, consider taking another job in a related industry—such as audio equipment design or sales—and record on weekends as a second job.

The usual way to join a commercial recording studio is to have an interview. Supply examples of your best work. Be willing to start as a gofer, and show that you can drum up business for the studio, perhaps bringing clients with you. Work your way up to operating recorders, placing mics, and copying/backing up files. Eventually, if you have personal and musical skills as well as technical skills, you can hope to sit at the console as a mixing engineer. Good luck!

AAX (Avid Audio Extension)   An Avid plug-in format used in Pro Tools 10. It's 64-bit ready and lets you use native or DSP processing.

ACCENT MICROPHONE   See SPOT MICROPHONE.

AES/EBU   Also called IEC 988 Type 1. An interface format for digital signals, using a balanced 110 ohm mic cable terminated with XLR-type connectors. See also S/PDIF.

AIFF (Audio Interchange File Format)   A standard Mac format for uncompressed digital audio files.

AMBIENCE   Room acoustics, early reflections, and reverberation. Also, the audible sense of a room or environment surrounding a recorded instrument.

AMBIENCE MICROPHONE   A microphone placed relatively far from its sound source to pick up ambience.

AMPLITUDE, PEAK   On a graph of a sound wave, the sound pressure of the waveform peak. On a graph of an electrical signal, the voltage of the waveform peak. The amplitude of a sound wave or signal as measured on a meter is 0.707 times the peak amplitude.

ANALOG-TO-DIGITAL (A/D) CONVERTER   A circuit that converts an analog audio signal into a stream of digital data (bitstream).

ANTI-ALIAS FILTER   In an A/D converter, a lowpass filter that removes all frequencies above 20 kHz before sampling in order to prevent audio artifacts called aliasing.

ANTI-IMAGING FILTER   In a D/A converter, a lowpass filter that smooths the voltage steps in the analog signal that were generated by translating digital numbers into analog voltages. The anti-imaging filter recovers the waveform of the original analog signal.

ASIO (Audio Stream Input/Output)   Steinberg's computer audio driver spec for Mac and Windows. ASIO has low latency because it interfaces directly between a sound card and the audio application software.

ASSIGN   To route or send an audio signal to one or more selected channels.

ATTACK   The beginning of a note. The first portion of a note's envelope in which a note rises from silence to its maximum volume.

ATTACK TIME   In a compressor, the time it takes for gain reduction to occur in response to a musical attack.

ATTENUATE   To reduce the level of a signal.

AUDIO INTERFACE   A device that connects to a computer and converts an audio signal into computer data for storage in memory or on hard disk. The interface also converts computer data into an audio signal. See Breakout Box, I/O Box, and Sound Card.

AUTOMATED MIXING   A system of mixing in which a computer remembers and updates mixer control settings and moves. With this system, a mix can be performed and refined in several stages and played back at a later date exactly as set up previously.

AUXILIARY BUS (AUX BUS)   A bus or channel that contains a mix of the aux-send signals of the input modules in a mixer. An aux bus is used to send signals to an effects unit or monitor system. See Effects Bus.

AUXILIARY SEND (AUX SEND)   A control in a mixer's input module used to send that module's signal to an aux bus. The aux-send level adjusts the amount of effects heard on an instrument, or adjusts the loudness of that instrument in the monitor system.

BAFFLED-OMNI   A stereo miking arrangement that uses two ear-spaced omnidirectional microphones separated by a hard padded baffle.

BALANCE   The relative volume levels of various tracks or instruments.

BALANCED LINE   A cable with two conductors surrounded by a shield, in which each conductor is at equal impedance to ground. With respect to ground, the conductors are at equal potential but opposite polarity; the signal flows through both conductors.

BANDPASS FILTER   In a crossover, a filter that passes a band or range of frequencies but sharply attenuates or rejects frequencies outside the band.

BASIC TRACKS   Recorded tracks of rhythm instruments (bass, guitar, drums, and sometimes keyboard).

BASS TRAP   An assembly that absorbs low-frequency sound waves in the studio.

BI-AMPLIFICATION (BI-AMPING)   Driving a woofer and tweeter with separate power amplifiers. An active crossover is connected ahead of these power amplifiers.

BIDIRECTIONAL MICROPHONE   A microphone that is most sensitive to sounds arriving from two directions—in front of and behind the microphone. It rejects sounds approaching either side of the microphone. Sometimes called a cosine or figure-eight microphone because of the shape of its polar pattern.

BINAURAL RECORDING   A 2-channel recording made with an omnidirectional microphone mounted near each ear of a human or a dummy head, for playback over headphones. The object is to duplicate the acoustic signal appearing at each ear.

BIT DEPTH (word length)   The number of bits (ones and zeros) making up a word in a digital signal (such as 16 or 24 bits). Each word is a binary number that is the value of each sample. A sample is a measurement of an analog waveform that is done several thousand times a second during analog-to-digital conversion. High bit depth=long word length=high resolution of the analog signal amplitude=high sound quality.

BOARD   See MIXING CONSOLE.

BOUNDARY MICROPHONE   A microphone designed to be used on a boundary (a hard reflective surface). The microphone capsule is mounted very close to the boundary so that direct and reflected sounds arrive at the microphone diaphragm in phase (or nearly so) for all frequencies in the audible band.

BREAKOUT BOX (I/O BOX)   A group of audio input and output connectors in a chassis, which is wired to a sound card (PCI audio interface) in a computer. Used to interface analog audio signals (and often MIDI and digital signals) with a computer.

BUS   A common connection of many different signals. An output of a mixer or submixer. A channel that feeds a tape track, signal processor, or power amplifier.

BUS IN   An input to a program bus, usually used for effects returns.

BUS MASTER   In the output section of a mixing console, a potentiometer (fader or volume control) that controls the output level of a bus. Also see Group Fader.

BUS OUT   The output connector of a bus.

BUZZ   An unwanted edgy tone that sometimes accompanies audio, containing high harmonics of 60 Hz.

CARDIOID MICROPHONE   A unidirectional microphone with side attenuation of 6 dB and maximum rejection of sound at the rear of the microphone (180 degrees off-axis). A microphone with a heart-shaped directional pattern.

CARVE   Apply an EQ dip.

CD   See Compact Disc.

CD-R (CD-Recordable)   A recordable compact disc that cannot be rewritten. Once recorded, it cannot be erased and reused.

CD-ROM DRIVE   A computer disk drive that plays computer data from CD-ROM disks. The disks are read optically by a laser, as in a compact disc player.

CD-RW (CD-ReWritable)   A recordable compact disc that can be rewritten. Once recorded it can be erased and reused.

CHANNEL   A single path of an audio signal. Usually, each channel contains a different signal.

CHANNEL ASSIGN   See ASSIGN.

CHORUS   1. A special effect in which a signal is delayed by 15 to 35 msec, the delayed signal is combined with the original signal, and the delay is varied randomly or periodically. This creates a wavy, shimmering effect. 2. The main portion of a song that is repeated several times throughout the song with the same lyrics.

CLEAN   Free of noise, distortion, overhang, leakage. Not muddy.

CLEAR   Easy to hear, easy to differentiate. Reproduced with sufficient high frequencies.

CLIP   To cut off or flatten the tops of a signal waveform, causing distortion. Clipping is caused by setting signal level too high in an audio device or audio software. Also see REGION.

COINCIDENT-PAIR   A stereo microphone, or two separate microphones, placed so that the microphone diaphragms occupy approximately the same point in space. They are angled apart and mounted one directly above the other.

COMB-FILTER EFFECT   The frequency response caused by combining a sound with its delayed replica. The frequency response has a series of peaks and dips caused by phase interference. The peaks and dips resemble the teeth of a comb.

COMPACT DISC (CD)   A read-only optical disc medium for storing digital audio programs up to 80 minutes long. The compact disc stores data in the form of a spiral groove of microscopic pits and is read optically by a laser. A CD's digital audio format is 44.1 kHz sampling rate and 16-bit word length.

COMPLEX WAVE   A wave with more than one frequency component.

COMPING   Recording several musical performances (takes) of a single instrument or vocal on different tracks, and selecting the best parts of each take to be played in order on a composite track during mixdown.

COMPOSITE TRACK   A track containing the best parts of several takes of a musical-instrument or vocal performance.

COMPRESSION   1. The portion of a sound wave in which air molecules are pushed together, forming a region with higher-than-normal atmospheric pressure. 2. In signal processing, the reduction in dynamic range or gain caused by a compressor. 3. Data compression or data reduction is an encoding scheme to reduce the size of a data file by throwing away audio data deemed inaudible because of masking. ATRAC, MP3, MLP, AAC, RealAudio, OGG, and Microsoft Media are examples of compressed data formats.

COMPRESSION RATIO (SLOPE) In a compressor, the ratio of the change in input level (in dB) to the change in output level (in dB). For example, a 2:1 ratio means that for every 2 dB change in input level, the output level changes 1 dB.

COMPRESSOR A signal processor that reduces dynamic range or gain by means of automatic volume control. An amplifier whose gain decreases as the input signal level increases above a preset point.

CONDENSER MICROPHONE A microphone that works on the principle of variable capacitance to generate an electrical signal. The microphone diaphragm and an adjacent metallic disk (called a backplate) are charged to form two plates of a capacitor. Incoming sound waves vibrate the diaphragm, varying its spacing to the backplate, which varies the capacitance, which in turn varies the voltage between the diaphragm and backplate.

CONNECTOR A device that makes electrical contact between a signal-carrying cable and an electronic device, or between two cables. A device used to connect or hold together a cable and an electronic component so that a signal can flow from one to the other.

CONSOLE See MIXING CONSOLE.

CONTACT PICKUP A transducer that contacts a musical instrument and converts its mechanical vibrations into a corresponding electrical signal.

CONTROL ROOM The room in which the engineer controls and monitors the recording. It houses most of the recording hardware.

CONTROLLER SURFACE A chassis with faders (and sometimes buttons and knobs) that resembles a mixer, used to adjust virtual controls that appear on-screen in computer editing software. Connected to the computer by USB or

FireWire, the controller surface might include analog and digital input/output connectors and MIDI connectors.

CONVOLUTION REVERB (SAMPLING REVERB)   A reverberation device or plug-in that creates the reverb from impulse-response samples (WAV files) of real acoustic spaces, rather than from algorithms. The resulting sound quality is very natural.

CUE, CUE SEND   In a mixing-console input module, a control that adjusts the level of the signal feeding the cue mixer that feeds a signal to headphones in the studio.

CUE MIXER   A submixer in a mixing console that takes signals from cue sends as inputs and mixes them into a composite signal that drives headphones in the studio.

CUE SYSTEM   A monitor system that allows musicians to hear themselves and previously recorded tracks through headphones.

DATA COMPRESSION   A data encoding scheme for reducing the amount of data storage on a medium. Also known as data reduction. See Compression, ATRAC, and MP3.

DAW   Abbreviation for digital audio workstation.

dB   Abbreviation for decibel.

DEAD   Having very little or no reverberation.

DECAY   The portion of the envelope of a note in which the envelope goes from maximum to some midrange level. Also, the decline in level of reverberation over time.

DECAY TIME   See REVERBERATION TIME.

DECIBEL   The unit of measurement of audio level. Ten times the logarithm of the ratio of two power levels. Twenty times the logarithm of the ratio of two voltages. dBV is decibels relative to 1 volt, dBu is decibels relative to 0.775 volt, dBv is decibels relative to 0.7746 volt, dBm is decibels relative to 1 milliwatt, dBFS is decibels, full scale (all bits on at a signal peak), dBSPL is decibels of sound pressure level, and dBA is decibels, A-weighted. See Weighted.

DE-ESSER   A signal processor or plug-in that removes excessive sibilance ("s" and "sh" sounds) by compressing high frequencies around 5 to 10 kHz.

DELAY   The time interval between a signal and its repetition. A digital delay or a delay line is a signal processor that delays a signal for a short time.

DELAY COMPENSATION   Adjusting the timing of a track that is processed by a plug-in, to make it in sync with nonprocessed tracks. Plug-ins add latency (delay) to a track.

DEPTH   The audible sense of nearness and farness of various instruments. Instruments recorded with a high ratio of direct-to-reverberant sound are perceived as being close; instruments recorded with a low ratio of direct-to-reverberant sound are perceived as being distant.

DESIGN CENTER   The portion of fader travel (usually shaded), about 10 to 15 dB from the top, in which console gain is distributed for optimum headroom and signal-to-noise ratio. During normal operation, each fader in use should be placed at or near design center.

DESIGNATION STRIP   A strip of paper taped near console faders to designate the instrument that each fader controls. Also called a scribble strip.

DESK   The British term for mixing console.

DI   Short for direct injection, recording with a direct box.

DIGITAL AUDIO   An encoding of an analog audio signal in the form of binary digits (ones and zeroes).

DIGITAL AUDIO WORKSTATION (DAW)   A computer, audio interface, and recording software that allows you to record, edit, and mix audio programs entirely in digital form. A standalone DAW is a digital multitrack recorder-mixer. Standalone DAWs include real mixer controls; computer DAWS have virtual controls on-screen.

DIGITAL RECORDING   A recording system in which the audio signal is stored as binary digits (ones and zeroes).

DIGITAL-TO-ANALOG (D/A) CONVERTER   A circuit that converts a digital audio signal into an analog audio signal.

DIRECT BOX   A device used for connecting an amplified instrument directly to a mixer mic input. The direct box converts a high-impedance unbalanced audio signal into a low-impedance balanced audio signal.

DIRECT INJECTION (DI)   Recording with a direct box.

DIRECTIONAL MICROPHONE   A microphone that has different sensitivity in different directions. A unidirectional or bidirectional microphone.

DIRECT OUTPUT, DIRECT OUT   An output connector following a mic preamplifier, fader, and equalizer, used to feed the signal of one instrument to one track of a multitrack recorder.

DIRECT SOUND   Sound traveling directly from the sound source to the microphone (or to the listener) without reflections. Also, DirectSound is an audio driver for the Windows operating system, intended as an enhancement to MultiMedia Extensions (MME).

DISTORTION   An unwanted change in the audio waveform, causing a raspy or gritty sound quality. The appearance of frequencies in a device's output signal that were not in the input signal. Distortion is caused by recording at too high a level, using improper mixer settings, components failing, or vacuum tubes distorting. (Distortion can be desirable, e.g., for an electric guitar.) In digital recording, distortion called quantization error can also occur at very low signal levels, where there are not enough bits to record the signal accurately. Also see Clip.

DITHER   Low-level noise added to a digital signal to reduce quantization distortion caused by truncating (removing) bits in a digital word. It's a good idea to add dither to a 24-bit program just before converting it to 16 bits for CD release.

DOUBLING   A special effect in which a signal is combined with its 15- to 35-msec-delayed replica. This process mimics the sound of two identical voices or instruments playing in unison. In another type of doubling, two identical performances are recorded and played back to thicken the sound.

DROP-OUT   During playback of a hard-disk recording, a momentary loss of signal caused by a buffer memory being emptied. Increasing buffer size usually prevents drop-outs.

DRUM MACHINE   A device—hardware or software—that plays samples of real drums and includes a sequencer to record rhythm patterns.

DRY   Having no echo or reverberation. Referring to a close-sounding signal that has not yet been processed by a reverberation or delay device or plug-in.

DSP (digital signal processing)   Modifying a signal in digital form by doing calculations on the numbers. DSP is used for level changes, EQ, effects, and so on.

DYNAMIC MICROPHONE   A microphone that generates electricity when sound waves cause a conductor to vibrate in a stationary magnetic field. The two types of dynamic microphones are moving coil and ribbon. A moving-coil microphone is usually called a dynamic microphone.

DYNAMIC RANGE   The range of volume levels in a program from softest to loudest.

ECHO   A delayed repetition of a signal or sound. A sound delayed 50 msec or more that is combined with the original sound.

EDITING   The cutting and rejoining of magnetic tape to delete unwanted material, to insert silent spaces, or to rearrange recorded material into the desired sequence. Also, the same actions performed on a digital recording with a DAW, hard-disk recorder, or MiniDisc recorder-mixer.

EFFECTS   Interesting sound phenomena created by signal processors, such as reverberation, echo, flanging, doubling, compression, or chorus. See SOUND EFFECTS.

EFFECTS BUS   The bus that feeds effects devices (signal processors).

EFFECTS LOOP   A set of connectors in a mixer for connecting an external effects unit, such as a reverb or delay device. The effects loop includes a send section and a receive section. See EFFECTS SEND, EFFECTS RETURN.

EFFECTS RETURN (AUX RETURN)   In the output section of a mixing console, a control that adjusts the amount of signal received from an effects unit. Also, the connectors in a mixer to which you connect the effects-unit output signal. They might be labeled "bus in" instead. The effects-return signal is mixed with the program bus signal.

EFFECTS SEND (AUX SEND)   In an input module of a mixing console, a control that adjusts the amount of signal sent to an effects device, such as a reverberation or delay unit. Also, the connector in a mixer that you connect to the input of an effects unit. The effects-send or aux-send control adjusts the amount of effects heard on each instrument.

ELECTRET-CONDENSER MICROPHONE   A condenser microphone in which the electrostatic field of the capacitor is generated by an electret—a material that permanently stores an electrostatic charge.

END-ADDRESSED   Referring to a microphone whose main axis of pickup is perpendicular to the front of the microphone. You aim the front of the mic at the sound source. See SIDE-ADDRESSED.

ENVELOPE   The rise and fall in volume of one note. The envelope connects successive peaks of the waves that make up the note. Each harmonic in the note might have a different envelope.

EQUALIZATION (EQ)   The adjustment of frequency response to alter the tonal balance or to attenuate unwanted frequencies.

EQUALIZER   A circuit (usually in each input module of a mixing console, or in a separate unit) that alters the frequency spectrum of a signal passed through it.

EXPANDER   1. A signal processor that increases the dynamic range of a signal passed through it. 2. An amplifier whose gain decreases as its input level decreases. When used as a noise gate, an expander reduces the gain of low-level signals to reduce noise between notes.

FADE-OUT   To gradually reduce the volume of the last several seconds of a recorded song, from full level down to silence, by slowly pulling down the master fader, or by selecting a fade-out process in a digital editing program.

FADER A linear or sliding potentiometer (volume control), used to adjust signal level.

FEED 1. To send an audio signal to some device or system. 2. An output signal sent to some device or system.

FEEDBACK 1. The return of some portion of an output signal to the system's input. 2. The squealing sound you hear when a PA system microphone picks up its own amplified signal through a loudspeaker.

FILTER 1. A circuit that sharply attenuates frequencies above or below a certain frequency. Used to reduce noise and leakage above or below the frequency range of an instrument or voice. 2. A MIDI filter removes selected note parameters.

FIREWIRE A standard protocol for high-speed transfer of data between digital devices. Also called IEEE 1394.

FLANGING A special effect in which a signal is combined with its delayed replica, and the delay is varied between 0 and 20 msec. A hollow, swishing, ethereal effect like a variable-length pipe, or like a jet plane passing overhead. A variable comb filter produces the flanging effect.

FLETCHER–MUNSON EFFECT Named after the two people who discovered it, the psychoacoustical phenomenon in which the subjective frequency response of the ear changes with program level. Because of this effect, a program played at a lower volume than the original level subjectively loses low- and high-frequency response.

FLUTTER ECHOES A rapid series of echoes that occurs between two parallel walls.

FOLDBACK (FB) See CUE SYSTEM.

FREQUENCY The number of cycles per second of a sound wave or an audio signal, measured in hertz (Hz). A low frequency (for example, 100 Hz) has a low pitch; a high frequency (for example, 10,000 Hz) has a high pitch.

FREQUENCY RESPONSE 1. The range of frequencies that an audio device will reproduce at an equal level (within a tolerance, such as ±3 dB). 2. The range of frequencies that a device (mic, human ear, etc.) can detect.

FUNDAMENTAL   The lowest frequency in a complex wave.

FX   Abbreviation for effects.

GAIN   Amplification. The ratio, expressed in decibels, between the output voltage and the input voltage, or between the output power and the input power.

GAIN STAGING   Setting all the gain controls in a signal path to get the cleanest signal. At any stage in the signal path, the signal level should be set well above the noise floor but below clipping.

GATE   1. To turn off a signal when its amplitude falls below a preset value. 2. The signal-processing device used for this purpose. See also NOISE GATE.

GATED REVERB   Reverberation with the reverberant "tail" cut off before it fades out.

GENERAL MIDI FILE (GM file)   A MIDI file containing a standard set of musical instrument sounds. A General MIDI file produces the same sounds on any MIDI-capable instrument that supports the GM spec.

GOBO   A moveable partition used to prevent the sound of an instrument from reaching another instrument's microphone. Short for go-between.

GRAPHIC EQUALIZER   An equalizer with a horizontal row of faders; the fader-knob positions indicate graphically the frequency response of the equalizer. Usually used to equalize monitor speakers for the room they are in. Sometimes used for complex EQ of a track.

GROUND LOOP   1. A loop or circuit formed of ground leads. 2. The loop formed when unbalanced components are connected together via two ground paths—the connecting-cable shield and the power ground. Ground loops cause hum and should be avoided.

GROUP   See SUBMIX. 1. To select several faders to make them act in unison. For example, select all the faders for the drum tracks to so that you can adjust the overall level of the drums by pushing one fader. 2. To assign the output of several input modules to a single group or bus, whose level is controlled by a single group fader. For example, assign all the input modules of the drum mics to a single "drums" group. 3. A bus or channel in a mixer that contains the signals from several input modules. For example, a drums group is a group of all the signals of the drum-set mics.

GROUP FADER (submaster fader) In the output section of a mixing console, a potentiometer (fader or volume control) that controls the output level of a bus or group.

GSIF (GigaSampler Interface) Nemesys' Windows sound card driver that achieves low latency by interfacing directly between the sound card and the audio application software.

HARD DISK A random-access storage medium for computer data. A hard-disk drive contains a stack of magnetically coated hard disks that are read and written to by an electromagnetic head.

HARD-DISK RECORDER (hard-drive recorder) A device dedicated to recording digital audio on a hard-disk drive. A hard-disk recorder-mixer includes a built-in mixer.

HARMONIC An overtone whose frequency is a whole-number multiple of the fundamental frequency.

HARMONIZER A signal processor that provides a wide variety of pitch-shifting and delay effects.

HD Abbreviation for hard-disk drive.

HEADPHONES A head-worn transducer that covers the ears and converts electrical audio signals into sound waves.

HERTZ (Hz) Cycles per second, the unit of measurement of frequency.

HIGHPASS FILTER A filter that passes frequencies above a certain frequency and attenuates frequencies below that same frequency. A low-cut filter.

HISS A noise signal containing all frequencies, but with greater energy at higher octaves. Hiss sounds like wind blowing through trees. It is usually caused by random signals generated by microphones, electronics, and magnetic tape.

HOST A DAW recording program that supports plug-ins. See PLUG-IN.

HOT   1. A high recording level causing slight distortion, may be used for special effect. 2. High average level on a CD making it relatively loud, produced by peak limiting and normalization, or by compression and normalization. 3. A condition in which a chassis or circuit has a potentially dangerous voltage on it. 4. Referring to the conductor in a microphone cable that has a positive voltage on it at the instant that sound pressure moves the diaphragm inward.

HUM   An unwanted low-pitched tone (60 Hz and its harmonics) heard in the monitors. The sound of interference generated in audio circuits and cables by AC power wiring. Hum pickup is caused by such things as placing audio cables near power cables or transformers, faulty grounding, poor shielding, and ground loops.

HYPERCARDIOID MICROPHONE   A directional microphone with a polar pattern that has 12 dB attenuation at the sides, 6 dB attenuation at the rear, and two nulls of maximum rejection at 110 degrees off-axis.

IMAGE   An illusory sound source located somewhere around the listener. An image is generated by two or more loudspeakers. In a typical stereo system, images are located between the two stereo speakers.

IMPEDANCE   The opposition of a circuit to the flow of alternating current. Impedance is the complex sum of resistance and reactance. Abbreviated as Z.

INPUT   The connection going into an audio device. In a mixer or mixing console, a connector for a microphone, line-level device, or other signal source.

INPUT MODULE   In a mixing console, the set of controls affecting a single input signal. An input module usually includes an attenuator (trim), fader, equalizer, aux sends, and channel-assign buttons.

INPUT/OUTPUT (I/O) CONSOLE (IN-LINE CONSOLE)   A mixing console arranged so that input and output sections are aligned vertically. Each module (other than the monitor section) contains one input channel and one output channel.

INPUT SECTION   The row of input modules in a mixing console.

INSERT JACKS   Two jacks (send and return) in a console input module or output module that allow access to points in the signal path, usually for connecting a compressor. Plugging into the access jacks breaks the signal flow and allows you to insert a signal processor or recorder in series with the signal. In many mixers, a single insert jack has both send and return terminals. Also called access jacks.

I/O   Referring to input and output connectors.

I/O BOX   A breakout box type of audio interface.

iOS   Apple's mobile operating system used in the iPad, iPod Touch, and iPhone.

JACK   A female or receptacle-type connector for audio signals into which a plug is inserted.

KEYBOARD WORKSTATION   Several MIDI components in one chassis—a keyboard, a sample player, a sequencer, and perhaps a synthesizer and disk drive.

KILO   A prefix meaning one thousand. Abbreviated k.

LATENCY   The signal delay through an A/D, D/A converter, through a software program, or through a computer operating system. Monitoring latency is the delay between the time when a musician plays a note and when she hears the monitored signal of that note. Latency can make you play out-of-sync during overdubs.

LEAKAGE   The overlap of an instrument's sound into another instrument's microphone. Also called bleed or spill.

LED INDICATOR   A recording-level indicator using one or more light-emitting diodes.

LEVEL   The degree of intensity of an audio signal—the voltage, power, or sound pressure level. The original definition of level is the power in watts.

LEVEL SETTING   In a recording system, the process of adjusting the input-signal level to obtain maximum level on the recording medium without distortion. A VU meter, LED meter, or other indicator shows recording level.

LIGHTPIPE   An Alesis connection protocol that transfers 8 digital audio channels at once over a Toslink fiber-optic cable.

LIMITER A signal processor whose output is constant above a preset input level. A compressor with a compression ratio of 10:1 or greater, with the threshold set just below the point of distortion of the following device. Used to prevent distortion of attack transients or peaks. Also used to decrease the level of signal peaks so that the average level of a program can be raised. See LOOK-AHEAD.

LINE LEVEL In balanced professional recording equipment, a signal whose level is approximately 1.23 volts (+4 dBm). In unbalanced equipment (most home hi-fi or semipro recording equipment), a signal whose level is approximately 0.316 volt (−10 dBV).

LIVE 1. Having audible reverberation. 2. Occurring in real time, in person.

LIVE RECORDING A recording made at a concert. Also, a recording made of a musical ensemble playing all at once, rather than overdubbing.

LOCALIZATION The ability of the human hearing system to tell the direction of a real or illusory sound source.

LOOP In a sampling program, to play the sustain portion of a sound's envelope repeatedly. Also, a repeated rhythmic or musical pattern.

LOUDSPEAKER A transducer that converts electrical energy (the signal) into acoustical energy (sound waves).

LOWPASS FILTER A filter that passes frequencies below a certain frequency and attenuates frequencies above that same frequency. A high-cut filter.

M Abbreviation for mega, or one million (as in megabytes). Small m (m) is an abbreviation for milli, or one thousandth.

MASK To hide or cover up one sound with another sound. To make a sound inaudible by playing another sound along with it. Masking is used in many data reduction schemes.

MASTER A completed tape or CD used to generate tape copies or compact discs.

MASTER FADER A volume control that affects the level of all program buses simultaneously. It is the last stage of gain adjustment before the 2-track recorder.

MEMORY   A group of integrated circuit chips used to store digital data temporarily or permanently (such as an audio signal in digital format).

MEMORY RECORDER   A device that records audio on a memory chip, such as CompactFlash. Usually the recorded audio can be uncompressed WAV files or compressed MP3 files.

METER   A device that indicates voltage, resistance, current, or signal level.

MIC   An abbreviation for microphone.

MIC LEVEL   The level or voltage of a signal produced by a microphone, typically 2 millivolts.

MIC PREAMP   See PREAMPLIFIER.

MICROPHONE   A transducer or device that converts an acoustical signal (sound) into a corresponding electrical signal.

MICROPHONE TECHNIQUES   The selection and placement of microphones to pick up sound sources.

MIDI   Abbreviation for Musical Instrument Digital Interface, a specification for a connection between synthesizers, drum machines, and computers that allows them to communicate with and/or control each other.

MIDI CHANNEL   A route for transmitting and receiving MIDI signals. Each channel controls a separate MIDI musical instrument or synth patch. Up to 16 channels can be sent on a single MIDI cable.

MIDI CLOCK   A timing reference that sets a common tempo for all the equipment. The MIDI clock is a series of bytes in the MIDI data stream that conveys timing information. The clock is like a conductor's baton movements, keeping all the performers in sync at the same tempo. The clock bytes are added to the MIDI performance information in the MIDI signal. The clock signal is 24, 48, or 96 pulses per quarter note (ppq). That is, for every quarter note of the performance, 24 or more clock pulses (bytes) are sent in the MIDI data stream.

MIDI CONTROLLER   A musical performance device (keyboard, drum pads, breath controller, etc.) that outputs a MIDI signal designating note numbers, note on, note off, and so on.

MIDI/DIGITAL AUDIO SOFTWARE   Software that combines a MIDI sequencer with a multitrack digital audio recorder/editor.

MIDI IN   A connector in a MIDI device that receives MIDI messages.

MIDI INTERFACE   A circuit that plugs into a computer and converts MIDI data into computer data for storage in memory or on hard disk. The interface also converts computer data into MIDI data.

MIDI OUT   A connector in a MIDI device that transmits MIDI messages.

MIDI THRU   A connector in a MIDI device that duplicates the MIDI information at the MIDI IN connector. Used to connect another MIDI device in the series.

MIKE   To pick up with a microphone.

MILLI   A prefix meaning one thousandth, abbreviated m.

MIX   1. To combine two or more different signals into a common signal. 2. A control on an effects processor that varies the ratio between the dry (unprocessed) signal and the processed signal.

MIXDOWN   The process of playing recorded tracks through a mixing console and mixing them to two stereo channels for recording on a two-track recorder. Also applies to a surround-sound mixdown to 6 or 8 channels.

MIXER   A device that mixes or combines audio signals and controls the relative levels of the signals.

MIXING CONSOLE   A large mixer with additional functions such as equalization or tone control, pan pots, monitoring controls, solo functions, channel assigns, and control of signals sent to external signal processors.

MONAURAL   Referring to listening with one ear. Often incorrectly used to mean monophonic.

MONITOR   A loudspeaker in a control room, or headphones, used for judging sound quality. Also, a video display screen used with a computer.

MONITORING   Listening to an audio signal with a monitor.

MONO, MONOPHONIC   1. Referring to a single channel of audio. A monophonic program can be played over one or more loudspeakers, or one or more headphones. 2. Describing a synthesizer that plays only one note at a time (not chords).

MONO-COMPATIBLE   A characteristic of a stereo program, in which the program channels can be combined to a mono program without altering the frequency response or balance. A mono-compatible stereo program has the same frequency response in stereo or mono because there is no delay or phase shift between channels to cause phase interference.

MOVING-COIL MICROPHONE   A dynamic microphone in which the conductor is a coil of wire moving in a fixed magnetic field. The coil is attached to a diaphragm that vibrates when struck with sound waves. Usually called a dynamic microphone.

MP3 (MPEG Level-1 Layer-3)   A data compression format for audio. In an MP3 file (.mp3), the data has been compressed or reduced to one-tenth of its original size or less. Compressed files take up less memory, so they download faster. You download MP3 files to your hard drive, then listen to them. MP3 audio quality at a 128 kbps rate is nearly the same as that of CDs (depending on source material).

MUDDY   Unclear sounding; having excessive leakage, reverberation, or overhang.

MULTIEFFECTS PROCESSOR   See MULTIPROCESSOR.

MULTIPROCESSOR   A signal processor that can perform several different signal-processing functions.

MULTITRACK   Referring to a recorder with more than two tracks.

MUTE   To turn off an input signal on a mixing console by disconnecting the input-module output from channel assign. During mixdown, the mute function is used to reduce tape noise and leakage during silent portions of tracks, or to turn off unused performances. During recording, mute is used to turn off mic signals.

NATIVE   Referring to DAW plug-ins that process audio with a computer's CPU rather than with DSP cards.

NEAR-COINCIDENT   A stereo microphone technique in which two directional microphones are angled apart symmetrically on either side of center and spaced a few inches apart horizontally.

NEARFIELD MONITORING   A monitor-speaker arrangement in which the speakers are placed very near the listener (usually just behind the mixing console) to reduce the audibility of control-room acoustics.

NOISE   Unwanted sound, such as hiss from electronics or tape. An audio signal with an irregular, nonperiodic waveform.

NOISE GATE   A gate used to reduce or eliminate noise between notes.

NONDESTRUCTIVE EDITING   In a digital audio workstation, editing done by changing pointers (location markers) to information on the hard disk. A nondestructive edit can be undone.

NONLINEAR   1. Referring to a storage medium in which any data point can be accessed or read almost instantly in a random fashion, rather than sequentially. Examples are a hard disk, compact disc, and MiniDisc. See Random Access. 2. Referring to an audio device that is distorting the signal.

NORMALIZE   To raise the level of a digital audio signal so that the highest peak in the program is at the highest level allowed by the recording. For example, in a normalized 16-bit recording, the highest peak in the program has all 16 bits on (the highest possible level in a 16-bit recording).

OCTAVE   The interval between any two frequencies where the upper frequency is twice the lower frequency.

OFF-AXIS   Not exactly in front of a microphone or loudspeaker.

OMNIDIRECTIONAL MICROPHONE   A microphone that is equally sensitive to sounds arriving from all directions.

ON-LOCATION RECORDING   A recording made outside the studio, in a room or hall where the music usually is performed or practiced.

ORTF   Named after the French broadcasting network (Office de Radio-diffusion Television Française), a near-coincident stereo mic technique that uses two cardioid mics angled 110 degrees apart and spaced 17 cm horizontally.

OUTBOARD EQUIPMENT Signal processors that are external to the mixing console.

OUTPUT A connector in an audio device from which the signal comes and feeds successive devices.

OUTTAKE A take, or section of a take, that is to be removed or not used.

OVERDUB To record a new musical part on an unused track in synchronization with previously recorded tracks.

OVERLOAD The distortion that occurs when an applied signal exceeds a system's maximum input level.

OVERTONE In a complex wave, a frequency component that is higher than the fundamental frequency.

PAD See ATTENUATE.

PAN POT Abbreviation for panoramic potentiometer. In each input module in a mixing console, a control that divides a signal between two channels in an adjustable ratio. By doing so, a pan pot controls the location of a sonic image between a stereo pair of loudspeakers.

PARAMETRIC EQUALIZER An equalizer with continuously variable parameters, such as frequency, bandwidth, and amount of boost or cut.

PATCH 1. To connect one piece of audio equipment to another with a cable. 2. A setting of synthesizer parameters to achieve a sound with a certain timbre.

PATCH BAY (PATCH PANEL) An array of connectors, usually in a rack, to which equipment inputs and outputs are wired. A patch bay makes it easy to interconnect various pieces of equipment in a central, accessible location.

PATCH CORD A short length of cable with a phone plug on each end, used for signal routing in a patch bay.

PCM Abbreviation for pulse-code modulation, a method of analog-to-digital conversion in which the instantaneous amplitude of an analog waveform is measured or sampled several thousand times a second, and each measurement is assigned a binary value of a certain number of bits (ones and zeroes).

PEAK   On a graph of a sound wave or signal, the highest point in the waveform. The point of greatest voltage or sound pressure in a cycle.

PEAK AMPLITUDE   See AMPLITUDE, PEAK.

PEAKING EQUALIZER   An equalizer that provides maximum cut or boost at one frequency, so that the resulting frequency response of a boost resembles a mountain peak.

PERSONAL STUDIO   A minimal group of recording equipment set up for one's personal use, usually using a 4-track cassette or memory recorder-mixer. Also, a simple 4-track recorder-mixer for one's personal use.

PERSPECTIVE   In the reproduction of a recording, the audible sense of distance to the musical ensemble, the point of view. A close perspective has a high ratio of direct sound to reverberant sound; a distant perspective has a low ratio of direct sound to reverberant sound.

PHANTOM POWER   A DC voltage (usually 12 to 48 volts) applied to microphone signal conductors to power condenser microphones.

PHASE   The degree of progression in the cycle of a wave, where one complete cycle is 360 degrees.

PHASE CANCELLATION, PHASE INTERFERENCE   The cancellation of certain frequency components of a signal that occurs when the signal is combined with its delayed replica. At certain frequencies, the direct and delayed signals are of equal level and opposite polarity (180 degrees out of phase), and when combined, they cancel out. The result is a comb-filter frequency response having a periodic series of peaks and dips. Phase interference can occur between the signals of two microphones picking up the same source at different distances, or can occur at a microphone picking up both a direct sound and its reflection from a nearby surface.

PHASING   A special effect in which a signal is combined with its phase-shifted replica to produce a variable comb-filter effect. See also FLANGING.

PHASE SHIFT The difference in degrees of phase angle between corresponding points on two waves. If one wave is delayed with respect to another, there is a phase shift between them of 2 piFT, where pi=3.14, F=frequency in Hz, and T=delay in seconds.

PHONE PLUG A cylindrical, coaxial plug (usually 1/4-inch diameter). An unbalanced phone plug has a tip for the hot signal and a sleeve for the shield, which connects to ground. A balanced phone plug has a tip for the signal hot signal, a ring for the return signal, and a sleeve for the shield.

PHONO PLUG A coaxial plug with a central pin for the hot signal and a ring of pressure-fit tabs for the shield or ground. Also called an RCA plug. Phono plugs are used on Tascam modular digital multitrack recorders and on consumer stereo equipment.

PICKUP A piezoelectric transducer that converts mechanical vibrations to an electrical signal. Used in acoustic guitars, acoustic basses, and fiddles. Also, a magnetic transducer in an electric guitar that converts string vibration to a corresponding electrical signal.

PITCH The subjective lowness or highness of a tone. The pitch of a tone usually correlates with the fundamental frequency.

PITCH SHIFTER A signal processor that changes the pitch of an instrument without changing its duration.

PLUG A male connector that inserts into a jack. Also, short for plug-in.

PLUG-IN Effects software that you load into your DAW recording program (called the host). The plug-in becomes part of the host program and can be called up from within the host. Some manufacturers make plug-in bundles, which are a variety of effects in a single package.

POLAR PATTERN The directional pickup pattern of a microphone. A plot of microphone sensitivity plotted versus angle of sound incidence. Examples of polar patterns are omnidirectional, bidirectional, and unidirectional. Subsets of unidirectional are cardioid, supercardioid, and hypercardioid.

POLARITY Referring to the positive or negative direction of an electrical, acoustical, or magnetic force. Two identical signals in opposite polarity are 180 degrees out of phase with each other at all frequencies.

POP 1. A thump or little explosion sound heard in a vocalist's microphone signal. Pop occurs when the user says words with "p," "t," or "b" so that a turbulent puff of air is forced from the mouth and strikes the microphone diaphragm. 2. A noise heard when a mic is plugged into a monitored channel, or when a switch is flipped.

POP FILTER A screen placed on a microphone grille that attenuates or filters out pop disturbances before they strike the microphone diaphragm. Usually made of open-cell plastic foam or silk, a pop filter reduces pop and wind noise.

PORTABLE STUDIO A combination recorder and mixer in one portable chassis.

POWER AMPLIFIER An electronic device that amplifies or increases the power level fed into it to a level sufficient to drive a loudspeaker.

POWER GROUND (SAFETY GROUND) A connection to the power company's earth ground through the U-shaped hole in a power outlet. In the power cable of an electronic component with a 3-prong plug, the U-shaped prong is wired to the component's chassis. This wire conducts electricity to power ground if the chassis becomes electrically hot, preventing shocks.

PREAMPLIFIER (PREAMP) In an audio system, the first stage of amplification that boosts a mic-level signal to line level. A preamp is a standalone device or a circuit in a mixer.

PREDELAY Short for prereverberation delay. The delay (about 30 to 150 msec) between the arrival of the direct sound and the onset of reverberation. Usually, the longer the predelay, the greater the perceived room size. Using predelay on an instrument's or vocal's reverb helps to clarify the sound.

PRE-FADER/POST-FADER SWITCH A switch that selects a signal either ahead of the fader (pre-fader) or following the fader (post-fader). The level of a pre-fader signal is independent of the fader position; the level of a post-fader signal follows the fader position.

PREPRODUCTION Planning in advance what you're going to do at a recording session, in terms of track assignments, overdubbing, studio layout, and microphone selection.

PRESENCE The audible sense that a reproduced instrument is present in the listening room. Some synonyms are closeness, definition, and punch. Presence is often created by an equalization boost in the midrange or upper midrange, and by a high direct-to-reverb ratio.

PRESSURE ZONE MICROPHONE   A boundary microphone constructed with the microphone diaphragm parallel with, and facing, a reflective surface.

PRINT   To record on tape or disc.

PRODUCTION   1. A recording that is enhanced by effects. 2. The supervision of a recording session to create a satisfactory recording. This involves getting musicians together for the session, making musical suggestions to the musicians to enhance their performance, and making suggestions to the engineer for sound balance and effects.

PRO TOOLS   A popular digital audio editing platform for professional use. It offers computer multitrack recording, overdubbing, mixing, editing, and a variety of plug-in effects.

PROXIMITY EFFECT   The bass boost that occurs with a single-D directional microphone when it is placed a few inches from a sound source. The closer the microphone, the greater the low-frequency boost due to proximity effect.

PULSE CODE MODULATION (PCM)   A method of analog-to-digital conversion. The analog signal voltage is measured several thousand times a second, and each measurement is a digital word (a string of 1 s and 0 s) of a certain word length or bit depth.

PUNCH-IN/OUT   A feature in a multitrack recorder that lets you insert a recording of a corrected musical part into a previously recorded track by going into and out of record mode as the tape or disk is rolling.

PURE WAVEFORM   A waveform of a single frequency; a sine wave. A pure tone is the perceived sound of such a wave.

Q   In a parametric equalizer, Q is the sharpness of the boost or cut. A Q of 1 is a broad boost or cut that covers a wide range of frequencies. A Q to 10 is a narrow boost or cut that covers a small range of frequencies. Q or quality factor is a filter's center frequency divided by the bandwidth.

RACK A 19-inch-wide wooden or metal cabinet used to hold audio equipment.

RADIO-FREQUENCY INTERFERENCE (RFI) Radio-frequency electromagnetic waves induced in audio cables or equipment, causing various noises in the audio signal.

RANDOM ACCESS Referring to a storage medium in which any data point can be accessed or read almost instantly. Examples are a hard disk, compact disc, and MiniDisc.

RAREFACTION The portion of a sound wave in which molecules are spread apart, forming a region with lower-than-normal atmospheric pressure. The opposite of compression.

REAMPING Recording a guitar amp that is fed the signal from a direct-recorded electric guitar track. This technique lets you work on the amp's sound during mixdown, rather than during recording.

RECIRCULATION (REGENERATION) Feeding the output of a delay device back into its input to create multiple echoes. Also, the control on a delay device that affects how much delayed signal is recycled to the input.

RECORD To store an event in permanent form. Usually, to store an audio signal in magnetic form on magnetic tape or disk, or to store an audio signal in optical form on a CD-R or CD-RW. Recording is also possible on magneto-optical disk, on MiniDisc, in RAM, and on memory cards.

RECORDER-MIXER A combination multitrack recorder and mixer in one chassis.

RECORDING/REPRODUCTION CHAIN The series of events and equipment that are involved in sound recording and playback.

REFLECTED SOUND Sound waves that reach the listener after being reflected from one or more surfaces.

REGENERATION See RECIRCULATION.

REGION In a digital audio editing program, a defined segment of the audio program. Also called clip or zone.

RELEASE   The final portion of a note's envelope in which the note falls from its sustain level back to silence.

RELEASE TIME   In a compressor, the time it takes for the gain to return to normal after the end of a loud passage.

REMIX   To mix again; to do another mixdown with different console settings or different editing.

REMOTE RECORDING   See ON-LOCATION RECORDING.

REMOVABLE HARD DRIVE   A hard-disk drive that can be removed and replaced with another, used in a DAW or hard-drive recorder to store a program temporarily.

RENDER   To convert one audio format to another in a DAW or hard-drive recorder. Examples include mixing a multitrack recording to a 2-channel WAV, AIFF, RM, or MPC file; converting a track with real-time effects to a track with embedded effects; converting MIDI tracks to WAV tracks; and combining multiple clips into a single track.

REVERBERATION   Natural reverberation in a room is a series of multiple sound reflections that make the original sound persist and gradually die away or decay. These reflections tell the ear that you're listening in a large or hard-surfaced room. For example, reverberation is the sound you hear just after you shout in an empty gymnasium. A reverb effect simulates the sound of a room—a club, auditorium, or concert hall—by generating random multiple echoes that are too numerous and rapid for the ear to resolve. The timing of the echoes is random, and the echoes increase in number with time as they decay. An echo is a discrete repetition of a sound; reverberation is a continuous fade-out of sound.

REVERBERATION TIME (RT60)   The time it takes for reverberation to decay to 60 dB below the original steady-state level.

RFI   See RADIO FREQUENCY INTERFERENCE.

RHYTHM TRACKS   The recorded tracks of the rhythm instruments (guitar, bass, drums, and sometimes keyboards).

RIBBON MICROPHONE   A dynamic microphone in which the conductor is a long metallic diaphragm (ribbon) suspended in a magnetic field.

RIDE GAIN   To turn down the volume of a microphone when the source gets louder, and turn up the volume when the source gets quieter, in an attempt to reduce dynamic range.

ROOM MODES   See STANDING WAVE.

RT60   See REVERBERATION TIME.

SAFETY COPY   A copy of the master tape or CD, to be used if the master is lost or damaged.

SAFETY GROUND   See POWER GROUND.

SAMPLE   1. To digitally record a short sound event, such as a single note or a musical phrase, into computer memory. 2. A recording of such an event. 3. A measurement of an analog waveform that is done several thousand times a second during a PCM A/D conversion.

SAMPLING   1. Recording a short sound event into computer memory. The audio signal is converted into digital data representing the signal waveform, and the data is stored in memory chips, tape, or disc for later playback. 2. In PCM digital recording, measuring an analog waveform periodically, several thousand times a second.

SAMPLING RATE   In PCM digital recording, the frequency at which an analog waveform is sampled or measured. The sampling rate of CD-quality audio is 44,100 samples per second. The higher the sampling rate, the higher the high-frequency response of the recording.

SCRATCH VOCAL   A vocal performance that is done simultaneously with the rhythm instruments so that the musicians can keep their place in the song and get a feel for the song. Because it contains leakage, the scratch-vocal recording is usually erased. Then the singer overdubs the vocal part that is to be used in the final recording.

SENSITIVITY   1. The output of a microphone in volts for a given input in sound pressure level. 2. The sound pressure level a loudspeaker produces at one meter when driven with 1 watt of pink noise. See also SOUND PRESSURE LEVEL.

SEQUENCE   A MIDI data file of musical-performance note parameters, recorded by a sequencer.

SEQUENCER   A device or computer program that records a musical performance done on a MIDI controller (in the form of note numbers, note on, note off, etc.) into computer memory or hard disk for later playback. During playback, the sequencer plays synthesizer sound generators or samples.

SESSION   1. A time period set aside for recording musical instruments, voices, or sound effects. 2. On a CD-R, a lead-in, program area, and lead-out.

SHELVING EQUALIZER   An equalizer that applies a constant boost or cut above or below a certain frequency, so that the shape of the frequency response resembles a shelf.

SHIELD   A conductive enclosure (usually metallic) around one or more signal conductors, used to keep out electrostatic fields that cause hum or buzz. A shield in a mic cable is a cylindrical mesh of fine wires.

SHOCK MOUNT   A suspension system that mechanically isolates a microphone from its stand or boom, preventing the transfer of mechanical vibrations.

SIBILANCE   In a speech recording, excessive frequency components in the 5 to 10 kHz range, which are heard as an overemphasis of "s" and "sh" sounds.

SIDE-ADDRESSED   Referring to a microphone whose main axis of pickup is perpendicular to the side of the microphone. You aim the side of the mic at the sound source. See also End-Addressed.

SIGNAL   A varying electrical voltage that represents information, such as a sound.

SIGNAL PATH   The path a signal takes from input to output in a piece of audio equipment.

SIGNAL PROCESSOR   A device that is used to alter a signal in a controlled way.

SIGNAL-TO-NOISE RATIO (S/N)   The ratio in decibels between signal voltage and noise voltage. An audio component with a high S/N has little background noise accompanying the signal; a component with a low S/N is noisy.

SINE WAVE A wave following the equation $y = \sin x$, where x is degrees and y is voltage or sound pressure level. The waveform of a single frequency. The waveform of a pure tone without harmonics.

SLAP, SLAP BACK An echo following the original sound by about 50 to 200 msec, sometimes with multiple repetitions.

SNAKE A multipair or multichannel mic cable. Also, a multipair mic cable attached to a connector junction box.

SOFT SYNTH A synthesizer in software form.

SOLO On an input module in a mixing console, a switch that lets you monitor that particular input signal by itself. The switch routes only that input signal to the monitor system.

SOUND Longitudinal vibrations in a medium (such as air) in the frequency range 20 to 20,000 Hz.

SOUND CARD A circuit card that plugs into a computer and converts an audio signal into computer data for storage in memory or on hard disk. The sound card also converts computer data into an audio signal. A type of audio interface (see AUDIO INTERFACE).

SOUND EFFECTS Recordings of nonmusical sounds—such as a door slam, gunfire, thunderstorm, car, or telephone—used in dramatic productions, radio spots and commercials. Not to be confused with EFFECTS.

SOUND MODULE (SOUND GENERATOR) 1. A synthesizer without a keyboard, containing several different timbres or voices. These sounds are triggered or played by MIDI signals from a sequencer program, or by a MIDI controller. A sound module can be a standalone device or a circuit on a sound card. 2. An oscillator.

SOUND PRESSURE LEVEL (SPL) The acoustic pressure of a sound wave, measured in decibels above the threshold of hearing. The higher the SPL of a sound, the louder it is. $dBSPL = 20 \log(P/P_{ref})$, where P = the measured acoustic pressure and $P_{ref} = 0.0002 \, dyne/cm^2$.

SOUND WAVE The periodic variations in air pressure radiating from an object vibrating between 20 and 20,000 Hz.

SPACED-PAIR A stereo microphone technique using two identical microphones spaced several feet apart horizontally, usually aiming straight ahead toward the sound source.

S/PDIF (Sony Philips Digital Interface; IEC 958 Type II) A 2-channel digital signal interface format that uses a 75-ohm coaxial cable with RCA connectors, or an optical cable with TOSLINK connectors. See also AES/EBU.

SPEAKER See LOUDSPEAKER.

SPECTRUM The output level versus frequency of a sound source, including the fundamental frequency and overtones.

SPL See SOUND PRESSURE LEVEL.

SPOT MICROPHONE In classical music recording, a close-placed microphone that is mixed with more distant microphones to add presence or to improve the balance.

STANDING WAVE An apparently stationary waveform created by multiple reflections between opposite room surfaces. At certain points along the standing wave, the direct and reflected waves cancel, and at other points the waves add together or reinforce each other.

STEM A submix in ProTools lingo—a drums submix, keys submix, left-front submix, etc.

STEREO, STEREOPHONIC An audio recording and reproduction system with correlated information between two channels (usually discrete channels), meant to be heard over two or more loudspeakers to give the illusion of sound-source localization and depth.

STEREO BAR, STEREO MICROPHONE ADAPTER A microphone stand adapter that mounts two microphones on a single stand for convenient stereo miking.

STEREO IMAGING The ability of a stereo recording or reproduction system to form clearly defined audio images at various locations between a stereo pair of loudspeakers.

STEREO MICROPHONE   A microphone containing two mic capsules in a single housing for convenient stereo recording. The capsules usually are coincident.

STREAMING AUDIO   Audio sent over the Internet in real time. A streaming file plays as soon as you click on its title. A downloaded file doesn't play until you copy the entire file to your hard disk. Streaming audio is heard almost instantly, but usually sounds muffled and can be interrupted by Internet congestion. With downloaded audio, you must wait up to several minutes to hear the song, but the sound is high-fidelity and continuous.

STUDIO   A room used or designed for sound recording.

SUBMASTER   1. A master volume control for an output bus. 2. A recorded tape that is used to form a master tape

SUBMIX   A small preset mix within a larger mix, such as a drum mix, keyboard mix, vocal mix, etc. Also a cue mix, monitor mix, or effects mix.

SUBMIXER   A smaller mixer within a mixing console (or standalone) that is used to set up a submix, a cue mix, an effects mix, or a monitor mix.

SUPERCARDIOID MICROPHONE   A unidirectional microphone that attenuates side-arriving sounds by 8.7 dB, attenuates rear-arriving sounds by 11.4 dB, and has two nulls of maximum sound rejection at 125 degrees off-axis.

SUSTAIN   The portion of the envelope of a note in which the level is constant. Also, the ability of a note to continue without noticeably decaying, often aided by compression.

SYNTHESIZER   A musical instrument (usually with a piano-style keyboard) that creates sounds electronically and allows control of the sound parameters to simulate a variety of conventional or unique instruments.

TAIL   The end of a reverberation signal where the reverb fades down to silence.

TAKE   A recorded performance of a song. Usually, several takes are done of the same song, and the best one—or the best parts of several—becomes the final product.

TAKE SHEET   A list of take numbers for each song, plus comments on each take.

TALKBACK An intercom in the mixing console for the engineer and producer to talk to the musicians in the studio.

3-PIN CONNECTOR A 3-pin professional audio connector used for balanced signals. Pin 1 is soldered to the cable shield, pin 2 is soldered to the signal hot or in-polarity lead, and pin 3 is soldered to the signal cold or opposite-polarity lead. See also XLR-TYPE CONNECTOR.

THREE-TO-ONE RULE A rule in microphone applications. When multiple mics are mixed to the same channel, the distance between mics should be at least three times the distance from each mic to its sound source. This prevents audible phase interference.

THRESHOLD In a compressor or limiter, the input level above which compression or limiting takes place. In an expander, the input level below which expansion takes place.

TIGHT 1. Having very little leakage or room reflections in the sound pickup. 2. Referring to well-synchronized playing of musical instruments. 3. Having a well-damped, rapid decay.

TIMBRE The subjective impression of spectrum and envelope. The quality of a sound that allows us to differentiate it from other sounds. For example, if you hear a trumpet, a piano, and a drum, each has a different timbre or tone quality that identifies it as a particular instrument.

TONAL BALANCE The balance or volume relationships among different regions of the frequency spectrum, such as bass, mid-bass, midrange, upper midrange, and highs.

TRACK A path on magnetic tape containing a single channel of audio. A group of bytes in a digital signal (on tape, on hard disk, on compact disc, or in a data stream) that represents a single channel of audio or MIDI. Usually one track contains a performance by one musical instrument. The verb "to track" means to record tracks.

TRANSDUCER A device that converts energy from one form to another, such as a microphone or loudspeaker.

TRANSIENT   A short signal with a rapid attack and decay, such as a drum stroke, cymbal hit, or acoustic-guitar pluck.

TRANSIENT RESPONSE   The ability of an audio component (usually a microphone or loudspeaker) to follow a transient accurately.

TRIM   1. In a mixing console, a control for fine adjustment of level, as in a Bus Trim control. 2. In a mixing console, a control that adjusts the gain of a mic preamp to accommodate various signal levels.

TRS (tip-ring-sleeve)   A phone-plug connector used for aux send/return, unbalanced stereo, or balanced mono connections.

TUBE   A vacuum tube, an amplifying component made of electrodes in an evacuated glass tube. Tube sound is characterized as being "warmer" than solid-state or transistor sound.

TWEETER   A high-frequency loudspeaker

UNBALANCED LINE   An audio cable having one conductor surrounded by a shield that carries the return signal. The shield is at ground voltage.

UNIDIRECTIONAL MICROPHONE   A microphone that is most sensitive to sounds arriving from one direction—in front of the microphone. Examples are cardioid, supercardioid, and hypercardioid.

UNITY GAIN   A condition in which the input and output levels of a device are equal.

USB (UNIVERSAL SERIAL BUS)   A Mac/PC computer serial interface for connecting external devices such as MIDI interfaces and audio interfaces to a computer. Faster than a standard serial port.

VALVE   British term for vacuum tube. See TUBE.

VBR (variable bitrate MP3 encoding)   The bitrate varies with the audio signal.

VIRTUAL CONTROLS   Audio-equipment controls that are simulated on a computer monitor screen. You adjust them with a mouse or with a controller surface.

VIRTUAL TRACK   A recording of a single take of one instrument or vocal on a random-access medium. The best parts of several virtual tracks can be bounced a single track, forming a complete, perfect performance. This process is called comping tracks.

VST (VIRTUAL STUDIO TECHNOLOGY)   Steinberg's virtual instrument integration standard format for plug-ins.

WAV (.wav)   A computer audio file format for Windows. It encodes sound without any data reduction, using pulse code modulation. Its audio resolution is 16 bit, 44.1 kHz, or higher.

WAVEFORM   A graph of a signal's sound pressure or voltage versus time. The waveform of a pure tone is a sine wave.

WAVELENGTH   The physical length between corresponding points of successive waves. Low frequencies have long wavelengths; high frequencies have short wavelengths.

WEIGHTED   Referring to a measurement made through a filter with a specified frequency response. An A-weighted measurement is taken through a filter that simulates the frequency response of the human ear.

WINDOWS MEDIA   See WMA.

WINDSCREEN   See POP FILTER.

WMA (WINDOWS MEDIA AUDIO)   A popular compressed audio file format for streaming audio and for downloads. Windows Media 8 promises performance similar to that of MP3PRO—near-CD quality at 48 kbps and CD quality at 64 kbps.

WOOFER   A low-frequency loudspeaker.

WORD   A binary number that is the value of each sample in an analog-to-digital conversion. A sample is a measurement of an analog waveform that is done several thousand times a second during a PCM A/D conversion.

WORD LENGTH   See BIT DEPTH.

WORKSTATION A system of MIDI- or computer-related equipment that works together to help you compose and record music. Usually, this system is small enough to fit on a desktop or equipment stand. See also KEYBOARD WORKSTATION and DIGITAL AUDIO WORKSTATION.

WRAPPER Software that converts a plug-in from an unsupported format to a supported format. For example, a DAW recording program (host) that cannot use VST plug-ins directly might include a wrapper that converts VST plug-ins to DirectX plug-ins.

XLR-TYPE CONNECTOR An ITT Cannon part number that has become the popular definition for a 3-pin professional audio connector. See also 3-PIN CONNECTOR.

X-Y See COINCIDENT-PAIR.

Y-ADAPTER A cable that divides into two cables in parallel to feed one signal to two destinations.

Z Abbreviation for impedance.

ZONE See REGION.

Index